Dynamics, Transport and Photochemistry in the Middle Atmosphere of the Southern Hemisphere

NATO ASI Series

Advanced Science Institutes Series

A Series presenting the results of activities sponsored by the NATO Science Committee, which aims at the dissemination of advanced scientific and technological knowledge, with a view to strengthening links between scientific communities.

The Series is published by an international board of publishers in conjunction with the NATO Scientific Affairs Division

A Life Sciences	Plenum Publishing Corporation
B Physics	London and New York
C Mathematical	Kluwer Academic Publishers
and Physical Sciences	Dordrecht, Boston and London
D Behavioural and Social Sciences	
E Applied Sciences	
F Computer and Systems Sciences	Springer-Verlag
G Ecological Sciences	Berlin, Heidelberg, New York, London,
H Cell Biology	Paris and Tokyo

Series C: Mathematical and Physical Sciences - Vol. 321

Dynamics, Transport and Photochemistry in the Middle Atmosphere of the Southern Hemisphere

edited by

Alan O'Neill
Hadley Centre for Climate Prediction and Research,
Meteorological Office, Bracknell, U.K.

Kluwer Academic Publishers

Dordrecht / Boston / London

Published in cooperation with NATO Scientific Affairs Division

Proceedings of the NATO Advanced Research Workshop on
Dynamics, Transport and Photochemistry in the Middle Atmosphere of the Soutern
Hemisphere
San Francisco, California, U.S.A.
April 15–17, 1989

Library of Congress Cataloging-in-Publication Data

```
NATO Advanced Research Workshop on Dynamics, Transport, and
  Photochemistry in the Middle Atmosphere of the Southern Hemisphere
  (1989 : San Francisco, Calif.)
    Dynamics, transport, and photochemistry in the middle atmosphere
  of the Southern Hemisphere : proceedings of the NATO Advanced
  Research Workshop on Dynamics, Transport, and Photochemistry in the
  Middle atmosphere of the Southern Hemisphere, San Francisco,
  California, U.S.A., 15-17 April, 1989 / edited by Alan O'Neill.
        p.   cm. -- (NATO ASI series. Series C, mathematical and
  physical sciences ; v. 321)
    Includes indexes.

    1. Middle atmosphere--Southern Hemisphere--Congresses.  2. Dynamic
  meteorology--Congresses.  3. Photochemistry--Congresses.
  I. O'Neil, Alan, 1950-   . II. Title. III. Series.
  QC881.2.M53N37  1989
  551.5'1--dc20                                              90-5315
```

ISBN-13: 978-94-010-6797-3   e-ISBN-13: 978-94-009-0693-8
DOI: 10.1007/ 978-94-009-0693-8

Published by Kluwer Academic Publishers,
P.O. Box 17, 3300 AA Dordrecht, The Netherlands.

Kluwer Academic Publishers incorporates the publishing programmes of
D. Reidel, Martinus Nijhoff, Dr W. Junk and MTP Press.

Sold and distributed in the U.S.A. and Canada
by Kluwer Academic Publishers,
101 Philip Drive, Norwell, MA 02061, U.S.A.

In all other countries, sold and distributed
by Kluwer Academic Publishers Group,
P.O. Box 322, 3300 AH Dordrecht, The Netherlands.

Printed on acid-free paper

TABLE OF CONTENTS

vi

PREFACE

The NATO Advanced Research Workshop on "Dynamics, Transport and Photochemistry in the Middle Atmosphere of the Southern Hemisphere" was held in San Francisco, California, U.S.A., 15-17 April 1989. In addition to NATO, the workshop was supported by the University of California, Los Angeles, and by the National Aeronautics and Space Administration, U.S.A. (NASA). The American Meteorological Society was a co-operating organization. The venue for the workshop was the Lone Mountain Conference Center of the University of San Francisco. The workshop was organized and directed by Dr A.O'Neill (Hadley Centre for Climate Prediction and Research, Meteorological Office, Bracknell, U.K.) and Prof C.R. Mechoso (Dept of Atmospheric Sciences, University of California, Los Angeles, U.S.A.).

The workshop was the third one held as part of the Middle Atmosphere in the Southern Hemisphere (MASH) project, an international effort (under the auspices of the Middle Atmosphere Program) to learn more about dynamics, transport and photochemistry in the middle atmosphere of the southern hemisphere. Before the discovery that, during recent years, a dramatic thinning of the ozone layer takes place over Antarctica in spring - the "ozone hole" - the middle atmosphere of the southern hemisphere had received much less attention than that of the northern hemisphere from meteorologists and atmospheric chemists. The MASH project was instituted to remedy this comparative lack of interest.

The workshop was organised around four sessions of invited papers given by leading atmospheric scientists. Ample time was allowed for in-depth discussions after the formal presentations. Most of the papers are included in these proceedings. I decided not to divide the volume into separate sections on dynamics, transport and photochemistry in order to encourage interdisciplinary reading, and also to mirror the interdisciplinary nature of the workshop. Several of the papers would not fit naturally under one heading.

Support was provided by NATO through Dr C Sinclair of the Scientific Affairs Division. His courteous advice is much appreciated. Support from UCLA came through Dr. C.A. Hall Jr., and from NASA (NAGW-1021) through Dr M.J. Prather. Dr J.D. Farrara helped with the organisation, and Dr V.D. Pope gave editorial assistance. Patient encouragement from Mrs Nel de Boer (Kluwer) stimulated my efforts to coax manuscripts from busy scientists. I am especially grateful to Prof. C.R. Mechoso, ably assisted by Mrs C. Wong, whose organisational skills were essential to the success of the workshop. Finally, I would like to thank the invited speakers and participants for making the workshop not only scientifically stimulating but also most enjoyable.

Bracknell, U.K.   Alan O'Neill
25 June 1990

MIDDLE ATMOSPHERIC DYNAMICS AND TRANSPORT: SOME CURRENT CHALLENGES TO OUR UNDERSTANDING

MICHAEL E. McINTYRE
Department of Applied Mathematics and Theoretical Physics
Silver Street, Cambridge CB3 9EW
U. K.

ABSTRACT. The fluid dynamics of wave propagation, wave breaking, and the resulting turbulence – be it the fully three-dimensional small-scale turbulence due to breaking internal gravity waves, or the layerwise two-dimensional turbulence due to breaking Rossby waves – poses three major challenges to research on middle atmospheric dynamics and chemical transport. These are, first, the unjustifiability of the eddy-diffusivity concept, under conditions often met with in the atmosphere, second, the ill-understood nature of the Rossby-wave-associated dynamical feedbacks on the global circulation and, third, an acute difficulty in parameterizing vertical mixing by convectively overturning gravity waves in the mesosphere and lower thermosphere.

1. Introduction

The three major challenges I have chosen to discuss are all related to a tendency for naturally-occurring turbulence to be spatially inhomogeneous. This often gives rise to situations very different from the situations assumed in classical turbulence theory.

One of the contributing factors seems to be the importance of various kinds of wave propagation mechanism. Besides leading to the propagation of information, angular momentum, etc., between different parts of the atmosphere, the dynamical restoring mechanisms to which the waves owe their existence can act locally to suppress turbulent motion, often in a spatially very selective way. Indeed one often has the impression of dealing with closely adjacent, strongly interacting regions of wavelike motion and turbulent motion, a kind of highly inhomogeneous 'wave-turbulence jigsaw puzzle' in which the waves strongly modify, indeed often give rise to, the turbulence, and in which the turbulence, in turn, modifies the local spatial distribution of the wave restoring mechanism, and also, after a propagation delay, the wave field at greater distances. The word 'turbulence' is being used here in the broad sense that includes not only three-dimensional turbulence – such as that due to cumulus convection, to boundary-layer friction, to Kelvin-Helmholtz instability or to breaking internal gravity waves – but also layerwise two-dimensional turbulence, such as that due to baroclinic instability or to breaking Rossby waves.

The first of the three challenges to be discussed is that posed by the concept of the 'eddy diffusivity' of materially conserved quantities such as long-lived tracer substances (§3 below). Although that concept correctly expresses the statistical tendency for such tracers to move downgradient, in some long time average sense – and will, inevitably, continue to be used *faute de mieux* as a modelling device for some time to come – we cannot be too careful about the use we make of the concept. Model predictions that depend on it too heavily need to be regarded with suspicion. This is because the conditions under which

1

A. O'Neill (ed.),
Dynamics, Transport and Photochemistry in the Middle Atmosphere of the Southern Hemisphere, 1–18.
© 1990 *Kluwer Academic Publishers.*

it can be strictly justified in terms of classical turbulence theories, namely spatial near-homogeneity of the turbulence, seldom appear to be met, for the reasons just mentioned. For some purposes the concept may actually be wrong qualitatively, in the sense that fluxes are not even roughly proportional to mean gradients. The Antarctic ozone hole provides one of the more conspicuous examples of this. Various other paradoxes arise, as will be pointed out. There is also the simple but potentially important fact that current models of chemical evolution assume that chemical constituents interdiffuse and react when, in reality, they may well be separate and non-reacting in a fully-resolved, fine-grain view (e.g. Tuck 1979).

The second challenge is how to parameterize global-scale dynamical feedbacks in simplified general circulation models, particularly the feedbacks associated with the Rossby-wave-associated or layerwise two-dimensional phenomena that appear to dominate events in the lower and middle stratosphere. It is widely recognized that there will be a continuing need for simplified circulation models to aid the assessment of multidecadal ozone–climate scenarios. Furthermore, research on the feedback mechanisms will be an important means, in its own right, of improving our understanding of what is robust, and what is sensitive, in the complicated web of causal links involved. There has been no spectacular progress over the past few years, as far as I know, and so the present discussion (§4 below) will confine itself to a very brief update on the lengthier discussion given in a previous informal review and forward look, written for the proceedings of the Erice Workshop (McIntyre 1987, hereafter E87). Progress on this extremely difficult set of problems, which are not unrelated to some of the problems to be discussed in §3, seems likely to depend on continued efforts to combine theoretical insights with ultra-high-resolution numerical modelling as well as with observational information. On the observational side, the proposed new generation of space-based wind sounders and high-resolution scanning limb sounders should in due course provide a particularly important source of new information.

It seems more certain than ever that the strange quantity known as potential vorticity will play a central role in the quest for an improved understanding of the global-scale dynamical feedbacks, and that it will also be important for future attempts at the simultaneous retrieval of dynamical and chemical information from future observing systems, and in the quality control of observational data processing. As a preliminary to §§3 and 4, therefore, §2 briefly recalls some of the fundamental properties of PV, including its relevance to wave–mean interaction theory and the differences between PV and chemical-tracer behaviour.

The third challenge (§5) concerns what may be an even tougher part of the global circulation and transport problem, albeit that its most critical importance appears likely, on present knowledge, to be confined to mesospheric and higher altitudes. This is the problem of how to parameterize the vertical turbulent mixing by convectively breaking internal gravity waves in the mesosphere and lower thermosphere. Such mixing seems likely to be important in summertime, for instance, in transporting atomic oxygen downward against the mean circulation (e.g. Thomas et al. 1984, Garcia and Solomon 1985). The turbulence is, again, highly inhomogeneous, but interestingly enough it seems possible that the notion of eddy diffusivity might be justifiable to some extent in a non-classical way, for this purpose. However, it can also be shown that there is still no such thing as a single eddy diffusivity that is applicable for all purposes, and that the tacit assumption that there is such a thing as 'the' eddy diffusivity, which one often encounters in the literature, is still highly dangerous.

Another feature of the problem seems to be a pathological sensitivity of the efficiency of vertical mixing to the precise degree of wave supersaturation. In particular, convectively-breaking gravity waves seem likely to be inefficient at vertical mixing unless they break very violently indeed, so as to give a very large supersaturation, or amplitude overshoot. Again, the discussion will be kept very brief since most of the ground has already been covered elsewhere (McIntyre 1989b). Breaking gravity waves provide, also, a particularly striking illustration of the aforementioned differences between PV and chemical-tracer behaviour – differences which can be shown to be of central importance, qualitatively as well as quantitatively, to our understanding of the global circulation.

2. The fundamental properties of potential vorticity

We recall these only briefly since they have been discussed extensively elsewhere (Hoskins *et al.* 1985; Haynes and McIntyre 1987, 1990; McIntyre and Norton, 1990). The most accurate and general version of the potential-vorticity concept, for a continuously stratified fluid, is that associated with the exact set of definitions given by Ertel (1942). A hydrostatic version was given slightly earlier by Rossby (1940). The difference between the exact and hydrostatic versions is usually unimportant in atmospheric dynamics, the main exception being when one seeks to understand in detail how the potential vorticity is affected by three-dimensional turbulent mixing. The exact (Ertel) definition is usually taken for convenience as

$$Q = \rho^{-1} \mathbf{q_a} \cdot \nabla \theta \ , \tag{1}$$

where ρ is the mass density, $\mathbf{q_a}$ the three-dimensional absolute vorticity, including the vorticity of the earth's rotation, ∇ the three-dimensional gradient operator with respect to geometric position $\mathbf{x}$, and θ the potential temperature. As Ertel pointed out, it would be equally valid in principle to adopt any of an infinite number of other definitions in which θ is replaced by some monotonic function of θ. For convenience we shall refer to Q, as defined by (1), as 'the' potential vorticity, hereafter 'PV'. There are three main points about Q.

2.1. MATERIAL TENDENCY, AND VISUALIZABILITY

The first is the well known fact that the PV is *materially conserved* if diabatic heating and nonconservative forces are negligible. This is a particular case of the general result

$$DQ/Dt = -\rho^{-1}\nabla \cdot \mathbf{N}_Q \ , \tag{2}$$

where the material derivative, and the nonadvective flux or transport, are defined respectively by

$$D/Dt = \partial/\partial t + \mathbf{u} \cdot \nabla \ , \qquad \mathbf{N}_Q = -H\mathbf{q_a} - \mathbf{F} \times \nabla \theta \ , \tag{3a, b}$$

$\mathbf{u}$ being the three-dimensional velocity field, $\mathbf{F}$ the viscous or other nonconservative body force per unit mass, and H the diabatic heating rate expressed as the material rate of change of θ. To the extent that the right-hand side of (2) is effectively small, the PV becomes approximately an air-mass marker, making its evolution easy to grasp conceptually, and easy to visualize in terms of isentropic distributions of PV, that is, in terms of layerwise two-dimensional distributions of PV on constant-θ surfaces.

4

2.2. BALANCE AND INVERTIBILITY

The second main point is the idea, which goes back to Charney (1948) and Kleinschmidt (1950a,b, 1951) – for further historical notes see the review by Hoskins *et al.* (1985) – that, as well as being easy to visualize, isentropic distributions of PV contain nearly all the dynamical information that is relevant to the stratification-constrained, layerwise two-dimensional part of the motion. In other words, isentropic distributions of PV contain nearly all the information about the dynamics of the air motion apart from any inertio-gravity oscillations that may be present. More precisely, there is an 'invertibility principle' saying that if a suitable balance condition is imposed, and a suitable reference state specified – for instance by specifying the mass under each isentropic surface as one does in the theory of available potential energy – then a knowledge of the distribution of PV on each isentropic surface, and of potential temperature at the lower boundary, is sufficient to deduce, diagnostically, all the other dynamical fields such as winds, temperatures, and geopotential heights.

The balance or 'slow-manifold' condition says that inertio-gravity oscillations are either absent, or can be averaged out and ignored. We know that in principle this can be true in general only as an approximation; but the approximation is often amazingly good, far better than might be supposed from the usual theories of balanced motion based on filtered equations (McIntyre and Norton 1990a,b). The succinctness of being able to represent so much dynamical information in the form of isentropic distributions of a single scalar field, the PV, whose evolution is so easy to visualize, is a powerful aid to understanding and quantitatively depicting the layerwise two-dimensional motion. This includes all the usual barotropic and baroclinic instabilities, and other Rossby-wave-associated phenomena. It provides in addition a very succinct way of understanding, and significantly generalizing, the main results of wave–mean interaction theory, and showing how they are relevant to understanding the global atmospheric circulation. For full details, including the relationship between the momentum viewpoint and the viewpoint in terms of isentropic distributions of PV, the reader is referred to a recent paper by McIntyre and Norton (1990b) and to the extensive set of references therein.

2.3. GENERAL CONSERVATION PROPERTIES

The third point has recently been the subject of some controversy. It concerns PV behaviour for general **F** and H. The flux form of (2), from which (2) itself can be recovered using the mass-conservation equation $\partial \rho / \partial t + \nabla \cdot \{\rho \mathbf{u}\} = 0$, is exactly

$$\frac{\partial}{\partial t}(\rho Q) + \nabla \cdot \mathbf{J} = 0 \qquad \text{where} \quad \mathbf{J} = \mathbf{u}\rho Q + \mathbf{N}_Q . \qquad (4a,b)$$

This is an equation in what is known as 'conservation form'. Here one does not, of course, mean conservation in the material or Lagrangian sense of §2.1, but in the traditional, general sense used in theoretical chemistry and physics. Material conservation is the special case $\mathbf{N}_Q = 0$. The exact conservation form of (4) is a direct consequence of the mathematical form of (1) and the fact that div curl of any vector field vanishes identically.

One way of talking about the general meaning of (2) and (4) that I find vivid and useful is to state it in terms of the analogy between PV and chemical mixing ratios. This is because the general notion of conservation can be put into correspondence with the notion of an indestructible chemical substance, or decay-corrected radioactive tracer, that is to say a chemical constituent that has zero source. Equations (2) and (4) can be interpreted as saying that the PV behaves like the mixing ratio of a peculiar chemical 'substance', or generalized tracer, that has zero source away from boundaries. The word 'source' is being used here in its standard chemical sense. A chemical substance with zero source means a chemical substance whose molecules are neither created nor destroyed. The *mixing ratio* of such a substance can of course change – for instance by dilution – and so, likewise, of course, can the PV.

One of the peculiarities of the generalized tracer substance whose mixing ratio is the PV is that one can have both positive and negative amounts of it, like electric charge. Another peculiarity is what might be called the 'impermeability property'. This expresses a strikingly simple fact about the flux or transport **J** relative to isentropic surfaces that promises useful simplifications in our thinking about the global circulation. It seems likely, in any case, that we shall tend more and more to think about everything relative to isentropic surfaces, both for dynamical (recall §2.2) and chemical reasons (e.g. Yang and Tung 1990, this Proceedings), taking advantage of the fact that radiative transfer tends to keep the middle atmosphere stably stratified.

The impermeability property is another direct consequence of the mathematical form of the definition (1). One can show quite straightforwardly, by manipulating the expressions (3b) and (4b) giving the total (advective plus nonadvective) flux or transport **J**, that

A point moving with velocity $\mathbf{J}/(\rho Q)$ always remains on exactly the same isentropic surface.

Various alternative proofs are given in Haynes and McIntyre (1987, 1990) and in McIntyre and Norton (1990a). This result says that isentropic surfaces behave exactly as if they were impermeable to the generalized tracer substance whose mixing ratio is the PV. This is true not just for adiabatic motion, but also when diabatic heating or cooling ($H \neq 0$) makes isentropic surfaces permeable to mass and chemical substances. In this respect isentropic surfaces can be said to act like semi-permeable membranes. These statements, being completely general, apply to the real global atmospheric 'wave-turbulence jigsaw puzzle' in all its enormous complexity.

The general conservation property is mathematically equivalent to an integral relation pointed out by Thorpe and Emanuel (1985), and the impermeability property was, as far as I know, first pointed out by Haynes and myself (1987, *q.v.* for further history). In publishing the latter paper we inadvertently got ourselves into some controversy over both properties. Part of the reason was insufficient care on our part over the wording of the introduction to that paper. (The interested reader is recommended to start at §2!) This was compounded by an interdisciplinary language-barrier problem of which we were unaware at the time. [It appears that, despite the analogy between PV and chemical mixing ratios, a separate convention has grown up in which PV behaviour is thought of in a manner not directly related to the traditional, general notions of conservation relation and conservable quantity, and that along with this has grown up a separate usage of words like flux, transport, creation, destruction, etc., when used in connection with the PV. I have not seen a systematic account of this separate convention, nor a set of explicit definitions of the associated vocabulary, but the convention appears to define the words 'flux' and 'transport' to mean the quantity $\mathbf{u}\rho Q$, in traditional language the *advective contribution* to the total transport **J**. There appears also, although I am not sure of this, to be a third convention in which the word 'transport' is used to mean the quantity $\mathbf{u}\cdot\nabla Q$, in traditional language the *advection*. It should be noted that each of the latter two conventions prohibit the use of a traditional physico-chemical idea such as 'molecular-diffusive transport' since they render such an idea self-contradictory.]

The other reason for the controversy is more substantial. There appears to be a genuine mistake in the literature, to the effect that PV behaves like a chemical mixing ratio in all essential respects, even when three-dimensionally turbulent mixing is taking place. The impermeability theorem makes it particularly clear, however, that no such behaviour is possible, inasmuch as there is nothing to stop chemicals being turbulently mixed across isentropic surfaces. The same conclusion can be drawn, albeit less directly, from other well known principles like the Kelvin-Bjerknes circulation theorem.

The origin of the mistake appears to be a tacit neglect of the strong diabatic heating or cooling that occurs on the Kolmogorov microscale during turbulent mixing. This is crucially important first for seeing how air and trace chemicals can cross isentropic surfaces, and second for seeing how (2) and (4) are satisfied in the presence of three-dimensional turbulence. The point is relevant irrespective of whether one is thinking of the ρ, θ, u and q_a fields as explicitly representing the detailed, fine-grain reality including the Kolomogorov microscales, so that the diabatic heating field H represents the effects of molecular conduction only, or whether one is taking the coarse-grain view necessary in practical observational work, in which the ρ, θ, u and q_a fields are considered to be coarse-grain approximations to reality, with corresponding adjustments to the fields H and F to include the eddy flux convergences from unresolved scales. The interested reader may consult our (1990) paper for further discussion, along with a forthcoming paper by Keyser and Rotunno (1990). An example that is relevant to the global circulation problem, and in which the fundamental difference between PV and chemical behaviour shows up especially strikingly, will be noted in §5.

The fact that the notional tracer substance whose mixing ratio is the PV is, in some ways, more like electric charge, ties in with the notion of PV invertibility. In the Boussinesq, quasi-geostrophic approximation, for a fluid of constant static stability, the mathematical process of PV inversion is the same as the mathematical process of finding the electrostatic potential from a given charge distribution (Obukhov 1962; Hoskins *et al.* 1965, §5).

3. Some remarks on the eddy diffusivity concept

One often reads about 'the' eddy diffusivity of the atmosphere, or 'the' small-scale mixing, as if it were something ubiquitous whose prior existence can be taken for granted, and which always acts to smooth out small-scale variations in any quantity in which we might be interested. Some models of atmospheric motion have even been criticized from time to time as 'unrealistic' for the *lack* of a powerful eddy smoothing term.

I have been worried about our dependence on the eddy-diffusivity concept for a long time, without knowing quite what to do about it. We depend on numerical models to help us grasp the complexities of atmospheric motion, and its implications for chemical evolution. We hope to develop them to the point where they can be useful predictors of, among other things, global environmental change. But nearly all our atmospheric numerical modelling, be it one, two, or three dimensional, depends on the notion of eddy diffusivity, in one version or another. To what extent does that notion ever make sense, except possibly as an artifice that we have been compelled to adopt *faute de mieux* in order to keep our numerical models stable, and as a very rough way of expressing the observed tendency of materially conserved tracer quantities to be transported down their gradients, in some average sense?

3.1. THE INHOMOGENEITY OF SMALL-SCALE TURBULENCE

Anyone who has observed what can be seen out of aeroplane windows must feel uncomfortable with the idea – which seems to be widely held – that the atmosphere *above* the planetary boundary layer is permeated by quasi-uniform, small-scale, three-dimensional turbulence giving rise to a ubiquitous, pre-existing small-scale eddy diffusivity. Looking at the variety of cloud forms, all the way from turbulently-convecting cumulus and Kelvin-Helmholtz billows to silky-smooth, laminar-looking wave clouds revealing thin layers in the humidity field, one gets the impression, rather, of extreme spatial inhomogeneity. For what it is worth, this impression is reinforced by the amazing smoothness of passenger-jet flight in the upper troposphere or lower stratosphere, interrupted only now and then by the more obvious encounters with clear-air turbulence. The sample may be biased but the point seems significant nonetheless. The impression one gets, in this and in other ways, is that very large portions of the atmosphere are actually in laminar motion, in the sense that small-scale, truly three-dimensional turbulence is altogether absent. This hypothesis is consistent with the fact that Richardson numbers are usually observed to be considerably larger than unity.

3.2. SMALL-SCALE INHOMOGENEITY, LARGE-SCALE INHOMOGENEITY AND WAVE BREAKING

One can make sense of the extreme spatial and temporal intermittency of small-scale turbulence in the real atmosphere, outside convective clouds and the planetary boundary layer, by adopting the time-honoured hypothesis that clear-air turbulence is often due to breaking waves of one kind or another. This notion can actually be generalized in a significant way, adumbrated long ago by Deem and Zabusky (1978). When suitably formulated (a careful discussion is given in McIntyre and Palmer 1985), it applies in various forms over a vast range of scales, from ordinary small-scale gravity waves up to planetary-scale Rossby waves.

Just as breaking gravity waves produce three-dimensional turbulence and mix potential temperature vertically, so do breaking Rossby waves produce layerwise two-dimensional 'turbulence' and mix PV isentropically, that is to say along isentropic surfaces. Recent satellite data studies and high-resolution numerical simulations have vividly revealed just how spatially inhomogeneous is the resulting wave-turbulence jigsaw puzzle when viewed on a global scale (Juckes and McIntyre 1987, hereafter JM, Haynes and Norton 1989, Salby *et al.* 1989, Juckes 1989, Juckes *et al.* 1990). Animated grayscale PV maps were shown at the workshop, vividly conveying this inhomogeneity in a numerical model of the wintertime stratosphere; see also figure 1 below.

3.3. THE ANTARCTIC OZONE HOLE, AND SUBPOLAR PV BARRIERS

A striking feature seen again and again in the high-resolution simulations is made clear by looking at the motion of a band of PV values embedded in the region of very steep PV gradients that mark the edge of the main polar vortex. Such a band of values was highlighted in one of the animated sequences shown at the workshop, and a frame from that sequence is shown here in figure 1 below. The band exhibits a peculiar resilience or elasticity that, in the case shown, altogether prevents it from being entrained into the more turbulent-looking regions on either side of it, especially on the tropical side, where the motion is arguably two-dimensionally turbulent by most accepted criteria and where it seems appropriate to speak of a Rossby-wave 'surf zone'. It is clear from the animated version of figure 1 that, by contrast, the band undulates reversibly: its motion is wavelike rather than turbulent. To very good approximation it is a material contour; and this means that *all the material it encloses is chemically isolated from its surroundings*. This vortex isolation phenomenon – JM called the polar vortex a 'containment vessel' – seems to be a crucial factor in the genesis of the Antarctic ozone hole, possibly just as crucial as the non-standard chemistry involved[†](e.g. Tuck *et al.* 1989, and other papers appearing in the two special JGR issues on the Airborne Antarctic Ozone Experiment). As is well known, neither factor was represented in the models on which the Montreal Protocol on chlorofluorocarbons was based.

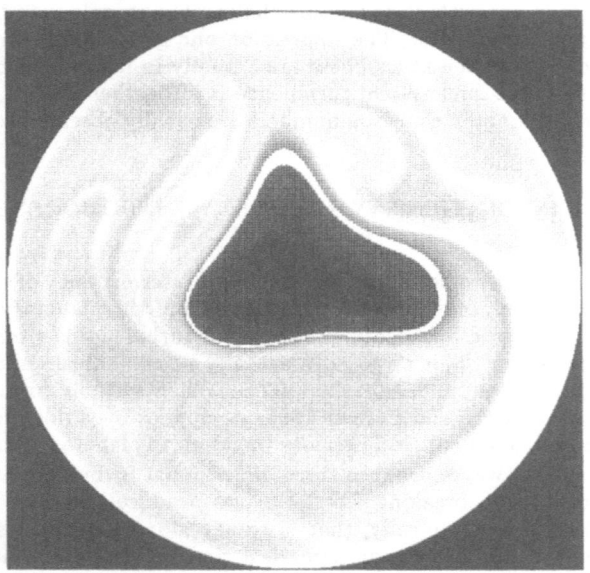

Figure 1. Grayscale representation of the PV in a hemispheric single-layer model of the strato-sphere (shallow water equations with mean equivalent depth 4 km), shown in a polar stereographic projection. The grayscale mapping is monotonic (dark cyclonic and light anticyclonic) *except* that a substantial band of PV values embedded in the main polar vortex edge is also made light, in order to make the edge, or region of steep PV gradients, more easily visible. In an animated version of this display shown at the Workshop, this white band was seen to undulate as if made of elastic, making the PV barrier effect due to the concentrated Rossby-wave restoring mechanism visually evident.

Part of the explanation for the contour's effectiveness as a dynamical barrier is that its undular motion is a case of Rossby-wave motion. As with other wavemotions there is a *restoring mechanism*, in this case a sideways restoring mechanism due to the large PV gradient concentrated in the vortex edge – somewhat analogous to the vertical, gravity-wave restoring mechanism due to the potential temperature gradient concentrated in a strong temperature-inversion layer. But that is only part of the story. The problem of why the contour presents such an effective barrier to material incursions even on the very smallest

†Also Solomon, S. (this Proceedings).

resolved scales, on which the Rossby restoring mechanism is relatively weak (Hoskins *et al.*, 1985, p.920), is in reality a highly nonlinear, highly scale-interactive problem – hence the need for very accurate, fine-resolution simulations (see JM and Juckes *et al.* 1990 for a demonstration of their numerical integrity) before claiming that this 'PV barrier' effect has really been shown to be implied by the dynamics. As already emphasized elsewhere (McIntyre 1989a), "It is possible, for all we can tell from linearized Rossby-wave theory, that material might be able to cross the putative PV barrier like water through a sieve". But in nonlinear reality this is not so: careful numerical simulations like those cited consistently show that the barrier is effective. Some very recent, multi-layer high resolution numerical experiments that add to this evidence will be described in a forthcoming paper by Haynes and Norton (1990). An independent laboratory demonstration of a PV barrier (in a single-layer system), confirming that it is a robust consequence of the fluid dynamics, has been given by Sommeria *et al.* (1989).

It hardly needs saying that the nonlinear PV barrier dynamics just described contradicts the eddy-diffusivity assumption, as it is usually applied. The transport of PV and long-lived chemical tracers is least, not greatest, where PV and chemical gradients are greatest. Of course one can to some extent mimic polar vortex isolation in a model based on the eddy-diffusivity assumption, by forcing the assumed eddy diffusivity, K_{yy} say, to be highly structured in space and time (e.g. Plumb and Mahlman 1987, Juckes 1989, Yang and Tung 1989), and in particular by forcing it to a very small value at some subpolar location (M.P. Chipperfield, J.A. Pyle, personal communication). Such devices will continue to be used until improved modelling concepts are developed. Yang and Tung (1990, this proceedings) have taken the use of spatially variable K_{yy} in height-latitude global chemical models to a considerable degree of sophistication; and this may well lead to a line of work that represents the best we can do with the K_{yy} concept. The approach depends crucially on the isentropic distributions of all the chemicals concerned being well correlated with the isentropic distributions of PV (in reality, as well as in the small-displacement *theory* used to motivate the modelling approach – which, incidentally, is the origin of the scale separation in that approach). It is clear that for the real atmosphere this does not add up to a fully justifiable procedure, and it remains to be seen how well it does in practice.

3.4. QBO DYNAMICS SUGGESTS THE EXISTENCE OF SUBTROPICAL PV BARRIERS

The observed quasi-biennial oscillation of the zonal winds in the equatorial lower strato-sphere (QBO) implies the existence of subtropical PV barriers that must, to a considerable extent, isolate the tropical region from the mid-latitude planetary wave activity of the winter hemisphere. These, too, will need to be taken into account in future chemical modelling.

The point is better appreciated by contrasting the real tropics with the tropics in the one-layer numerical experiments already cited. An *unrealistic* feature of these experiments, seen for instance in figure 1 above, is a tendency for the planetary-scale Rossby waves to break all the way to the equator. This produces unrealistically strong easterly accelerations in the tropics, about an order of magnitude greater than those observed in the easterly acceleration phase of the QBO in the real atmosphere. We are forced to the conclusion that the planetary-scale Rossby wave 'breakers' in the real atmosphere usually reach no further than the subtropics, and that the deep tropics is to a large extent isolated laterally from middle latitudes, albeit possibly not completely. Satellite observations of volcanic aerosols lend independent support to this idea (G.S. Kent, M.P. McCormick, personal communication).

Stop press: Drs P.H. Haynes and W.A. Norton (personal communication) have very recently taken single-layer modelling to a higher level of sophistication, including a promising way of relating single-layer model runs more closely to multi-layer model runs and, it is hoped, ultimately to observational data. One result has been the discovery of single-layer cases (not shown here) that appear to have a more realistic tropics. This has opened the way to some key experiments on the interaction between nonlinear eddy transport with photochemically different types of chemical constituent, which we believe will have a bearing on recent arguments about extravortical ozone depletion (Proffitt *et al.* 1989).

3.5. WHEN CAN ONE JUSTIFY THE NOTION OF EDDY DIFFUSIVITY?

Why does the classical justification of the eddy-diffusive or flux-gradient hypothesis involve assuming that the turbulence is nearly homogeneous? The reason is that there has to be a separation of spatial scales: it has to be assumed *inter alia* that the mean gradients are characterized by much larger spatial scales than the largest turbulent eddies (e.g. Batchelor and Townsend, 1956). Many examples, including the foregoing, suggest that this scale-separation assumption may often be one of the worst modeling assumptions one can make. The real atmosphere often seems to prefer a diametrically opposite state of affairs in which 'mean gradients' develop scales as small as the definition of 'mean' allows them to be. The dynamical PV barrier that permits the Antarctic ozone hole to form provides merely the most conspicuous example.

It is not just a question of fluxes generally failing to be proportional to gradients, as the existence of PV barriers shows. It is also a question of non-localness. The flux-gradient hypothesis, as usually applied in practice, carries not only the assumption that fluxes are at least roughly proportional to gradients, with a proportionality 'constant' K say (scalar or tensor), but also the assumption that K can be preassigned or computed *locally*, e.g. from some measure of local turbulent intensity. This is another way of seeing that a separation of scales is required. (In the case of molecular diffusion, molecular mean free paths or interaction distances provide the inner scale, and temperatures provide the preassigned intensity).

I do not want to overdo my critique of the eddy-diffusivity concept; and the concluding §6 will mention two naturally occurring cases where the concept does appear, in fact, to have some partial *prima facie* justifiability.

3.6. A 'HOMOGENEOUS TURBULENCE' PARADOX

It is interesting to recall how drastically the eddy-diffusivity idea can fail even for passive-tracer dispersion in a classical homogeneous-turbulence model, if the scale-separation assumption is violated by the *tracer* distribution. One might be tempted for instance to take the classical enstrophy-cascading inertial subrange model of 2D turbulence (k^{-3} energy spectrum) as a model of a stratospheric Rossby-wave 'surf zone'. A well known property of this model is the scale-independence of eddy turnaround times (e.g. the enstrophy or mean-square vorticity in an octave from wavenumber k_0 to $2k_0$ is proportional to $\int_{k_0}^{2k_0} k^2 k^{-3} dk = \ln 2$, independent of k_0). Take a tracer blob whose initial spatial distribution has characteristic scale L. The ensemble-mean concentration can certainly be expected to spread out as time goes on, the tracer in other words being turbulently transported down its gradient. But can we characterize this process by an eddy diffusivity K homogeneous in space? Certainly not, since if we could, then the time for the mean tracer distribution to double its spatial scale would $\propto \bar{K}L^2$. But this immediately contradicts the basic property of the homogeneous turbulence model, that characteristic eddy times are independent of spatial scale. The model clearly predicts, therefore, that if we repeated the thought-experiment with L half as big, then the size-doubling time would be the same as before, not four times faster.

Some other, very intriguing, paradoxes associated with the idea of eddy diffusivity are described by Kraichnan (1976; see also Corrsin 1974). I am indebted to T.G. Shepherd for drawing my attention to these two references.

4. Dynamical feedbacks on the global circulation: Rossby-wave aspects

Besides those just discussed, another basic difficulty is that Rossby-wave propagation is not easily quantified. Wavelengths are comparable to basic-state variations, phase speeds are comparable to mean flow speeds, and amplitudes are typically large, in the appropriate dimensionless sense, this of course being the reason for the relevance of the generalized 'wave breaking' concept. Unresolved questions include the longstanding question of how to quantify the forcing of stratospheric planetary-scale Rossby waves by their putative tropospheric sources. Another – an inhomogeneous-turbulence question par excellence – concerns the (perhaps highly variable) reflectivity of Rossby-wave surf zones. Such questions were aired extensively in E87 (q.v., & references therein), and still need to be answered before final warmings, QBO modulations, and other possible manifestations of planetary-wave coupling can be said to be fully understood.

One of the key challenges here will, I think, be that of understanding the Antarctic final warming. Not only is this of great potential interest in its own right, as suggested recently by Farman et al. (1988), for understanding the interannual variability being observed in ozone-hole chemistry; it also seems to me to be one of the litmus tests of our understanding of the range of purely dynamical questions just mentioned. I can do no better than to quote from an earlier paper (McIntyre 1989a), regarding scientific opportunities arising from the recently increased supply of observational information about the Antarctic:

"One of these [opportunities] might be the long-awaited chance to test theoretical ideas on vortex erosion as a self-tuning Rossby-wave resonance mechanism (Haynes 1985). Self-tuning resonance, albeit with a different tuning mechanism, was originally suggested by Plumb (1981) as a contributing factor in the dynamics of northern hemispheric stratospheric warmings. There are good theoretical grounds (Haynes, op. cit.) for believing that the poleward-downward vortex erosion suggested by Antarctic observational data for late winter and early spring (e.g. Farrara and Mechoso 1986, Shiotani and Gille 1987) will tune the vortex in such a way that the speed of long Rossby waves systematically becomes less retrograde. Diabatic heating as well as erosion is involved, of course, especially at the higher stratospheric altitudes, and so it would be more accurate to think in terms of a highly interactive combination of self-tuning and externally imposed tuning.

"Suppose, for instance, that the gravest wave-1 free mode of the vortex were to become stationary as a result of the poleward-downward attrition of the vortex. The physical implication is that the vortex would become easy to push off the pole. Any forcing mechanism tied to geography, such as tropospheric storm-track activity or blocking, could do the pushing. Numerical experiments are being planned to test this idea. [I can add that these are now under way (P.D. Clark, personal communication).] They not only promise to give us a better understanding of Antarctic final warmings but, because stratospheric warmings in the northern hemisphere pose a far more complicated modelling problem, which has yet to

12

be surmounted, the Antarctic might also turn out to give us the first convincing test of the relevance of Rossby-wave resonance in the real atmosphere."

Other updates to the discussion of E87 are:

1. Progress has been made on the fundamental theory of wave-activity conservation relations (Haynes 1988). As a result, we now know not only how quantities like the Eliassen-Palm flux are related to the underlying Hamiltonian dynamics, but also how to fit it into a coherent wave-mean interaction theory, taking account of wave dissipation. However, the details are dauntingly complicated, and there is a deeper problem of knowing what choice to make to define the basic state. It is not clear to me that any analytically simple averaging operator gives a good way of doing this; the concept of PV rearrangement, subject to the fundamental conservation properties summarized in §2.3, seems likely to be more fruitful, but there is still an enormous range of apparently arbitrary choices to be made.

2. I now think that spontaneous inertio-gravity-wave emission is not very likely to be a powerful contributor, at least not directly, to the middle-atmospheric angular momentum balance. Kelvin-Helmholtz envelope radiation (Fritts 1984) still seems to be a strong contender.

3. Signficant progress on assessing the mathematical status, and ultimate accuracy, of the PV inversion and balance concepts has been made (Egger 1990; McIntyre and Norton 1990b).

5. Further remarks on breaking gravity waves

The observed structure of breaking gravity waves both in the laboratory (e.g. Koop and McGee 1986) and in the atmosphere (e.g. Kopp et al. 1985) provides a small-scale three-dimensional example of the extreme spatial inhomogeneity that seems to characterize many naturally-occurring turbulent fluid flows. There are further small-scale examples in the book by Turner (1973). In the case of breaking gravity waves, an important consequence is that vertical mixing of tracers is often far less efficient than one might guess from the requirements of wave dissipation, or from observed turbulent dissipation rates (Chao and Schoeberl 1984, Fritts and Dunkerton 1985, Coy and Fritts 1988, McIntyre 1989b, Walterscheid and Schubert 1989). This seems true at any rate unless the waves break extremely suddenly and violently, as may occur in the winter mesosphere and the summer lower thermosphere. The latest conclusions on this problem seem to be:

1. Vertical mixing seems likely to be even more sensitive to the wave supersaturation $(a-1)$ than was deduced in earlier work; and

2. There is no such thing as 'the' eddy diffusivity for all purposes. For instance, values required to account for wave dissipation can be quite different from values that might be relevant to the vertical transport of chemical constituents. In particular, the idea of 'the' turbulent Prandtl number does not appear to be well defined.

In the gravity-wave case, it is worth noting that breaking not only mixes entropy and chemicals vertically, but also transports PV sideways, and often upgradient; a detailed analysis is given in the paper by McIntyre and Norton (1990a). This is consistent with the basic theorems of §2.3: the notional tracer substance whose mixing ratio is the PV can be, and in this case is, transported exactly along isentropic surfaces, no approximations being involved. In both the gravity and the Rossby case, and for more complicated wave types such as the various equatorial waveguide modes, the irreversible effect of breaking or otherwise-dissipating waves on PV and vorticity distributions is perhaps the most general way of describing the wave-induced mean effects associated with breakdown of the nonacceleration theorem of wave-mean interaction theory; again, the reader may consult the recent paper by McIntyre and Norton (1990a) for a detailed discussion. The high-resolution Rossby wave-breaking simulations cited in §3 provide one very clear illustration of these wave-induced mean flows.

6. Concluding remarks

Every time one looks at the evidence about real fluid motion, or pushes up the resolution in one's numerical experiments, one sees an increasingly striking departure from the conditions of near-homogeneous turbulence that comprise the classical justification of the eddy-diffusivity concept. There is also the problem of finite chemical 'mixdown time' discussed in JM, stemming from the remarks in Tuck (1979). It is difficult to escape the conclusion that we should be looking for radically new modelling concepts in order to represent correctly both the dynamics, and the interactions between dynamics and chemistry.

As already mentioned, I don't want to overdo my criticism of the eddy diffusivity concept. It may somehow manage to be qualitatively correct in some instances where it is not justifiable. There are even a few cases in which a partial justification seems possible – mainly concerning situations where it makes sense to consider only fluxes and mean states averaged over long times, and where problems like mixdown are not important. For instance Holton (1986) gives an interesting argument as to why globally averaged one-dimensional models might actually make sense as along as attention is confined to very long-lived tracers, and to global-average mixing ratios on isentropic surfaces. The requirement of very long chemical timescales (much longer than global circulation turnover times, i.e. several years) is stringent. Also, PV barriers might vitiate the argument!

There are two other naturally occurring turbulent cases I know of at present where the eddy-diffusivity concept appears to have some chance of a *prima facie* justifiability. One is the ocean gyre case, where the Rossby length is small in comparison with gyre scales and gives at least the possibility of a scale separation (Rhines 1977). But in this case it is still difficult to believe in the relevance of preassigning the eddy intensity.

The other case was recently pointed out to me by T. VanZandt, who drew my attention to an interesting paper by Woodman and Rastogi (1984) on the thin turbulent layers observed by the Arecibo radar in the tropical lower stratosphere. They argue cogently that, according to the radar evidence, there is some degree of statistical homogeneity in the altitude distribution of the occurrence of these layers. In that case one *can* get a scale separation, and hence eddy-diffusive behaviour, for long-lived tracers with smooth vertical profiles, despite the undoubtedly extreme inhomogeneity of the turbulence within each turbulent layer. The scale separation is based on the thinness of the complete turbulent layers relative to the scale height of the tracer.

In this last case, the idea of a preassigned eddy intensity might make sense also, up to a point. If one assumes that the turbulence is due to an input of inertio-gravity waves generated by topography, or for instance by spontaneous wave emission from jet streaks, with the waves subsequently breaking via Kelvin-Helmholtz instability as can be expected for such waves, then one can imagine sensible modelling experiments in which the wave input is preassigned. Such a scenario seems self-consistent in that waves that propagate vertically while breaking could well give approximately homogeneous statistics of turbulent layer occurrence, after averaging over many breaking events and detailed profile changes. It would probably be pushing the idea too far, however, to suppose that the resulting eddy diffusivity is a true diffusivity that acts most strongly to smooth the *smaller* vertical

14

scales, since the problem of statistical inhomogeneity will reappear at such scales, This is presumably why balloon and aircraft data (e.g. Proffitt and McLaughlin 1983; Danielsen *et al.* 1987, Tuck *et al.* 1989) often show fine vertical scale features, much as one might also expect from casual observations of lenticular clouds.

Modelling studies to examine such situations more closely might well be of great interest, aiming for an improved understanding of the breaking of lower-stratospheric inertio-gravity waves and the detailed way in which this interacts with the vertical profile changes due to individual breaking events, perhaps leading to a more accurate model to which to fit the radar results. Such studies might also throw some light on an unresolved disagreement between radar results like those of Woodman and Rastogi (also references therein) and an interpretation of aircraft data by Lilly *et al.* (1974). The radar results give values of the order of 0.2 to $0.3 \mathrm{m}^2\mathrm{s}^{-1}$, but the aircraft data an order of magnitude less, at $0.012 \mathrm{m}^2\mathrm{s}^{-1}$. Of course the measurements were made at different times and places, and might merely be reminding us that we should never fall into the trap of thinking in terms of 'the' eddy diffusivity.

Corresponding results might well hold also for the mesosphere, apart from the different predominant mode of wave breaking (and also, it must be presumed, in the ocean thermocline), although in the case of the mesosphere this will not take away the problem of sensitivity to supersaturation noted in §5. In the mesosphere it is interesting to note that the inefficiency of vertical mixing, which is associated with the fact that we expect these waves to be breaking more by convective overturning than by Kelvin-Helmholtz instability, can be shown to mean that the effective thickness of the turbulent layers is likely to be a great deal less than their instantaneous thickness as seen in rocket profiles such as those presented in Kopp *et al.* (1985). This is because of the geometrical nature of the distortion of the isentropic surfaces by the convectively overturning gravity waves, in relation to the phase of the waves.

Thus, in the upper mesosphere and lower thermosphere, there might also be some justification to the notion that vertical mixing of long-lived tracers could be described by an eddy diffusivity. The difficulty remains that, even if so, its value appears certain to be highly sensitive to wave parameters, particularly wave intermittency. This again has been discussed at greater length elsewhere (McIntyre 1989b). In particular, a given long-time-mean momentum flux, such as would be required to account for a given global mean circulation strength, could well be associated with very different intensities of vertical mixing, being smaller or greater, in fact, according to whether the wave flux is more steady or more intermittent.

The foregoing implies a correction to the mixdown-time estimate given in JM (penultimate section). If the suggestion of a vertical eddy diffusivity of the order of 0.2 to $0.3 \mathrm{m}^2\mathrm{s}^{-1}$ is correct, then the mixdown time of a month becomes more like a week, and the corresponding vertical scale a few hundred rather than a few tens of metres. In this connection an examination of the aircraft data from the recent airborne polar stratospheric ozone expeditions would be of the greatest interest, and comprises yet another interesting challenge, in which I should like to get involved very soon.

Acknowledgements. T.G. Shepherd kindly drew my attention to the papers by Corrsin and Kraichnan, and T. VanZandt to that by Woodman and Rastogi. I am indebted to them, and also to the following, for stimulating conversations and correspondence: D.G. Andrews, E.F. Danielsen, D.G. Dritschel, D.W. Fahey, D. Fairlie, J.C. Farman, D.C. Fritts, S.E. Gaines, W.L. Grose, P.H. Haynes, R.S. Hipskind, J.R. Holton, B.J. Hoskins, I.N. James, M.N. Juckes, D. Keyser, D.M. Murphy, A. O'Neill, W.A. Norton, T.N. Palmer, L.

Pfister, C.R. Philbrick, M.H. Proffitt, P.B. Rhines, J. Röttger, R. Rotunno, P.G. Saffman, K.P. Shine, S. Solomon, A.J. Thorpe, J. Tribbia, M.R. Schoeberl, W.H. Schubert, K.P. Shine, G.J. Shutts, A.F. Tuck, W.E. Ward, and A. A. White. Support from the UK Department of the Environment, the Meteorological Office, the Science and Engineering Research Council, the Natural Environment Research Council through the UK Universities' Global Atmospheric Modelling Project, and the US Office of Naval Research, is gratefully acknowledged.

REFERENCES

Batchelor, G.K., 1952: The effect of homogeneous turbulence on material lines and surfaces. Proc. Roy. Soc. A **213**, 349-366.

Batchelor, G.K., 1959: Small-scale variation of convected quantities like temperature in turbulent fluid. Part 1. General discussion and the case of small conductivity. J. Fluid Mech. **5**, 113-133.

Batchelor, G.K., and Townsend, A.A., 1956: Turbulent diffusion. In *Surveys in Mechanics* (G.K.Batchelor and R.M.Davies, eds.) Cambridge University Press, pp 352-399. See also Batchelor, 1950: Q. J. Roy. Meteorol. Soc. **76**, 133.

Chao, W.C., and Schoeberl, M.R., 1984: On the linear approximation of gravity wave saturation in the mesosphere. J. Atmos. Sci., **41**, 1893-1898.

Corrsin, S., 1974: Limitations of gradient transport models in random walks and in turbulence. Adv. Geophys. **18A**, 25-71.

Coy, L. and Fritts, D.C., 1988: Gravity wave heat fluxes: a Lagrangian approach. J. Atmos. Sci., **45**, 1770-1780.

Danielsen, E. F., Hipskind, R. S., Gaines, S. E., Sachse, G. W., Gregory, G. L., and Hill, G. F., 1987: Three-dimensional analysis of potential vorticity associated with tropopause folds and observed variations of ozone and carbon monoxide. J. Geophys. Res., **92**, 2103-2111.

Deem, G. S. , and Zabusky, N. J., 1978: Vortex waves: stationary "V-states", interactions, recurrence, and breaking. Phys. Rev. Lett. **40**, 859-862.

Egger, J., 1990: Some aspects of the invertibility principle. J. Atmos. Sci., in press.

Ertel, H., 1942: Ein neuer hydrodynamischer Wirbelsatz. Met. Z., **59**, 271-281.

Farman, J. C., Gardiner, B. G., and Shanklin, J. D., 1988: How deep is an 'ozone hole'? Nature, **336**, 198.

Farrara, J. D., and Mechoso, C. R., 1986: An observational study of the final warming in the southern hemisphere stratosphere. Geophys. Res. Letters, **13**, 1232-1235.

Fritts, D.C., and Dunkerton, T.J., 1985: Fluxes of heat and constituents due to convectively unstable gravity waves. J. Atmos. Sci., **42**, 549-556.

Fritts, D. C., 1984: Shear excitation of atmospheric gravity waves. Part II: Nonlinear radiation from a free shear layer. J. Atmos. Sci., **41**, 524-537.

Garcia, R. R., and Solomon, S., 1985: The effect of breaking gravity waves on the dynamics and chemical composition of the mesosphere and lower thermosphere. J. Geophys. Res. **90 D**, 3850-3868.

Haynes, P. H., 1985: A new model of resonance in the winter stratosphere. Handbook for MAP, vol. 18: Extended abstracts of papers presented at the 1984 MAP Sympo-

Haynes, P. H., 1988: J. Atmos. Sci., **45**, 2352-2362.

sium, Kyoto. S. Kato, ed., 126-131. Available from SCOSTEP secretariat, University of Illinois, 1406 W. Green St, Urbana, Ill. 61801, U.S.A.

Haynes, P.H. and McIntyre, M.E., 1987: On the evolution of vorticity and potential vorticity in the presence of diabatic heating and frictional or other forces. J. Atmos. Sci. 44, 828-841.

Haynes, P.H., and McIntyre, M.E., 1990: On the conservation and impermeability theorems for potential vorticity. J. Atmos. Sci., 47, in press for June 1990.

Haynes, P. H., Marks, C. J., McIntyre, M. E., Shepherd, T. G., and Shine, K. P., 1990: On the downward control of extratropical diabatic circulations by eddy-induced mean zonal forces. J. Atmos. Sci., 47, in press.

Haynes, P.H., and Norton, W. A., 1990. On mass and chemical tracer transports through the edge of a stratospheric polar vortex. Manuscript in preparation.

Hitchman, M. H., Gille, J. C., Rodgers, C. D., and Brasseur, G., 1989: The separated polar winter stratopause: a gravity wave driven climatological feature. J. Atmos. Sci., 46, 410-422.

Holton, J.R., 1986: A dynamically based transport parameterization for one-dimensional photochemical models of the stratosphere. J. Geophys. Res., 91 D, 2681-2686.

Hoskins, B.J., McIntyre, M.E. and Robertson, A.W. 1985: On the use and significance of isentropic potential-vorticity maps. Q. J. Roy. Meteorol. Soc. 111, 877-946. Also 113, 402-404.

Juckes, M.N. and McIntyre, M.E., 1987: A high resolution, one-layer model of breaking planetary waves in the stratosphere. Nature, 328, 590-596. [Referred to as 'JM'.]

Juckes, M. N., McIntyre, M. E., and Norton, W. A., 1990: High-resolution barotropic simulations of breaking stratospheric planetary waves and polar-vortex edge formation. Q. J. Roy. Meteorol. Soc., to be submitted.

Juckes, M. N., 1989: A shallow water model of the winter stratosphere. J. Atmos. Sci., 46, 2934-2955.

Kanzawa, H., 1989: Warm stratopause in the Antarctic winter. J. Atmos. Sci., 46, 435-438.

Keyser, D., and R. Rotunno, 1990: On the formation of potential-vorticity anomalies in upper-level jet front systems. Mon. Wea. Rev., submitted.

Koop, C. G., and B. McGee, 1986: Measurements of internal gravity waves in a continuously stratified shear flow. J. Fluid Mech., 172, 453-480.

Kopp, E., Bertin, F., Björn, L.G., Dickinson, P.H.G., Philbrick, C.R. and Witt, G., 1985: The 'CAMP' campaign 1982. Proc. 7th ESA Symp. on European Rocket & Balloon Programmes & Related Res., (Paris, Europ. Space Agency SP-229, July 1985).

Kraichnan, R.H., 1976: Eddy viscosity in two and three dimensions. J. Atmos. Sci., 33, 1521-1536.

McIntyre, M.E., 1987: Dynamics and tracer transport in the middle atmosphere: an overview of some recent developments. In Transport Processes in the Middle Atmosphere, ed. G. Visconti and R.R. Garcia. Dordrecht, Reidel, 267-296. (Proc. NATO workshop, Erice.)

McIntyre, M.E., 1989a: On the Antarctic ozone hole. J. Atmos. Terrest. Phys., 51, 29-43. [See Haynes et al., 1990, for a correction regarding the 'downward control' principle]

McIntyre, M. E., 1989b: On dynamics and transport near the polar mesopause in summer. J. Geophys. Res., 94, 14617-14628. [Special Issue for the International Workshop on

Noctilucent Clouds.]

McIntyre, M.E. and Palmer, T.N. 1985: A note on the general concept of wave breaking for Rossby and gravity waves. Pure Appl. Geophys., 123, 964-975.

McIntyre, M.E. and Norton, W.A., 1990a: Dissipative wave-mean interactions and the transport of vorticity or potential vorticity. J. Fluid Mech., 212, 403-435 (G.K. Batchelor Festschrift Issue).

McIntyre, M. E., and Norton, W. A., 1990b: Potential-vorticity inversion on a hemisphere. J. Atmos. Sci., to be submitted.

Obukhov, A.M., 1962: On the dynamics of a stratified liquid. Dokl. Akad. Nauk SSSR, 145 (6), 1239-1242. English transl. in Soviet Physics – Doklady, 7, 682-684.

O'Neill, A., and Pope, V. D., 1988: Simulations of linear and nonlinear disturbances in the stratosphere. Q. J. Roy. Met. Soc., 114, 1063-1110.

Proffitt, M. H., and McLaughlin, R. J., 1983: Fast-response dual-beam UV-absorption ozone photometer suitable for use on stratospheric balloons. Rev. Sci. Instrum., 54, 1719-1728.

Proffitt, M. H., Fahey, D. W., Kelly, K. K. and Tuck, A. F., 1989: High latitude ozone loss outside the Antarctic ozone hole. Nature, 342, 233-237.

Plumb, R. A., 1981: Instability of the distorted polar night vortex: a theory of stratospheric warmings. J. Atmos. Sci., 38, 2514-2531.

Plumb, R.A. and Mahlman, J.D. 1987: The zonally averaged transport characteristics of the GFDL general circulation/transport model. J. Atmos. Sci., 44, 298-327.

Reiter, E.R., 1975: Stratospheric-tropospheric exchange processes. Revs. Geophys. Space Phys, 13, 459-474.

Rhines, P.B., 1977: The dynamics of unsteady currents. In The Sea, vol. 6 (E.D. Goldberg et al., eds.), 189-318. New York: Wiley.

Rossby, C. G., 1940: Planetary flows in the atmosphere. Q. J. Roy. Meteorol. Soc., 66, Suppl., 68-87.

Salby, M.L., Garcia, R.R., O'Sullivan, D., and Tribbia, J., 1989: Global transport calculations with an equivalent barotropic system. J. Atmos. Sci., in press.

Schröder, W., 1988: An additional bibliographical sketch on the development of Ertel's potential vorticity theorem. Q. J. Roy. Meteorol. Soc., 114, 1563-1567.

Shiotani, M., and Gille, J.C., 1987: Dynamical factors affecting ozone mixing ratios in the Antarctic lower stratosphere. J. Geophys. Res., 92 D, 9811-9824.

Sommeria, J., Meyers, S.D., and Swinney, H.L., 1989: Laboratory model of a planetary eastward jet. Nature, 337, 58-61.

Thomas, R. J., Barth, C. A., and Solomon, S., 1984: Seasonal variations of ozone in the upper mesosphere and gravity waves. Geophys. Res. Lett. 11, 673-676.

Thorpe, A. J., and Emanuel, K. A., 1985: Frontogenesis in the presence of small stability to slantwise convection. J. Atmos. Sci. 42, 1809-1824.

Tuck, A.F., 1979: A comparison of one-, two-, and three-dimensional model representations of stratospheric gases. Phil. Trans. Roy. Soc. Lond. A 290, 477-494.

Tuck, A. F., Watson, R. T., Condon, E. P., Margitan, J. J., and Toon, O. B., 1989: The planning and execution of ER-2 and DC-8 aircraft flights over Antarctica, August and September 1987. J. Geophys, Res., 94, 11181-11222. (The first of two special issues on the Airborne Antarctic Ozone Experiment; see also the other papers therein.)

Turner, J.S. 1973: Buoyancy effects in fluids. Cambridge University Press, 367pp.

Walterscheid, R.L. and Schubert, G., 1989: Nonlinear evolution of an upward propagating gravity wave: overturning, convection and turbulence. J. Atmos. Sci., in press.

WMO 1985: Atmospheric ozone 1985: Assessment of our understanding of the processes controlling its present distribution and change. Geneva, World Meteorological Organization, Global Ozone Research and Monitoring Report No. 16. Available from Global Ozone Research and Monitoring Project, World Meteorological Organization, Case Postale 5, CH 1211, Geneva 20, Switzerland. In 3 volumes, 1095pp + 86pp refs. See chap. 6.

Woodman, R. F. and Rastogi, P. K., 1984: Evaluation of effective eddy diffusive coefficients using radar observations of turbulence in the stratosphere. Geophys. Res. Lett., 11, 243-246.

Yang, H., and Tung, K.K., 1990: In *Dynamics, Transport and Photochemistry in the Middle Atmosphere* (Proc. San Francisco NATO Workshop on the Middle Atmosphere in the Southern Hemisphere), ed. A. O'Neill. Dordrecht, Kluwer (this Proceedings).

ON DATA SOURCES AND QUALITY FOR THE SOUTHERN HEMISPHERE
STRATOSPHERE

DAVID J. KAROLY*
Atmospheric and Oceanic Sciences Program
Princeton University
P. O. Box 308
Princeton, NJ 08542
U.S.A.

DENISE S. GRAVES
AT&T Bell Laboratories
Whippany, NJ 07981
U.S.A.

ABSTRACT. A brief review is presented of current operational analysis systems for the Southern
Hemisphere stratosphere, concentrating on the systems used at the National Meteorological Center
(USA) and the British Meteorological Office. An assessment is made of two major sources of error
in these current analyses; the tropospheric analyses used at the base-level and the thickness analyses
derived from satellite radiance data. Some results on the impact of different base-level analyses on
derived stratospheric circulation statistics in the Southern Hemisphere are described. The reliability
of analyses obtained from satellite data is assessed by sampling a numerical model simulation of
the stratosphere as if by satellite and comparing the sampled statistics with those from the model.
Both these areas are shown to lead to problems with the circulation statistics at high latitudes and
for differentiated quantities.

A possible improved stratospheric analysis system is described, based on modern data assimilation
and analysis systems used in the troposphere and using the operational system at the European
Centre for Medium Range Weather Forecasts as an example.

1. INTRODUCTION

Operational geopotential height analyses of the stratosphere in the Southern Hemisphere
(SH) have been made at several centres since about 1979. These analyses are possible
because of the availability of satellite observations of the SH stratosphere and the advent of
global tropospheric analysis systems during and after 1979, the year of the Global Weather
Experiment. These analyses have been used in a number of studies which have refined the
climatology of the SH stratosphere and led to an improved understanding of the differences
in the stratosphere between the two hemispheres.

---

* Permanent affiliation: *Centre for Dynamical Meteorology, Monash University,*
Clayton, Vic. 3168, Australia

A. O'Neill (ed.),
Dynamics, Transport and Photochemistry in the Middle Atmosphere of the Southern Hemisphere, 19–32.
© 1990 *Kluwer Academic Publishers.*

These height analyses have generally been prepared in a similar way at the different centres. The height field is built up from a base-level analysis for a level in the lower stratosphere, obtained from an operational tropospheric analysis system, and inter-level thickness analyses derived from satellite radiance data for the stratosphere. This method is described in more detail in the next section, using the operational SH stratospheric analysis systems at the National Meteorological Center (NMC) in the United States and the British Meteorological Office (BMO) as examples.

A comparison of satellite-derived dynamical quantities for the SH stratosphere was made at a MASH Workshop in Williamsburg in April, 1986 and the results are described by Miles and O'Neill (1989). The reliability of satellite-based temperature analyses in the SH from different instruments was considered as well as the impact of different base-level analyses. The horizontal and vertical resolution of the satellite instruments limits the reliability of the temperature analyses but they have similar accuracy in the SH as in the NH. Any errors in the temperatures can lead to larger errors in derived statistics. In the SH, there are larger differences between tropospheric analyses from different centres than in the NH. This leads to differences in base-level analyses and some sensitivity of the derived statistics to the particular base-level analysis used.

The two independent sources of information in stratospheric height analyses; the base-level analysis and the satellite-based thickness analysis, both may lead to errors. In section 3, the quality of operational SH tropospheric analyses and the impact of base-level analyses on stratospheric circulation statistics are discussed. In section 4, some results are presented from a study of the reliability of satellite-based analyses using a numerical model of the stratosphere.

A possible improved future analysis system is described in section 5, making use of data assimilation techniques which are used in operational tropospheric analyses. The analysis system at the European Centre for Medium Range Weather Forecasts (ECMWF) is used as an example of such a system.

2. CONVENTIONAL STRATOSPHERIC ANALYSIS SYSTEMS

Operational analyses of geopotential height in the SH stratosphere have been made routinely at several centres since about 1979 using similar analysis techniques. These involve building up the height fields from a base-level height analysis in the lower stratosphere, generally obtained from an operational tropospheric analysis system, and inter-level thickness analyses derived from stratospheric radiance data.

The operational stratospheric analyses from NMC and BMO will be used as examples of this type of conventional stratospheric analysis system. Both centres use radiance data from polar-orbiting satellites from the TOVS (TIROS-N Operational Vertical Sounder) observing system, which consists of the HIRS (High Resolution Vertical Sounder), MSU (Microwave Sounding Unit) and SSU (Stratospheric Sounding Unit) multi-channel radiometers. Each of these instruments has several channels in the stratosphere which sample the radiance over a deep layer (Smith et al., 1979). These channels have fairly coarse vertical resolution of around 10 km or more. The TOVS soundings use side-scanning as well as nadir viewing so the horizontal coverage is nearly global (about 87°S - 87°N) with resolution of around 5°. Since these instruments are mounted on polar-orbiting satellites, the zonal resolution is limited by the mapping of the asynoptic radiance data for an analysis at a specified time. Thus, the effective zonal resolution of the analyses is only about 30°. The TOVS radiance data is used in a different way by NMC and BMO to produce stratospheric thickness analyses.

2.1   NMC Stratospheric Analyses

The NMC stratospheric analysis system for the SH has been described by Geller et al. (1981) and Gelman et al. (1986). These analyses have been prepared operationally in the Climate Analysis Center of NMC, separately from the operational tropospheric analyses, since September, 1978. The base-level analysis is the 100 hPa height field from the operational tropospheric analyses at NMC. Temperatures at 40 pressure levels prepared at the National Environmetal Satellite Data Information Service (NESDIS) from TOVS radiances are used to obtain layer-mean temperatures for layers in the stratosphere. NESDIS uses a statistical inversion technique to obtain vertical profiles of temperature from the satellite radiance data. The layer-mean temperatures are used to obtain inter-level thicknesses and then heights at the pressure levels 70, 50, 30, 10, 5, 2, 1 and 0.4 hPa. Temperatures at these levels are obtained by linear interpolation of adjacent layer-mean temperatures. Since October, 1980, these NMC stratospheric analyses have used only TOVS temperature data at all SH analysis levels and have not used any of the SH radiosonde data which are available operationally in the lower stratosphere from a limited number of stations.

2.2   BMO Stratospheric Analyses

The BMO stratospheric analyses for the SH have been prepared by their stratospheric group and are available from 1979. TOVS radiance data from NESDIS are converted at the BMO directly to deep layer-mean thicknesses using a statistical regression procedure for the layers 100-20, 100-10, 100-5, 100-2 and 100-1 hPa (Clough et al., 1985). These thickness profiles are used to build up height analyses on a 5°× 5° grid at these levels in the stratosphere using an operational 100 hPa base-level height analysis. Temperatures are derived hydrostatically by interpolation from the thickness profiles. Operational analyses from several tropospheric systems have been used for the heights at the base-level, 100 hPa, and at 50 hPa. They were obtained from ECMWF during 1979, NMC during 1980-83 and BMO since 1984.

These conventional analysis systems for the stratosphere provide gridpoint fields of height and temperature at a number of pressure levels. The horizontal wind field can be derived from the height analyses using geostrophic, gradient or balance relationships.

There are two sources of error associated with these conventional stratospheric analysis systems, the base-level analyses and the satellite data. Each of these areas and possible errors associated with them are discussed in the next two sections.

3.   BASE-LEVEL ANALYSES

In the conventional stratospheric analysis systems described above, the base-level analyses are obtained from operational analyses for a level in the lower stratosphere (usually 100 hPa) from tropospheric analysis systems. The different centres have local constraints on their tropospheric analyses, such as data cut-off times, availability of synoptic and asynoptic data and forecast model first-guess fields, which lead to differences in the analyses. There have been marked improvements in global tropospheric data assimilation and analysis systems since 1979, the year of the Global Weather Experiment (Bengtsson and Shukla, 1988). However, the SH tropospheric analyses are generally regarded as being of poorer quality compared with NH analyses (Hollingsworth et al., 1985; Trenberth and Olson, 1988). This is associated with the very sparse and irregular synoptic data network in the SH, communication problems for data from Antarctica and forecast model problems in the region around Antarctica. Thus, there are still large differences between tropospheric analyses for the SH at times (Trenberth and Olson, 1988). Changes in the operational analysis systems at NMC and ECMWF and their impacts on the analyses have been described recently by Trenberth and Olson (1988) and references therein.

(a) $\bar{z}$, NMC

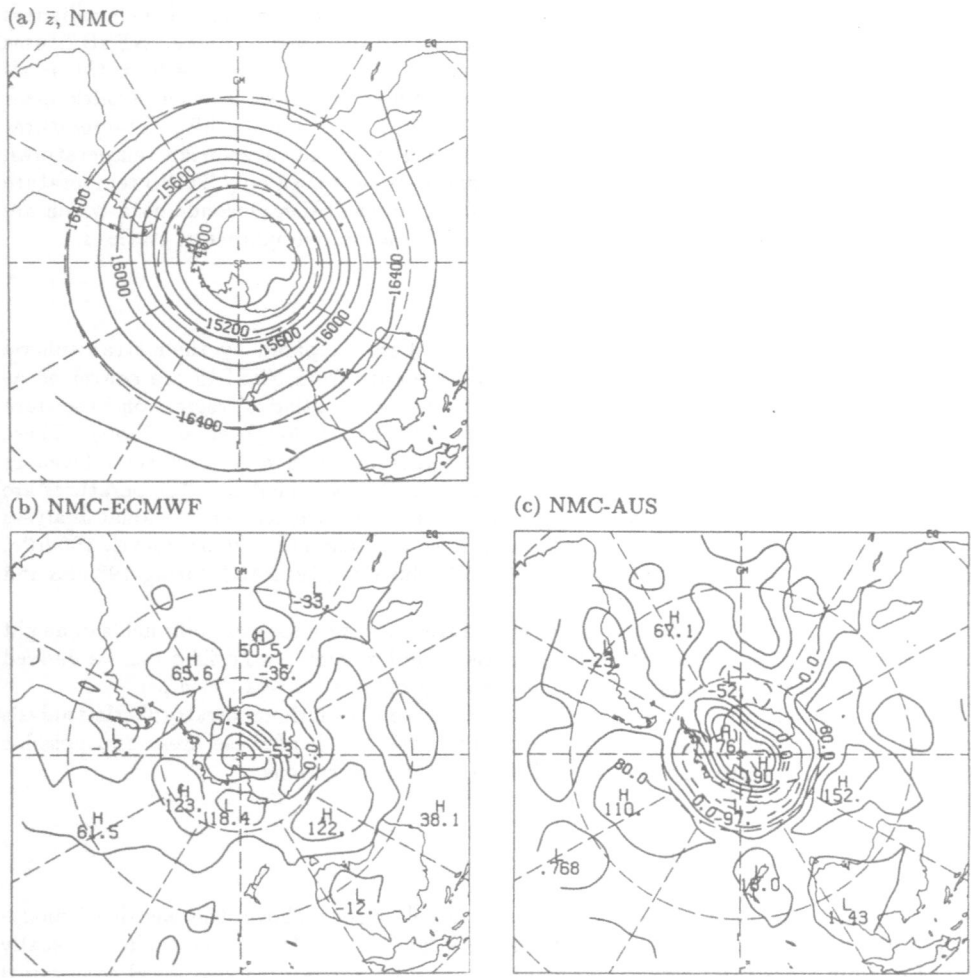

(b) NMC-ECMWF

(c) NMC-AUS

Figure 1. (a) Monthly mean height (in m) at 100 hPa from the NMC analyses for September, 1981 and the differences from the other analyses; (b) NMC - ECMWF and (c) NMC - AUS. Negative contours are dashed. (From Karoly, 1989).

Any differences between the base-level analyses from different centres will lead to differences in the stratospheric analyses and uncertainties in the circulation statistics. The impact of these differences between base-level analyses on derived stratospheric circulation statistics in the SH has been considered by Karoly (1989) and Grose and O'Neill (1989). To illustrate this impact, some of the results from Karoly (1989) are presented here.

Three sets of daily SH height analyses at 100 hPa for September, 1981 from NMC, ECMWF and the Australian Bureau of Meteorology (AUS) have been used as base-level analyses with a single set of satellite-based inter-level thickness analyses for the SH stratosphere from the BMO. Identical horizontal resolutions were used for the base-level analyses and the same techniques were used for computing stratospheric circulation statistics. Geostrophic winds derived from the height fields were used for calculating heat and momentum fluxes. All differences in these statistics arose from differences in the base-level analyses. Further details can be found in Karoly (1989). Comparisons of mean fields, standing eddy and transient eddy statistics, Eliassen-Palm diagnostics and vorticity fields were carried out but only a few of these will be presented here.

The differences of the monthly mean fields for September, 1981 between the base-level analyses (shown in Fig. 1) are of order 100 m, with the largest differences at high latitudes. This illustrates some of the problems with the SH analyses in the data-sparse regions of the SH. Since the height fields in the stratosphere are built up from the base-level analyses, these height differences are the same at all levels (barotropic). Differences between the second-order statistics obtained using the different base level analyses are not barotropic and are shown in Figs. 2 and 3. The standing eddy statistics in Fig. 2 show that there were large amplitude standing waves in September, 1981 in the SH. The differences are relatively small for the standing wave amplitudes and meridional heat flux but are larger for the momentum flux, for which there are differences of both zonal and meridional wind arising from the different base-level analyses. The monthly mean daily transient eddy statistics in Fig. 3 show that the transient eddy amplitudes were about 50% larger than the standing wave amplitudes. As for the standing eddy statistics, there are larger differences for the momentum flux than for the heat flux and the largest differences are at high latitudes.

These results indicate that the use of different base-level analyses does not affect the qualitative description of these monthly stratospheric circulation statistics. However, there are quantitative differences, which are important for derived quantities such as the Eliassen-Palm flux divergence. Also, the differences due to the base-level analyses are larger when considering individual days, particularly for the derived quantities (Karoly, 1989, Grose and O'Neill, 1989).

There are also systematic biases between the analyses from the different centres, as shown by the differences of the monthly mean 100 hPa analyses in Fig. 1. The systematic differences are largest over the data sparse regions of the SH at high latitudes and depend on the first-guess fields used for the operational analyses. In general, these systematic differences have decreased with time. Trenberth and Olson (1988) have compared the ECMWF and NMC analyses with data from the South Pole radiosonde station for the period 1979-86. The mean NMC bias of the 200 hPa height at the South Pole was more than 300 m in the winter months of 1983-85 but decreased to less than 50 m after May, 1986. In the ECMWF analyses, mean biases of more than -200 m occurred in the spring months of 1980-81 but there was a marked improvement in 1982, with biases of 50 m or less after this. It is likely that these differences arise because of communication problems with Antarctica so that the station data were not available in time for the operational analyses.

24

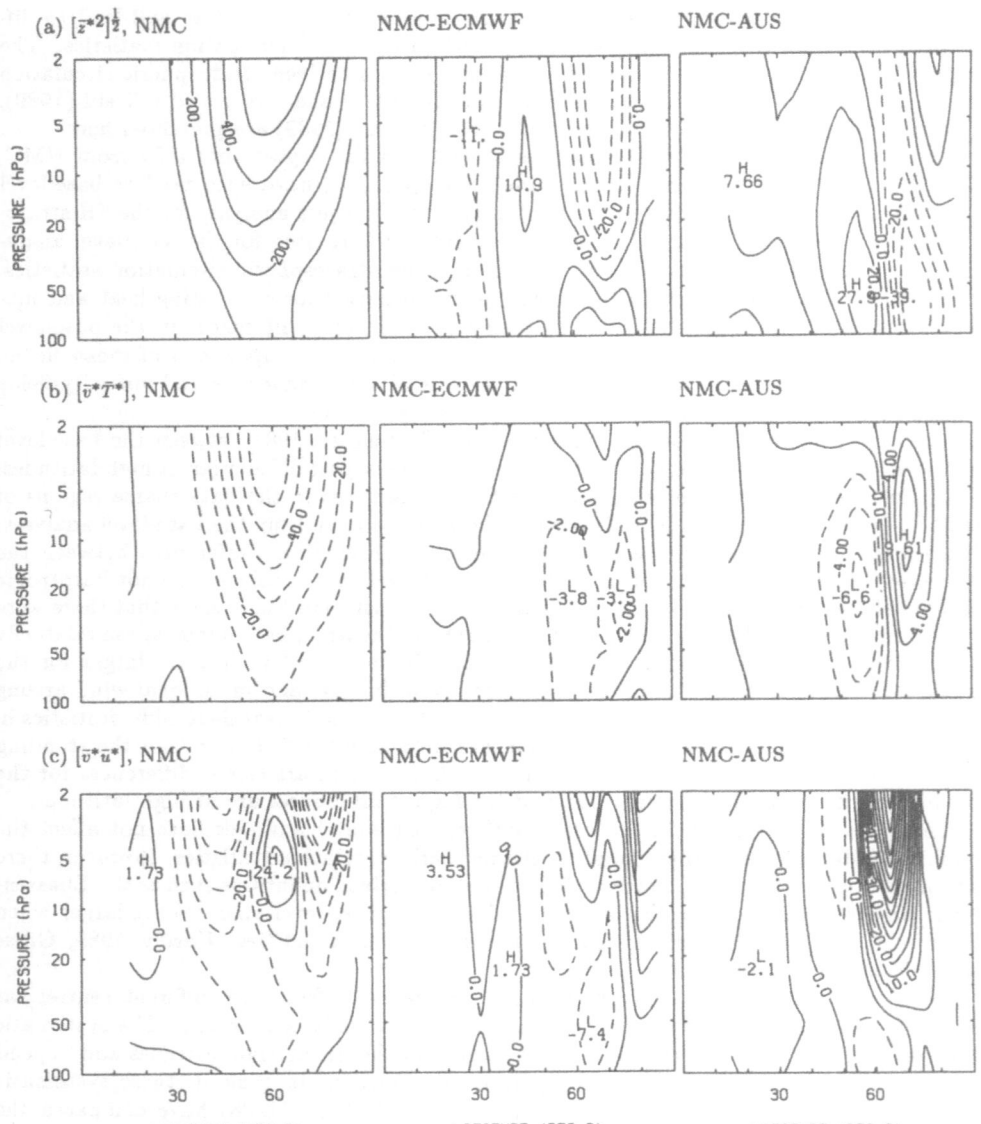

Figure 2. Zonal mean standing eddy statistics for September, 1981:
(a) root-mean-square zonal variations of height (in m),
(b) heat flux (in °Cms^{-1}) and
(c) momentum flux (in m^{2}s^{-2}).
The left panel is using the monthly mean NMC base-level analyses and the centre and right panels are the differences of the statistics, NMC - ECMWF and NMC - AUS, respectively. (From Karoly, 1989).

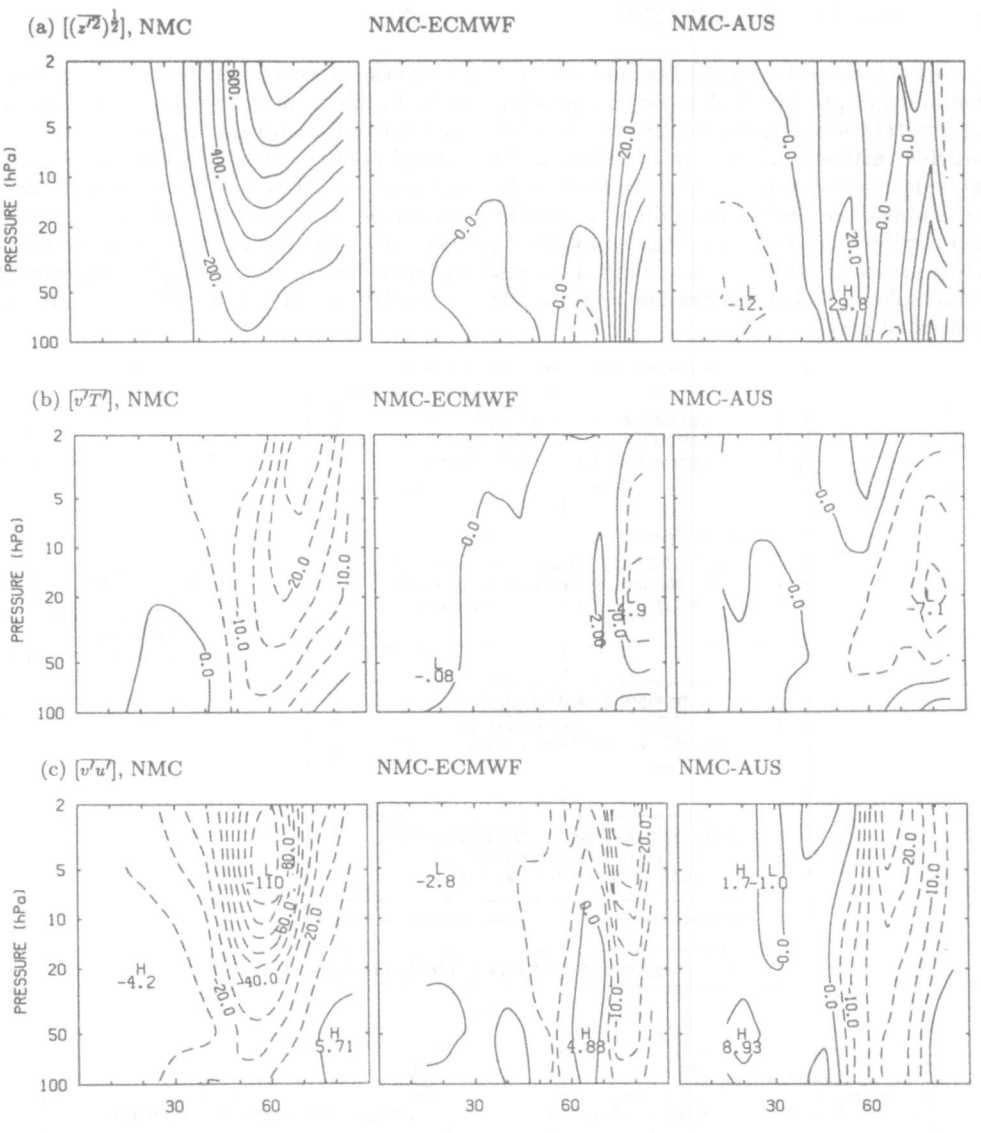

Figure 3. Zonal mean daily transient eddy statistics for September, 1981:
(a) Root-mean-square daily variations of height (in m),
(b) heat flux (in °Cms^{-1}) and
(c) momentum flux (in m^{2}s^{-2}).
The left panel is the monthly mean using the NMC base-level analyses and the centre
and right panels are the differences of the statistics, NMC - ECMWF and NMC - AUS,
respectively. (From Karoly, 1989).

4. SATELLITE DATA ANALYSIS

Errors in stratospheric analyses may also arise through the inversion and analysis of satellite radiances to provide thickness or temperature fields. Problems may be associated with the horizontal and vertical resolution of the satellite radiances, the inversion of the radiances to provide vertical profiles of temperature and the mapping of the asynoptic satellite data to a single analysis time. In general, these problems have been addressed by intercomparison of temperature profiles from different satellite instruments, radiosondes and rocketsondes, as described for the SH by Miles and O'Neill (1989). An alternative approach is to evaluate these problems of satellite sampling of the stratosphere using simulated sampling of a model stratosphere. This approach has been used by Graves (1986) and a few of those results are presented here.

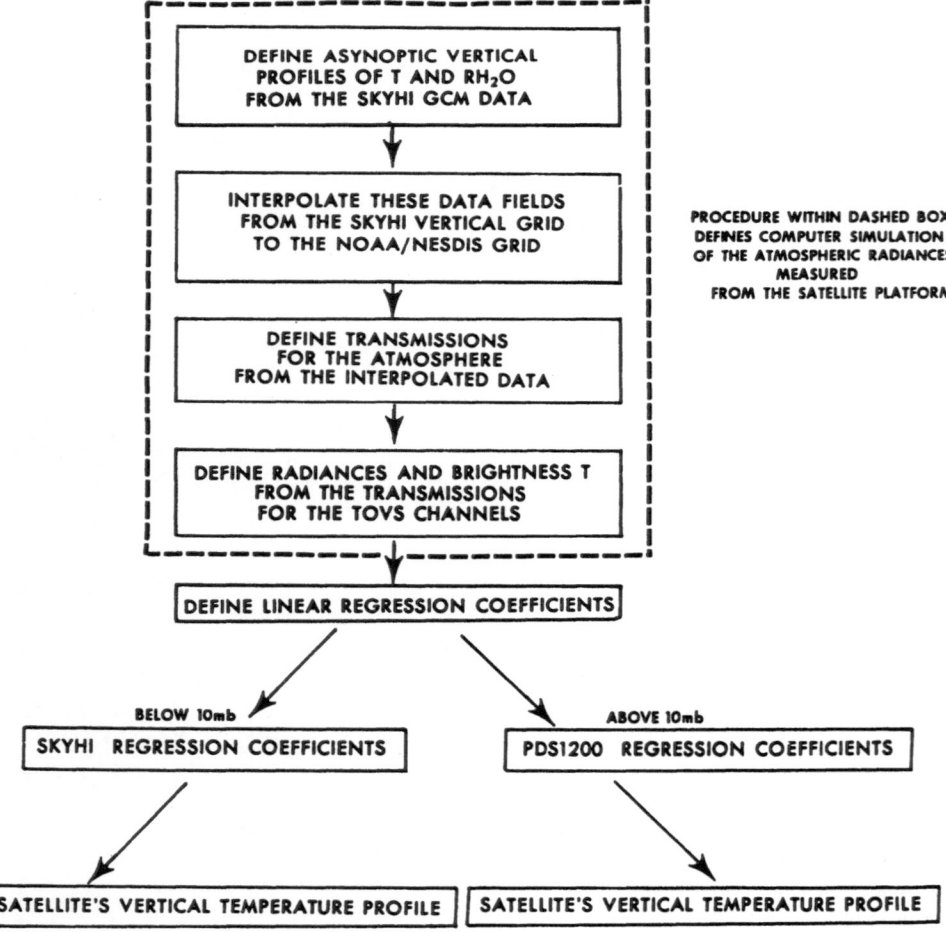

Figure 4. Flow chart of the procedure for obtaining satellite-sampled temperature profiles from the GFDL SKYHI general circulation model. (From Graves, 1986)

Graves (1986) used the GFDL SKYHI general circulation model (GCM) of the atmosphere as a representation of the "true" stratospheric circulation. The version of the SKYHI GCM used had horizontal resolution of 3° of latitude by 3.6° of longitude (N30 version) with 40 vertical levels from the surface to about 80 km (Fels et al., 1980, Andrews et al., 1983). The GCM provides a complete, consistent set of global data with known spatial and temporal resolution. The model atmosphere was sampled using a technique which matched the satellite sampling. Asynoptic sampled temperatures from the NH of the model for January were used to obtain transmittances and radiances corresponding to the TOVS radiometers. These radiances were then inverted using the statistical regression procedure from NESDIS to provide "satellite-based" vertical temperature profiles. This procedure is outlined by the flowchart in Fig. 4. Height fields in the stratosphere were built up from the model base-level field at 100 hPa and inter-level thickness analyses obtained from the temperature profiles, as at NMC. Circulation statistics computed directly from the model fields were compared with those obtained from the height analyses based on the satellite sampling.

The GCM contained a quiescent wave period during early January in the NH and a sudden warming in the second half of January. The results here will consider only the first half of January when the NH wave amplitudes were comparable to those in the SH late winter. Graves did not present any results for simulated satellite sampling of the SH of the GCM. Fig. 5 shows the temperature field from the GCM on 1 January at 5 hPa and that obtained from the satellite sampling of the model. There is a westward phase lag of about 15° longitude for the satellite temperature analysis but, overall, a similar pattern, apart from the smaller scale features in the model temperature field. Vertical profiles of temperature from the model and the satellite sampling are shown in Fig. 6 for longitude 46° W at latitudes 19.5° N and 55.5° N. The differences are typically of order 5 K except in the mesosphere, above 1 hPa, and near the tropopause, around 100 hPa, where the

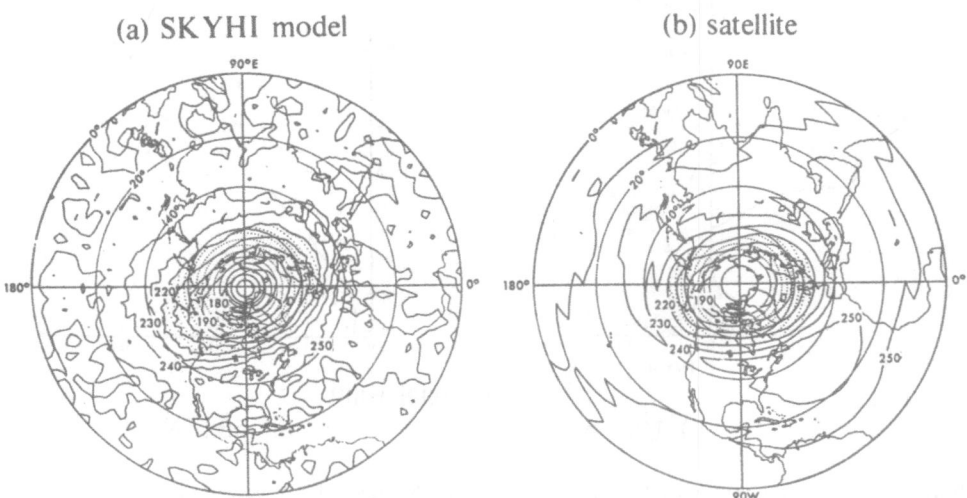

Figure 5. Temperature (in K) at 5 hPa on 1 January from (a) SKYHI model ("truth") and (b) satellite sampling of the model. (From Graves, 1986).

28

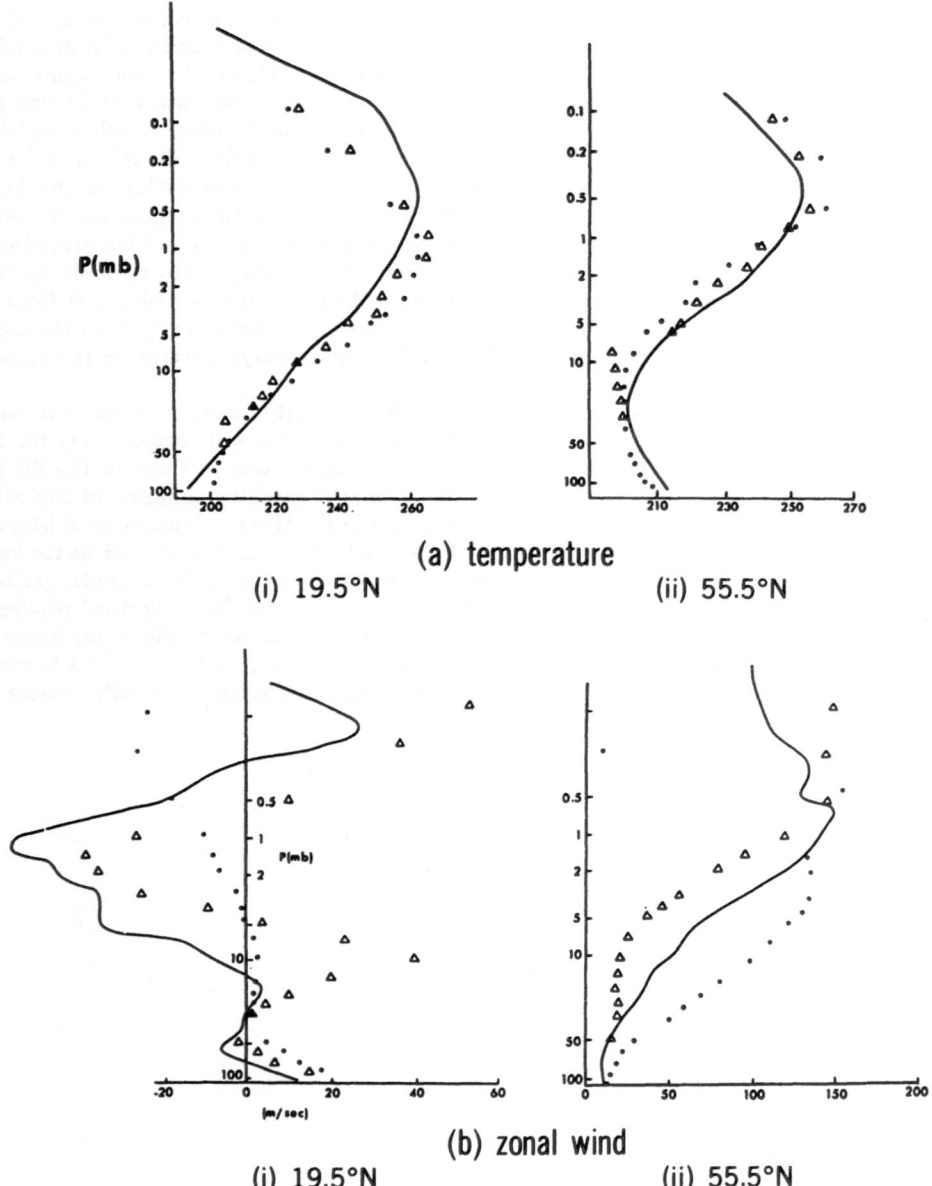

(a) temperature

(i) 19.5°N   (ii) 55.5°N

(b) zonal wind

(i) 19.5°N   (ii) 55.5°N

Figure 6. Vertical profiles of (a) temperature (in K) and (b) zonal wind (in ms^{-1}) on 1 January at 46°W and (i) 19.5°N (on the left) and (ii) 55.5°N (on the right). The solid line is from the SKYHI model and the dots and triangles are two versions of the satellite retrievals. (From Graves, 1986)

coarse vertical resolution of the satellite radiances prevents representing the temperature variations accurately. The phase lag shown in Fig. 5 may have contibuted to some of these temperature differences. Much larger differences are found between the gradient wind computed from the satellite height fields and the model winds, as shown by the vertical profiles of zonal wind in Fig. 6. The zonal wind differences typically increase with height and are larger than 30 ms^{-1} in much of the upper stratosphere. The large differences of the temperatures and winds above 1 hPa are due to the coarse vertical resolution of the highest SSU channel, which has its maximum response at 2 hPa (Smith et al., 1979).

The net effect of these differences between the satellite-sampled data and that from the model is summarised by the eddy meridional fluxes of momentum and heat shown in Fig. 7 for the first half of January. The satellite data correctly represents the sign and latitude of the maximum momentum flux, but exceeds the model amplitude by 25% and places the maximum at a much higher level. For the heat flux, again the satellite data correctly represents the sign and the latitude of the maximum flux and places the location of the maximum at a higher level, but now the magnitude is 25% smaller than in the model. The largest errors occur above 1 hPa, where the satellite data are unreliable. The Eliassen-Palm flux divergence is a very sensitive diagnostic obtained from differentiation of the momentum and heat fluxes. Comparison of this diagnostic between the model and satellite data shows large differences arising from the differences in the eddy fluxes.

In summary, during this quiescent period of the model stratosphere, the satellite estimates of the basic meteorological variables such as temperature and height are quite accurate, within 5% of the model values, in the stratosphere below 1 hPa. The amplitudes of the planetary scale disturbances are represented well, although there is a phase lag of the satellite estimates behind the model. However, for differentiated quantities such as the wind field or the Eliassen-Palm flux divergence, the differences are much larger. The differences between the satellite fields and those from the model are smaller the more averaging is involved in computing the field, either zonal- or time-averaging. This suggests that a substantial part of the errors associated with the satellite sampling involve the representation of the transient waves. A similar conclusion can be made about the errors due to the different base level analyses.

5. FUTURE STRATOSPHERIC ANALYSIS SYSTEMS

Many of the problems that have been described above arising from the base-level analyses and the satellite data analysis may be alleviated by the use of a modern data assimilation system. This type of system uses a numerical model to provide the first-guess for the analysis, allows for the insertion of data from different sources with different error variances and ensures dynamical consistency between the fields. Data assimilation and analysis systems have become normal for tropospheric analyses but there have been limited applications in the stratosphere (Takano et al., 1987, Rood et al., 1989).

Errors in the base-level analyses are often associated with relatively small-scale features in the upper troposphere which cannot be resolved by the satellite radiances. Hence, they lead to erroneous signals throughout the stratosphere arising from errors in the base-level. A problem associated with temperature retrievals from operational nadir-sounding satellites is their poor representation of features at the tropopause and in the mesosphere due to coarse vertical resolution. The statistical regression methods which are used to invert the

30

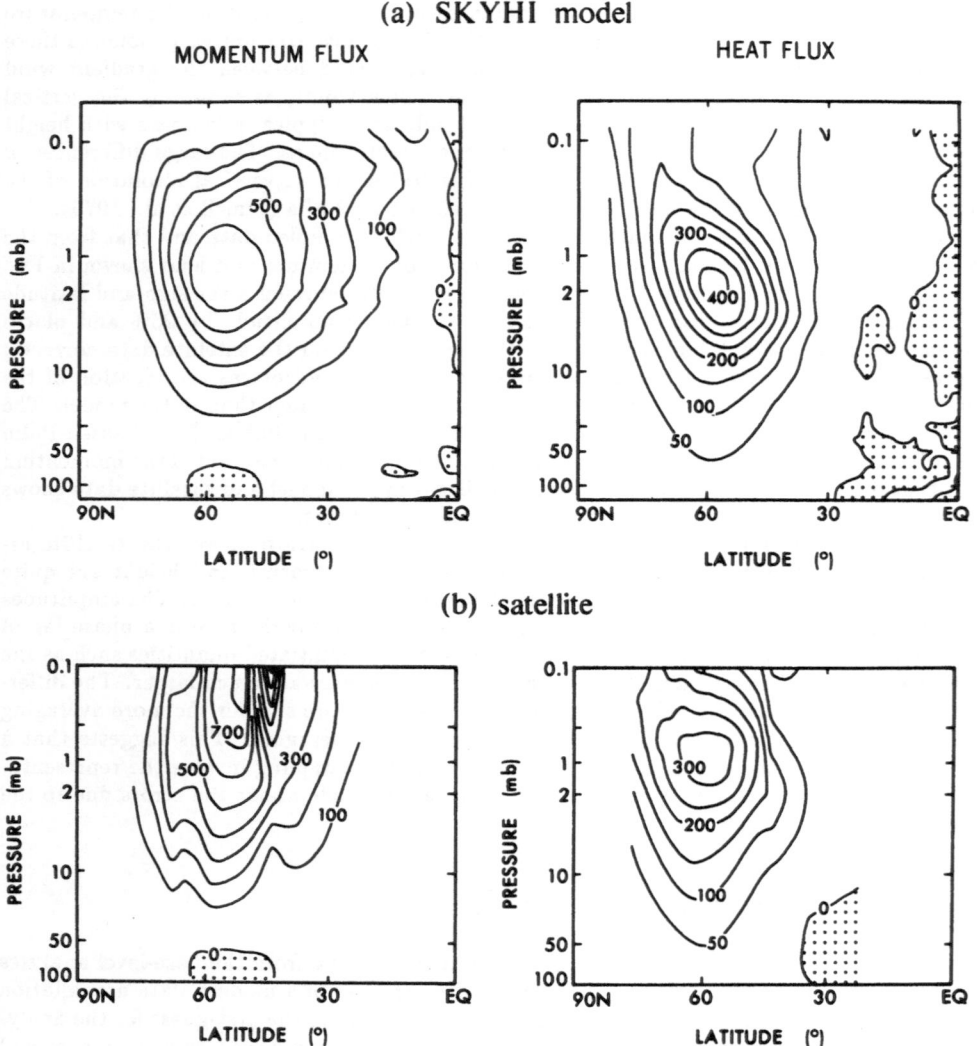

Figure 7. Eddy meridional flux of momentum (in m²s⁻², on the left) and heat (in Kms⁻¹, on the right) for the period 1-14 January from (a) SKYHI model and (b) satellite sampling. (From Graves, 1986)

radiances to provide temperature profiles depend on climatological or previous two-week average temperature profiles as the basis for the inversion. These problems associated with the resolution of the satellite radiances and the discontinuous treatment of the stratosphere at the base-level are likely to be reduced when a reliable model first-guess field is used as the basis of the data analysis.

The operational data assimilation system used at ECMWF for tropospheric analyses is used here as an example of a possible system for the stratosphere. The ECMWF system has been described recently by Shaw et al. (1987). A 6-hour forecast from a high-resolution global numerical weather prediction model is used to provide the first-guess fields for the analysis. All available observations (surface, radiosonde, satellite, aircraft, etc.) are used in a multi-variate optimum interpolation analysis scheme, with appropriate error variance estimates for the different data types. Since May, 1986, the forecast model has had 19 levels in the vertical, with the uppermost levels at about 10, 30, 50, 75, 100, 130,... hPa being used for analyses at levels 10, 30, 50, 70, 100,... hPa. All variables; height, temperature and wind, are analysed. The analysis scheme and the model first-guess ensure some dynamical consistency between the analysed variables. Operational twice-daily analyses at these levels in the lower stratosphere are available from ECMWF on a 2.5° grid from May, 1986. Although this data assimilation method is likely to ensure better vertical consistency between the tropospheric and stratospheric analyses, it is possible that biases in the forecast model may lead to errors in the analyses. Such a bias has existed in the ECMWF model, with the forecasts too warm in the lower stratosphere (Shaw et al., 1987).

The problems associated with the statistical inversion of satellite radiances to provide temperature profiles in the stratosphere may be avoided to some extent through the assimilation of the satellite radiances directly into the analysis system. This method uses matching of radiances computed using the forecast model with those from the satellite to modify the first-guess temperature profiles. It is a more physically-based method for inverting the satellite radiances to provide realistic temperature profiles at the high vertical resolution of the model. However, this method may have its own problems associated with the close relationship between the forecast model, the observed radiances and the analysed temperatures if there are model biases.

Although there are no plans for improved vertical resolution in stratospheric nadir soundings in the next series of operational polar-orbiting satellites, proposed research satellites using limb-scanning instruments, including the Upper Atmosphere Research Satellite (UARS) and the Earth Observing System (EOS) satellites, are planned to have better vertical resolution than available from the TOVS system.

ACKNOWLEDGMENTS

We wish to thank Tom Miles, Alan O'Neill, Bill Heckley and Mel Gelman for providing information about current operational stratospheric analysis systems. This manuscript was prepared while the first author was on leave from Monash University and he is grateful to the AOS Program, Princeton University and GFDL for their generous hospitality. This research was supported in part under NOAA Grant NA87EAD00039.

32

REFERENCES

Andrews, D. G., J. D. Mahlman and R. W. Sinclair, 1983: Eliassen-Palm diagnostics of wave-mean flow interaction in the GFDL "SKYHI" general circulation model. *J. Atmos. Sci.*, 40, 2768-2784.

Bengtsson, L., and J. Shukla, 1988: Integration of satellite and in situ observations to study global climate change. *Bull. Am. Meteorol. Soc.*, 69, 1130-1143.

Clough, S. A., N. S. Grahame and A. O'Neill, 1985: Potential vorticity in the stratosphere derived using data from satellites. *Quart. J. Roy. Meteorol. Soc.*, 111, 335-358.

Fels, S. B., J. D. Mahlman, M. D. Schwarzkopf and R. W. Sinclair, 1980: Stratospheric sensitivity to perturbations in ozone and carbon dioxide:radiative and dynamical response. *J. Atmos. Sci.*, 37, 2265-2297.

Geller, M. A., M. -F. Wu and M. E. Gelman, 1983: Troposphere-stratosphere (surface-55 km) monthly winter general circulation statistics for the northern hemisphere - four year averages. *Mon. Wea. Rev.*, 40, 1334-1352.

Gelman, M. E., A. J. Miller, K. W. Johnson and R. M. Nagatani, 1986: Detection of long-term trends in global stratospheric temperature from NMC analyses derived from NOAA satellite data. *Adv. Space Res.*, 6, 17-26.

Graves, D., 1986: *Evaluation of satellite sampling of the middle atmosphere using the GFDL SKYHI general circulation model*, Ph.D. Thesis, GFD Program,Princeton University, Princeton, NJ, 314pp.

Grose, W. L., and A. O'Neill, 1989: Comparison of data and derived quantities for the middle atmosphere of the southern hemisphere. *PAGEOPH*, 130, 195-212.

Hollingsworth, A., A. C. Lorenc, M. S. Tracton, K. Arpe, G. Cats, S. Uppala and P. Kallberg, 1985: The response of numerical weather prediction systems to FGGE level IIb data. Part I: Analyses. *Quart. J. Roy. Meteorol. Soc.*, 111, 1-66.

Karoly, D. J., 1989: The impact of base-level analyses on stratospheric circulation statistics for the Southern Hemisphere. *PAGEOPH*, 130, 181-194.

Miles, T., and A. O'Neill, 1989: *Comparison of satellite-derived dynamical quantities in the stratosphere of the Southern Hemisphere*, Proceedings of MASH Workshop, Williamsburg, VA.

Rood, R. B., D. J. Allen, W. E. Baker, D. J. Lamich and J. A. Kaye, 1989: The use of assimilated stratospheric data in constituent transport calculations. *J. Atmos. Sci.*, 46, 687-701.

Shaw, D. B., P. Lonnberg, A. Hollingsworth and P. Unden, 1987: Data assimilation: The 1984/85 revisions of the ECMWF mass and wind analysis. *Quart. J. Roy. Meteorol. Soc.*, 113, 533-566.

Smith, W. L., H. M. Woolf, C. M. Hayden, D. Q. Wark and L. M. McMillin, 1979: The TIROS-N operational vertical sounder. *Bull. Am. Meteorol. Soc.*, 60, 1177-1187.

Takano, K., W. E. Baker, E. Kalnay, D. J. Lamich, J. E. Rosenfield and M. A. Geller, 1987: Forecast experiments with the NASA/GLA stratospheric/tropospheric data assimilation system. *J. Meteorol. Soc. Japan*, 67, 83-89.

Trenberth, K. E., and J. G. Olson, 1988: An evaluation and intercomparison of global analyses from the National Meteorological Center and the European Centre for Medium Range Weather Forecasts. *Bull. Am. Meteorol. Soc.*, 69, 1047-1057.

THE SEASONAL EVOLUTION OF THE EXTRA-TROPICAL STRATOSPHERE IN THE
SOUTHERN AND NORTHERN HEMISPHERES: SYSTEMATIC CHANGES IN POTENTIAL
VORTICITY AND THE NON-CONSERVATIVE EFFECTS OF RADIATION.

A. O'NEILL AND V. D. POPE
Hadley Centre for Climate Prediction and Research
Meteorological Office, London Road
Bracknell, Berks
U. K.

ABSTRACT. Satellite data are used to summarize the main features of
the seasonal cycle of the extra-tropical stratosphere in the two
hemispheres during one year. Systematic changes in the overall
structure of the circulation are identified, and the contribution of
the non-conservative effects of radiation to these changes is
outlined.

1. INTRODUCTION

Over the past 15 years or so, radiometric measurements from polar
orbiting satellites have furnished a wealth of information on the
global structure of the circulation in the middle atmosphere. Until
recently, most studies using these data focused on the northern
hemisphere for several reasons, not least being that the middle
atmosphere of the northern hemisphere in winter is usually much more
dynamically active than that of the southern hemisphere, with
dramatic phenomena such as major mid-winter warmings. The MASH
project was designed to remedy this comparative lack of interest in
the southern hemisphere. Studies were advocated which compared
dynamics, transport and chemistry in the two hemispheres, on the
grounds that the two hemispheres showed marked differences whose
elucidation would give insight into dynamical and photochemical
processes.
    In this paper, we use satellite data to summarize and contrast
the main features of the seasonal cycle of the extra-tropical
stratosphere in the two hemispheres during one year. We focus on the
principal dynamical phenomena found in the extra-tropical
stratosphere: in the southern hemisphere the final warming, and in
the northern hemisphere the Canadian warming and the mid-winter major
warming. We point out that systematic changes in the overall
structure of the circulation stem from some warmings, and to
summarize these changes we use, following Butchart and Remsberg
(1986), a co-ordinate independent diagnostic - the area A(Q) enclosed

A. O'Neill (ed.),
Dynamics, Transport and Photochemistry in the Middle Atmosphere of the Southern Hemisphere, 33–54.
© 1990 Kluwer Academic Publishers.

by an isopleth of Ertel's potential vorticity, Q, on an isentropic
surface.

On the basis of an equation for the time rate of change of A(Q),
we then discuss some aspects of the interaction between radiation and
dynamics in the stratosphere of both hemispheres. We point out that
radiative effects connected with dynamically induced departures from
radiative equilibrium play a central role in producing systematic
changes in the overall structure of the stratospheric circulation. In
particular, we suggest that radiation contributes significantly to
the signature of a so-called main vortex/surf zone in the evolution
of the areas A(Q).

2. DATA

The data used were obtained mainly from nadir sounding radiometers on
board the satellite NOAA-6 of the Tiros-N series. The features and
performance of the stratospheric sounding unit (SSU), which provided
most information about fields in the middle and upper stratosphere,
were summarized by Pick and Brownscombe (1981). Clough et al. (1985)
gave details of the data and its analysis relevant to the calculation
of isentropic fields of potential vorticity. The analysed fields
were interpolated in time and space onto a 5°x5° latitude-longitude
grid at 12Z. Fields were smoothed by Fourier analysis with truncation
at wavenumber 12 in the meridional (pole to pole) and zonal
directions. This value is a compromise which reduces the resolution
of fields at high latitudes but over-estimates the resolution at low
latitudes.

Analyses were produced from radiances measured by the SSU, the
microwave sounding unit (MSU) and the high resolution infrared
sounder (HIRS-2). The SSU weighting functions have half widths of
about 12-15 km, but statistical tests suggest that the effective
resolution is probably about 10 km because of the overlap of several
weighting functions. Thicknesses of the 100-20, 100-10, 100-5, 100-2
and 100-1 mb intervals were retrieved from the equivalent brightness
temperatures (temperature weighted by the instrumental weighting
function) by statistical regression (Pick and Brownscombe, 1981). The
regression coefficients were determined using thicknesses derived
from a set of 1200 rocket soundings grouped into seven latitude and
seasonal zones.

Fields of geopotential height were produced by adding the
thicknesses to an objective analysis of geopotential height at 100
mb. In 1981 these analyses were provided by the National
Meteorological Center (Washington). NMC analyses were also used for
the 50 mb level. Winds and temperatures were derived from
geopotential heights using the geostrophic and hydrostatic
approximations.

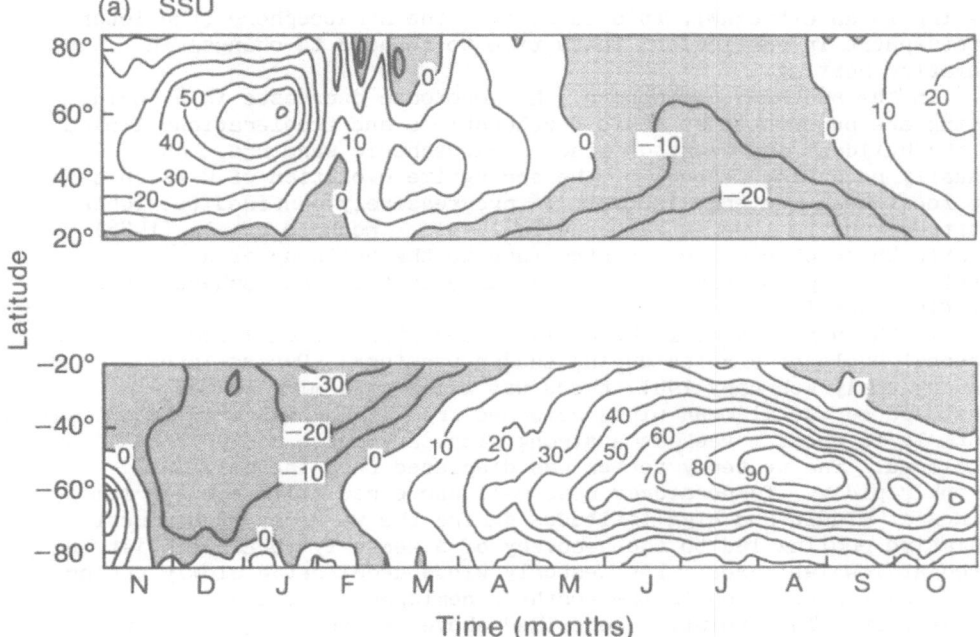

Fig. 1 Latitude-time section of zonal-mean geostrophic winds obtained from SSU observations at 10 mb for November 1980 to October 1981. Fields are smoothed in time with a nine-day running mean. Units are m s^{-1}. Negative values are shaded.

3. OVERVIEW OF THE SEASONAL EVOLUTION IN THE EXTRATROPICAL STRATOSPHERE

3.1 Zonal-mean wind

We now present a brief survey of the seasonal evolution of the extratropical stratosphere from November 1980 to October 1981. To focus the discussion, we refer to the evolution of the zonal-mean wind, $\bar{u}$, at 10 mb displayed in Fig. 1.

There is a pronounced seasonal cycle in $\bar{u}$ in both hemispheres. Zonal-mean winds change from easterly to westerly as solar heating decreases during autumn and early winter, and from westerly to easterly as solar heating increases during spring and early summer. Notice that the strongest westerlies in winter and easterlies in summer are attained a few weeks after the solstices (when solar heating is an extremum). This is because the stratosphere (the lower stratosphere in particular) takes time to respond to changes in radiative heating.

In the southern hemisphere, the monotonic decreases in $\bar{u}$ during spring are punctuated by rapid decelerations and accelerations linked to the build-up and decay of minor stratospheric warmings. An annually recurring feature of the springtime evolution of $\bar{u}$ is that the zonal-mean westerly jet becomes progressively confined to higher latitudes and to lower stratospheric levels. Mechoso et al. (1988) related these changes in jet structure to the build-up of a quasi-steady, planetary-scale disturbance in the stratosphere during the final warming.

In the northern hemisphere, large departures from a smooth seasonal cycle of $\bar{u}$ arise during sudden warmings. During late January/early February 1981, there was a major warming in the stratosphere. Zonal-mean winds reversed from strong westerly to easterly in about a week as a strong planetary-scale disturbance developed. The westerly vortex was displaced from the pole and shrank rapidly. Temperatures rose well above radiative equilibrium values at middle and high latitudes during the warming. Subsequent radiative cooling led to the recovery of a westerly zonal-mean jet in March at low latitudes. The westerly winds decelerated slowly during spring as temperatures in the northern hemisphere increased radiatively. The evolution described above is typical of years when there is a major warming during winter followed by a late winter cooling and a late final warming (Labitzke, 1982). The evolution is not typical of every winter in the northern hemisphere, however. In some years there is no major warming, whereas in others there is no significant recovery of the westerly flow after a warming (see for example Labitzke, 1982; Geller et al., 1984).

3.2 Synoptic maps

We now turn to maps of potential vorticity, Q, on the 850 K isentropic surface (near 10 mb in the mid stratosphere) to illustrate the main synoptic features in the seasonal evolution of the extra-tropical stratosphere, and in particular to identify systematic changes in the overall structure of the circulation, which we later relate to non-conservative processes.

The Southern Hemisphere

The flow in the southern hemisphere is generally almost zonally symmetric through most of the winter. Fig. 2 shows synoptic maps of Q during the southern winter and spring. Planetary-scale disturbances grow on the westerly vortex mainly during early winter and spring. In early winter 1981 (Fig. 2(a)), there was a minor warming similar to a special type of warming in the northern hemisphere, called a Canadian warming (see below). The cyclone moved away from the pole when an anticyclone formed near 180°E. This

(a) 1 May 1981

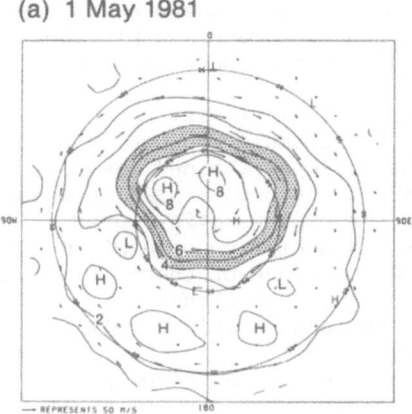

(b) 1 July 1981

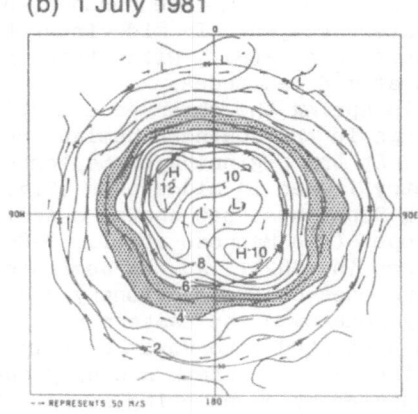

(c) 1 September 1981

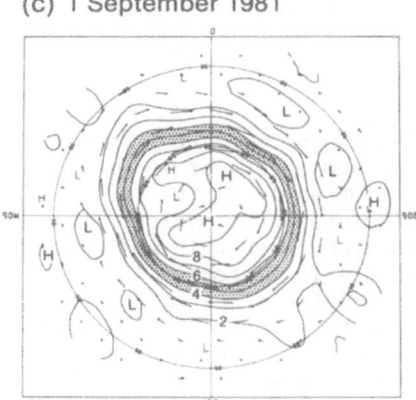

(d) 12 October 1981

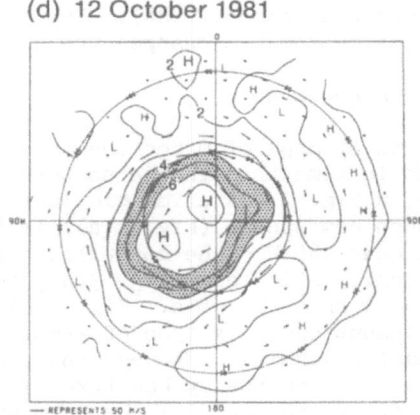

(e) 1 November 1981

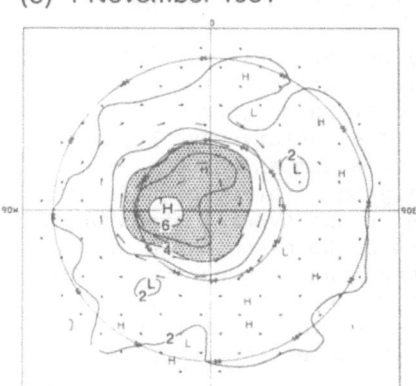

Fig. 2 Polar-stereographic maps of Ertel's potential vorticity Q and wind arrows obtained from SSU observations for the southern hemisphere on the 850K isentropic surface. The units are 10^{-4} m^2 K kg^{-1} s^{-1} and the contour interval is 1 in these units. Areas between Q=4 and Q=6 are shaded. All values of Q for the southern hemisphere are multiplied by -1.

anticyclone was a mobile feature, which travelled eastward as it grew
and later decayed. In mid-winter the cyclone became symmetric about
the pole (Fig. 2(b)), with strong meridional Q gradients and zonal
winds at mid-latitudes.

In spring the observed evolution was typical of many spring
warmings in the southern hemisphere (see for example Mechoso et al.,
1988). At the beginning of September, the cyclone retained the
symmetry it had in mid-winter (Fig. 2(c)) but covered a smaller area.
During September (not shown) a series of travelling anticyclones
developed (see Mechoso et al., 1988, for a description of similar
features in 1982). At the beginning of October one of these
anticyclones became stationary and grew, forming an elongated
anticyclone over the eastern half of the hemisphere (Fig. 2(d)),
which persisted into November (Fig. 2(e)).

The cyclone shrank steadily during spring. This shrinkage was
accompanied by a weakening of potential vorticity gradients outside
the cyclone (examine contours labelled 2 and 3 in Fig. 2(c) to (e)),
leading to what has been termed a main vortex/surf zone structure by
McIntyre and Palmer (1984). The non-conservative processes that led
to these systematic changes are considered below.

Northern Hemisphere

A series of maps of Q for the 850K isentropic surface, roughly evenly
spaced in time through the northern winter and spring of 1980/81, is
shown in Fig. 3. A climatological feature of early winter in the
northern hemisphere (Fig. 3(a)) is the persistent anticyclone that
forms near 180°E - the so-called Aleutian High. When this anticyclone
is strong, as it is in Fig. 3(a), the cyclone is distorted and
displaced from the pole. Fig. 3(b) shows a later stage of
development during a type of warming that Labitzke (1982) has termed
a Canadian warming. The salient features of this type of warming
have been described by Labitzke (1977) and by Clough et al. (1985).
In brief, these are: (1) the warming occurs in early winter; (2) it
involves the advection of warm air over Canada, but temperature rises
are much smaller than during strong warmings in late winter; and (3)
the cyclone is strongly distorted but not broken down (as happens in
major warmings).

Notice that Q contours in Fig. 3(b) are strongly distorted, with
an intrusion of low Q air from low latitudes adjacent to an extrusion
of high Q air from the cyclone. Such a pattern has been taken, by
McIntyre and Palmer (1983), as evidence of planetary-wave breaking.
They proposed that the phenomenon would be accompanied by
irreversible mixing of potential vorticity on isentropic surfaces,
which could lead to a significant change in the overall structure of
the stratospheric circulation, in particular to a shrinkage of the
cyclone and to the formation of weak Q gradients around it. In the
Canadian warming of 1980, however, there was no evidence of strong
mixing: the cyclone was actually bigger after the event (when the
flow was more nearly axisymmetric) than it was before (we give

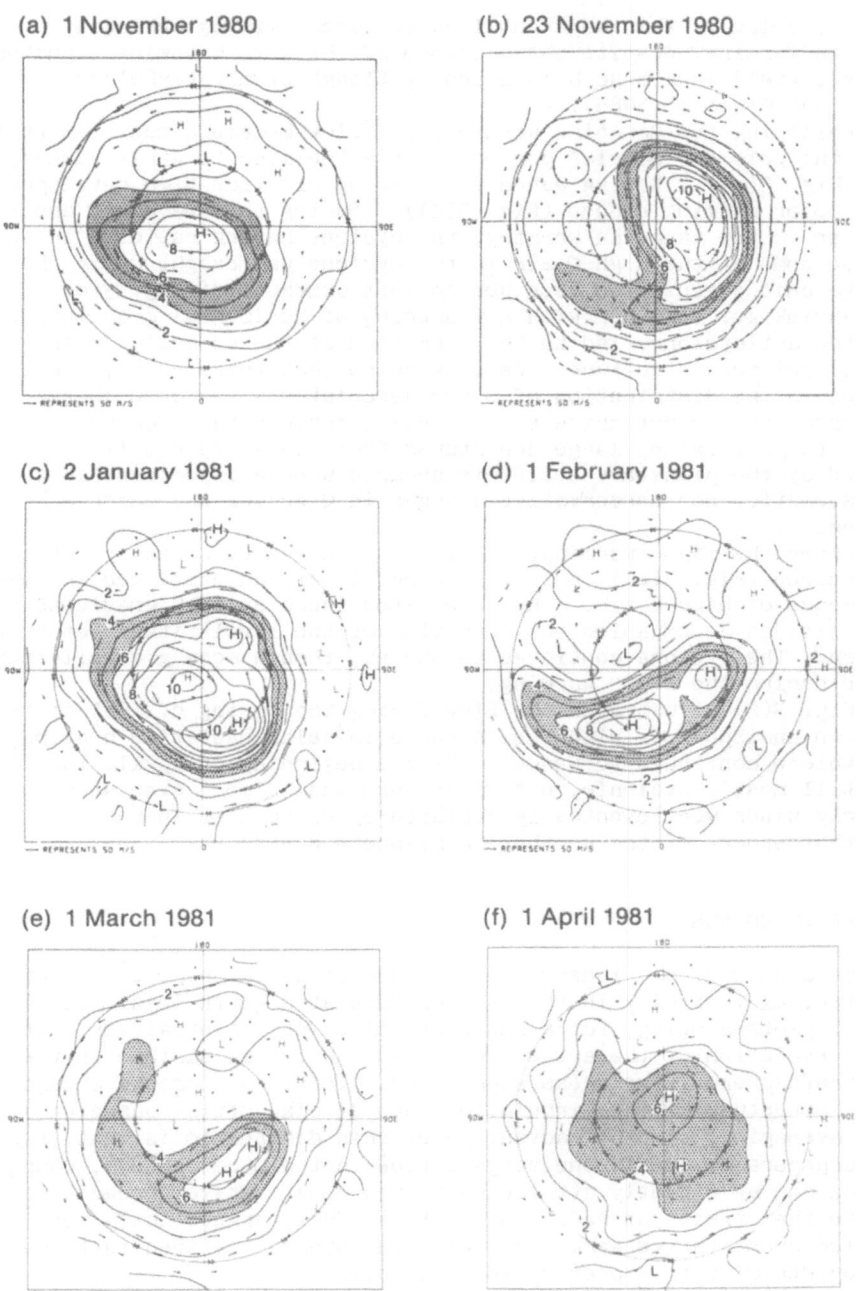

(a) 1 November 1980   (b) 23 November 1980

(c) 2 January 1981   (d) 1 February 1981

(e) 1 March 1981   (f) 1 April 1981

Fig. 3   As Fig. 2 but for the northern hemisphere.

evidence below). This apparent lack of strong mixing during a Canadian warming, despite the presence of the wave-breaking signature in the Q field, has also been noted by Clough et al. (1985) and by Juckes and O'Neill (1988).

Following the Canadian warming, the flow became almost symmetric about the pole for a brief period in late December to early January 1981 (Fig. 3(c)). During January, a strong Aleutian High developed and a major warming ensued (Fig. 3(d)). During this warming, and in contrast to the Canadian warming, the cyclone shrank rapidly, in part because some of the high-Q air in the cyclone was caught up in the anticyclonic circulation (see bottom left quadrant of Fig. 3(d)). This shrinkage, together with a weakening of horizontal gradients of Q in the anticyclone, led to the formation of the so-called main vortex/surf zone structure. We show below that these systematic changes in the distribution of Q are associated with non-conservative processes, and do not arise solely from a conservative rearrangement of Q. In particular, large departures from radiative equilibrium induced by the planetary-scale disturbance were a major contributor to systematic, non-conservative changes in Q during the major warming.

After the major warming, values of Q increased steadily outside the cyclone (Fig. 3(e)), mainly as a result of radiative cooling and subsidence of high-Q air. The associated increase in Q gradients at low latitudes accompanied the formation of the low-latitude jet noted earlier. The cyclone continued to shrink, though more gradually than it did during the major warming.

Fig. 3(f) illustrates the flow during the spring of 1981 in the northern hemisphere. With the increase in solar radiative heating, the cyclone continued to weaken. By the beginning of April, the flow was still mostly cyclonic, but winds were weak almost everywhere. Westerly winds were eventually replaced by easterlies in the mid-stratosphere of the northern hemisphere during May.

4. AREA INTEGRALS

To give a co-ordinate independent picture of the average distribution of potential vorticity in the stratosphere at a given level, we present plots showing, for a range of values of potential vorticity, Q, the evolution of the areas A(Q) on an isentropic surface in the mid stratosphere. The areas expose more clearly systematic changes in the structure of the circulation than do diagnostics based on zonal averaging. Another advantage of this diagnostic is that, for the large-scale, quasi-nondivergent flow in the stratosphere, changes in area can be directly attributed to non-conservative processes, e.g. to the effects of radiation. (Pope, 1989, argues that, for detailed quantitative work a circulation integral defined in terms of an isopleth of Q is a preferable diagnostic.)

4.1 Results

Area integrals, A(Q), calculated for various values of Q on the 850 K surface in the southern and northern hemispheres, are shown in Fig. 4. The spacing of the curves at a particular time gives an indication of the average spacing of Q contours on synoptic maps. For instance, the curves for the southern hemisphere in mid-winter are close together over most of the hemisphere, which corresponds to the strong Q gradients and associated westerly winds in the huge cyclone shown in Fig. 2(b).

A pronounced seasonal cycle is evident in the evolution of A(Q). Increases in area in autumn and early winter coincide with the growth of the westerly vortex, and decreases in late winter and spring with its decay. In the subsequent discussion, we shall ignore changes in area associated with the short-term (shorter than a week), noisy fluctuations shown in Fig. 4. We shall concentrate instead on the systematic changes in area associated with the formation of the main vortex/surf zone structure in late winter and spring, and on the non-conservative processes that led to these changes.

The signature in A(Q) of the formation of the main vortex/surf-zone structure is a decrease in A(Q) for high values of Q, and an increase in A(Q) for lower values. Fig. 4(a) illustrates this behaviour for the southern hemisphere in spring 1981. Notice that A(4) decreases while A(2) increases in October, so that, on average, a region of weak gradients of Q forms around the shrinking cyclone.

In the northern hemisphere, the Canadian Warming in November (Fig. 4(b)) had no clear effect on the build-up of the westerly vortex, despite apparent planetary-wave breaking. Fig. 4(b) shows that the areas A(Q) for Q contours that were irreversibly buckled during the event (e.g. Q=4 in Fig. 3(b)) increased steadily (apart from short-term noise). Although some mixing and apparent loss of high Q must have occurred, increases in A(Q) due to radiative cooling were large enough to mask any mixing. During the major warming (Fig. 3(d)), on the other hand, radiation augmented irreversible mixing in breaking down the westerly vortex. The vortex broke down much more rapidly than it did during the corresponding season in the southern hemisphere (compare the rate of decline of areas in Figs 4(a) and 4(b) during August and February).

The shrinkage of the vortex during late January and early February was accompanied, on average, by an increase in gradients of Q for both high and low values of Q, and by a decrease for intermediate values of Q (Fig. 4(b)). As values of Q increased outside the westerly vortex, contours with successively higher values of Q peeled off its edge. This peeling off extended much further into the centre of the westerly vortex in the northern hemisphere than it did in the southern hemisphere, because planetary-scale disturbances were (and usually are) stronger and more persistent in the northern hemisphere.

(a) Southern hemisphere

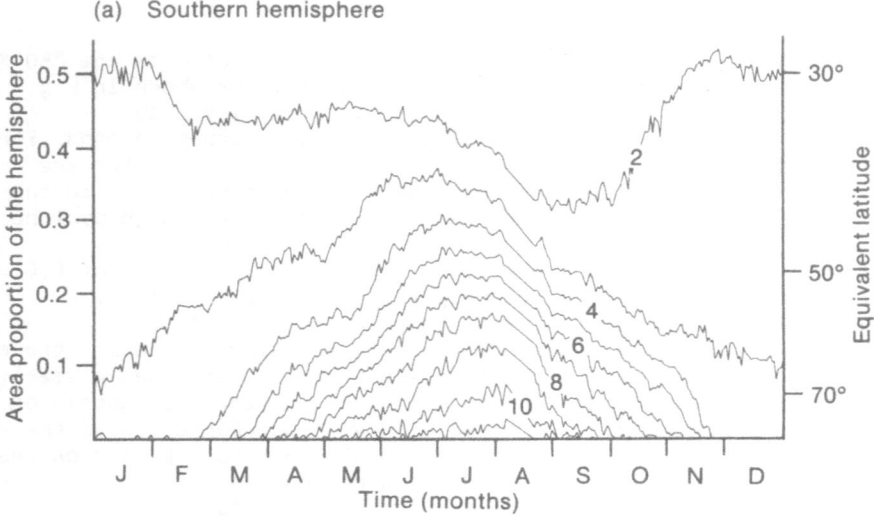

(b) Northern hemisphere

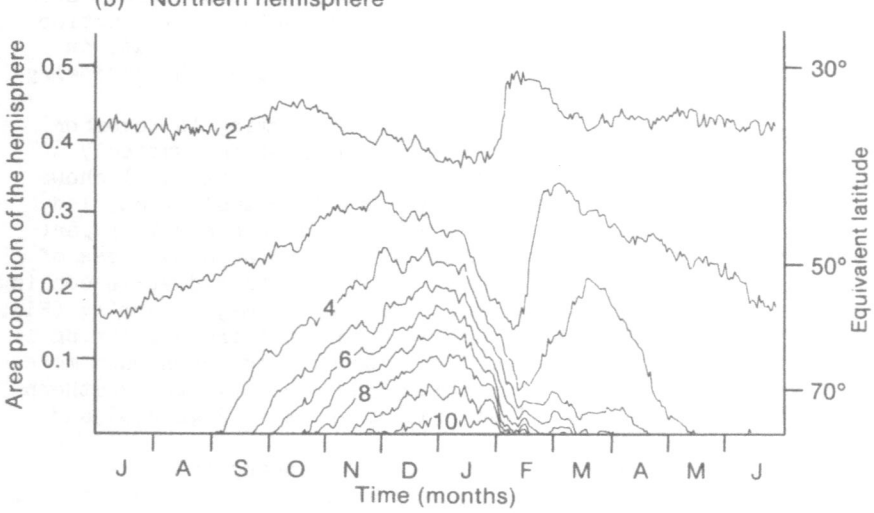

Fig. 4 The evolution of the area A(Q) enclosed by contours
Q=2 to 11 (in standard units of 10^{-4} m^2 K kg^{-1} s^{-1},
multiplied by -1 in the southern hemisphere) on the 850K
isentropic surface. (a) the southern hemisphere for 1 January
1981 to 31 December 1981. (b) The northern hemisphere for
1 July 1980 to 30 June 1981. The equivalent latitude for a given
Q contour is the latitude of a circular contour centred on the
pole enclosing an area equal to that enclosed by the Q contour.

These changes in area A(Q) are the hallmarks of the formation of the main vortex/surf zone structure noted by McIntyre and Palmer (1984) for the major warming of February 1979. They adduced such changes in area as evidence of apparent non-conservation due to isentropic mixing of Q. A strong hint that mixing was not be the only important process in 1981 comes from the evolution of areas in late February and March. Notice that area A(4) increased significantly in March. If this increase were attributable solely to isentropic mixing, one would expect to find substantial decreases in A(Q) for high values of Q, as portrayed in McIntyre and Palmer's Fig. 12 (1984). The decreases in area were small, however, and apparently not enough to account for the increases in A(4). Radiative effects must be invoked.

4.2 Theory

For frictionless flow with small Rossby number and large Richardson number (that of the extra-tropical stratosphere), the rate of change of the area A(Q) enclosed by a given isopleth of Q on a given isentropic surface, θ, is given by (Butchart and Remsberg, 1986, Pope, 1989)

$$\frac{\partial}{\partial t} A(Q) = \int \nabla_\theta \cdot u \, dA + \int S \, dl / \left| \nabla_\theta Q \right| \tag{1}$$

The first term on the right of this equation is an area integral of the divergence of the horizontal wind (with derivatives calculated on the isentropic surface). For the large-scale, quasi-nondivergent flow of the winter stratosphere, this term is generally small compared with the second term on the right of the equation (Pope op. cit.), and will not be considered further here. The second term is a contour integral, and is the term of interest. It gives the rate of change of A(Q) due to radiation. The quantity S appears in the expression for the isentropic rate of change of Q due to radiation

$$\frac{\partial Q}{\partial t} + u \frac{\partial Q}{\partial x} + v \frac{\partial Q}{\partial y} = S \tag{2}$$

where all partial derivatives are evaluated on the given isentropic surface. In terms of the radiative heating rate $\dot{\theta}$, S is given by

$$S = Q \frac{\partial \dot{\theta}}{\partial \theta} - \dot{\theta} \frac{\partial Q}{\partial \theta} \tag{3}$$

The two terms on the right of Eq. (3) have straightforward
interpretations. The first represents changes in potential vorticity
associated with radiatively induced changes in atmospheric stability.
For instance, where radiative cooling increases with height
(increasing potential temperature), the atmospheric stability is
reduced, and so is the potential vorticity following the isentropic
motion. The second term represents radiatively induced changes in
potential vorticity due to (near) vertical motion through the
isentropic surface. For instance, where Q increases with height (as
it generally does), radiative cooling will increase values of Q on an
isentropic surface by bringing down higher values from above.

In Eq. (1), S occurs in a contour integral around an isopleth of
Q (dl is an infinitesimal increment of length). It is readily seen
that the contour integral is proportional to the area integral of S
over a thin strip enclosed by the isopleths Q and Q + ΔQ. Hence, if
S is positive (or negative) on average on this thin strip (i.e. if Q
is increasing on average following the isentropic motion of air
parcels in the strip), then the area enclosed by a given isopleth of
Q increases (or decreases) with time.

Now Eq. (1) applies to continuous fields, whereas observational
(e.g. satellite) data have finite resolution in time and space. They
are "coarse-grain" measurements. Therefore, Eq. (1) will not be
exactly satisfied when terms are computed from observational data.
The discrepancy will in general have random and systematic causes -in
particular, random instrumental error and systematic misrepresenta-
tion of the effects of unresolved scales of motion. A possibly
important systematic effect in the stratosphere is irreversible,
isentropic mixing of potential vorticity during planetary wave
breaking. Apparent non-conservation of Q may lead to systematic
increases or decreases in A(Q) which cannot be accounted for
radiatively. Moreover, radiative effects may be misrepresented
systematically if, for example, the vertical structure of the
atmosphere is not fully resolved. The drag exerted by unresolved
breaking gravity waves may also lead to systematic discrepancies in
Eq. (1). As yet, there is no evidence that such drag is significant
in the middle stratosphere, the domain of interest here.

To calculate the radiative term in Eq. (1) for the middle
stratosphere, we use a sophisticated radiation scheme for solar
heating and infrared cooling (developed by Shine, 1987). To
calculate solar heating, a (zonal-mean) climatological distribution
of ozone is assumed (that used by Klenk et al., 1983). Pope (1989)
has shown that approximating the actual distribution of ozone by a
climatological distribution does not lead to serious error in the
calculations reported here. Because the vertical resolution of SSU
data is only about 10 km, radiative effects associated with vertical
structure on a smaller scale will not be adequately represented in
these calculations.

4.3 Systematic changes in A(Q) due to radiation

We used the diagnostic A(Q), shown in Fig. 4, to summarize the systematic changes in the average distribution of Q during the seasonal cycle of the stratosphere. Now we shall study the radiative contribution to those changes by examining the radiative field S that appears in the equation for the time rate of change of area, Eq. (1). We focus on two times of year: early winter when the stratospheric vortex is building up, and late winter or spring when the vortex is breaking down. Only a brief, mainly qualitative, outline is possible here; a much more extensive quantitative analysis, with a careful discussion of errors, is given by Pope (1989).

Before mid-winter 1981, the stratospheric flow in the southern hemisphere was roughly symmetric about the pole; planetary-scale disturbances were weak. The build-up of the westerly vortex, which is reflected by the increases in the areas shown in Fig. 4(a), is consistent with the field S. This is shown in Fig. 5(a) for the 850 K isentropic surface on 1 June 1981. Also plotted are three isopleths of Q, which allow the vortex to be located (see figure caption). S is positive over most of the vortex, which, from Eq. (1), corresponds to increases in the areas A(Q) for all isopleths of Q associated with the vortex. Pope (1989) shows that there is good quantitative agreement at this time between the observed rates of change of A(Q) and those given by the radiative term in Eq. (1), though the radiative term generally overestimates the rate of change of A(Q) when SSU data are used (see below).

By late spring, the final stratospheric warming was well underway, and the flow was highly asymmetric (see Fig. 2(d)). A large-amplitude, persistent planetary-scale disturbance induced departures from radiative equilibrium, and the resulting field of radiative cooling led to systematic changes in the average distribution of Q. Figure 5(b) shows the field S (together with a few isopleths of Q) for the 850 K isentropic surface on 12 October 1981. Broadly speaking, S is negative over most of the westerly vortex (where Q is greater than about 4 units), and positive over the region of anticyclonic circulation in the eastern hemisphere (where Q is less than 2 units). Thus, dynamically induced departures from radiative equilibrium led to an asymmetric distribution of S which reduced high values of Q in the cyclone and increased low values of Q in the anticyclone. Variations in Q around latitude circles were thereby reduced. (Note, however, that radiation does not relax Q instantaneously to a radiative equilibrium value (Pope 1989).)

From the field S shown in Fig. 5(b), which is typical for the final warming, one would expect on the basis of Eq. (1) that, during October, A(Q) should decrease systematically for values of Q greater than about 4 units, and increase systematically for smaller values. Such changes would, on average, correspond to a shrinkage of the westerly vortex and to the formation of weak Q gradients around it. A similar evolution is expected, on average, as a result of vortex erosion and mixing of Q during planetary-wave breaking (McIntyre and

46

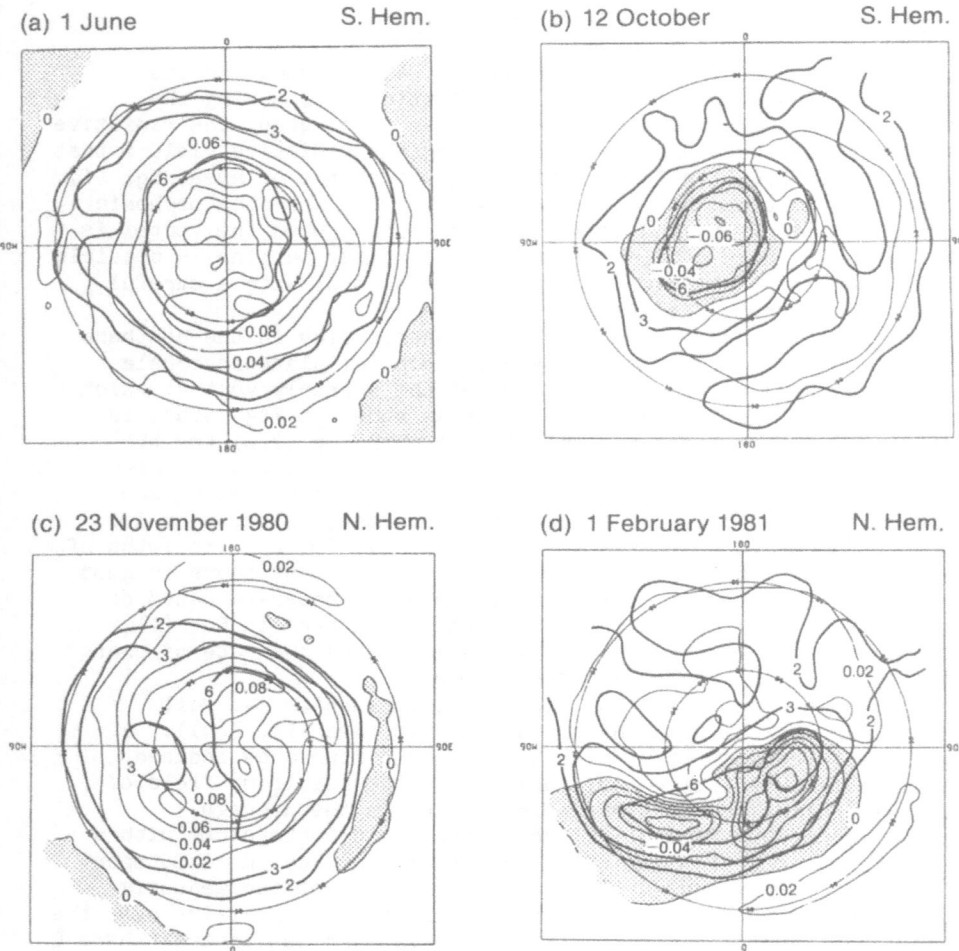

Fig. 5   Polar-stereographic maps of the radiative term S in the potential
        vorticity equation, for the 850K isentropic surface. The units for S are
standard Q units per day. The contour interval is 0.02 in these units and negative
values are shaded. Three Q contours are shown for reference, Q=2 positioned at
low latitudes, Q=3 on the edge of the cyclone, and Q=6 inside the cyclone. For
the southern hemisphere S and Q have been multiplied by −1.

Palmer 1983, 1984). Thus radiation and mixing may lead to similar
signatures in the evolution of A(Q), in particular to the formation
of a so-called main vortex/surf zone structure. Detailed
calculations by Pope (1989), who used both SSU data and fields from
simulations with a numerical model, have shown that radiation is a
major contributor to the changes in areas A(Q) that are said to
characterize the formation of the main vortex/surf zone.

Pope showed that the limited vertical resolution of the SSU
introduces a positive bias into the calculated value of S. This is
because increases in temperature with height are not fully resolved,
especially during major warmings, so that the first term on the right
of Eq. (3) is not as negative as it should be. In Fig. 4(a), the
area curves A(3) and A(2) separate with time during October as A(3)
gets smaller and A(2) bigger. As noted above, however, separation
would be expected at a value of Q of about 4 units. The positive
bias in S is a likely explanation of this discrepancy.

Before mid winter 1980/81, the stratospheric flow in the
northern hemisphere was much more disturbed than at the corresponding
time of year in the southern hemisphere. Fig. 3(b) shows that
isopleths of Q were highly distorted by a planetary-scale disturbance
- the Canadian warming discussed in section 3. The corresponding
field S, shown in Fig. 5(c), is much more symmetric than the field Q.
The reason is that the temperature perturbation during the Canadian
warming was weak, even though air parcel displacements were large.
The stratosphere has a quasi-barotropic structure during Canadian
warmings (as noted by Clough et al., 1985, and by Juckes and O'Neill,
1988). In Fig. 5(c), S is positive over most of the extra-tropical
northern hemisphere. Consistent with this, the areas A(Q) increase
with time during November for values of Q bigger than about 3 units
(Fig. 4(b)), but the rate of increase is overestimated by the
radiative term in Eq. (1). A possible explanation for this
discrepancy is the positive bias in S noted above.

Figure 5(d) shows the field S in the northern hemisphere during
the major stratospheric warming of 1981. As a result of a
planetary-scale disturbance, the flow is markedly baroclinic. Strong
downward motion leads to adiabatic warming and sharp rises in
temperature locally, i.e. to a highly asymmetric temperature field
(e.g. Fairlie et al. 1990). The field S is also highly asymmetric,
with S negative over most of the cyclone, reducing high values of Q,
and positive over the anticyclone, increasing low values of Q, thus
weakening both the cyclonic and anticyclonic circulations. As shown
in Fig. 4(b), areas A(Q) decrease rapidly for high values of Q in
early February, whereas areas increase rapidly for low values of Q in
the post-warming period in late February. In mid-February, Fig. 4(b)
shows that the classical signature of the so-called main vortex/surf
zone has developed: the area curves are bunched close together for
values of Q of 4 and above, corresponding to strong gradients of Q in
the cyclone, whereas they are far apart for values of Q less than 4,
corresponding on average to weak gradients around the cyclone.

48

Detailed calculations with SSU data by Pope (1989) show that
radiation can broadly account for the increases in area in the
post-warming period. Strong radiative cooling of very warm air over
middle and high latitudes in the stratosphere leads to downward
motion, which replaces the low values of Q in the Aleutian High by
higher values of Q from higher altitudes. Hence, the double-peak
structure in the areas A(Q) in Fig. 4(b) mentioned earlier can be
accounted for by radiative cooling in early winter as solar heating
decreases, and by radiative cooling in the post-warming period. The
calculations indicate, however, that radiation is not the dominant
contribution to the observed reduction in areas associated with the
shrinkage of the cyclone during the major warming. However, Pope
(1989) has shown that the radiative contribution may be seriously
underestimated when calculated from SSU data because of limited
vertical resolution. In view of the strong, irreversible deformation
of isopleths of Q witnessed during the warming (Fig. 3(d)), another
major factor is probably apparent loss of Q by quasi-horizontal
mixing.

5. INTERANNUAL VARIABILITY

Figures 6 and 7 show plots of A(Q) for December 1978 to December 1988
for the southern and northern hemispheres. Broadly speaking, the
areas exhibit much less interannual variability in the southern
hemisphere than they do in the northern hemisphere. Our discussion
of the systematic changes in the distribution of Q in the southern
hemisphere during 1981 applies to other years with little
modification. Notice, in particular, that each year the springtime
increase in A(Q) mentioned above applies only to the outermost of the
curves in Fig. 6. On synoptic maps, the contour Q=2 peels away from
the edge of the westerly vortex, but contours Q≳3 do not.
Disturbances and associated nonconservative effects are confined to
the outer portion of the intense westerly vortex. Consequently, the
area curves for Q≳3 are approximately mirror symmetric about their
maxima just after mid-winter, in sharp contrast to the curves for the
northern hemisphere (Fig. 7). Some interannual variability is
evident in the southern hemisphere, however. For instance, the
southern stratosphere during late winter and spring was much more
disturbed in 1988 than it was in 1987, and areas A(Q) decreased much
more rapidly for a time in 1988.

Despite the stronger interannual variability, the stratosphere
of the northern hemisphere also exhibits some annually recurring
features in the evolution of the areas A(Q), and hence in the gross
structure of the circulation. (Such regularity tends to be masked by
zonally averaged diagnostics, which is one reason for focusing on
co-ordinate independent diagnostics such as A(Q).) The area curves
in Fig. 7 are manifestly not mirror symmetric about mid-winter.
During years with particularly strong winter warmings (1978/79.
1980/81, 1981/82, 1983/84, 1984/85, 1986/87 and 1987/88), the curves

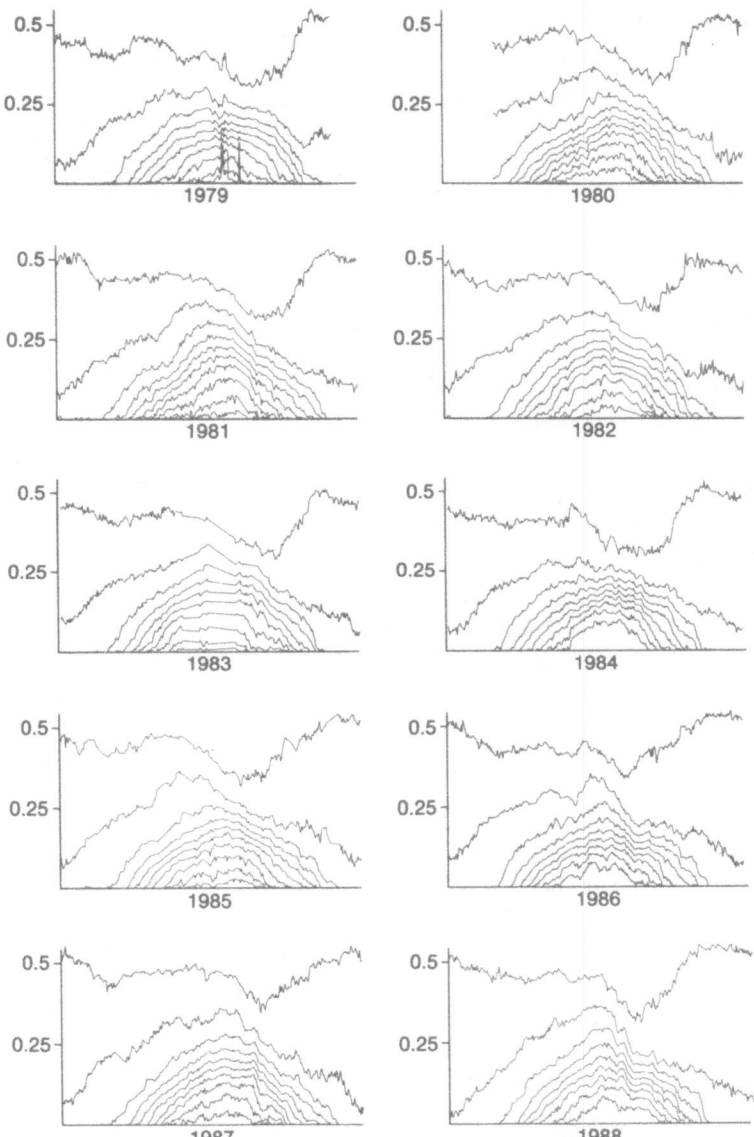

Fig. 6   Evolution of A(Q) in the southern hemisphere for
10 Q contours Q=2 to 11 (standard Q units) from the
top to the bottom of each plot. Fields are calculated for the
850K isentropic surface using SSU observations from
December 1978 to December 1988. Each plot shows data for
one year from 1 January to 31 December. The y axis is the
area proportion of the hemisphere.

50

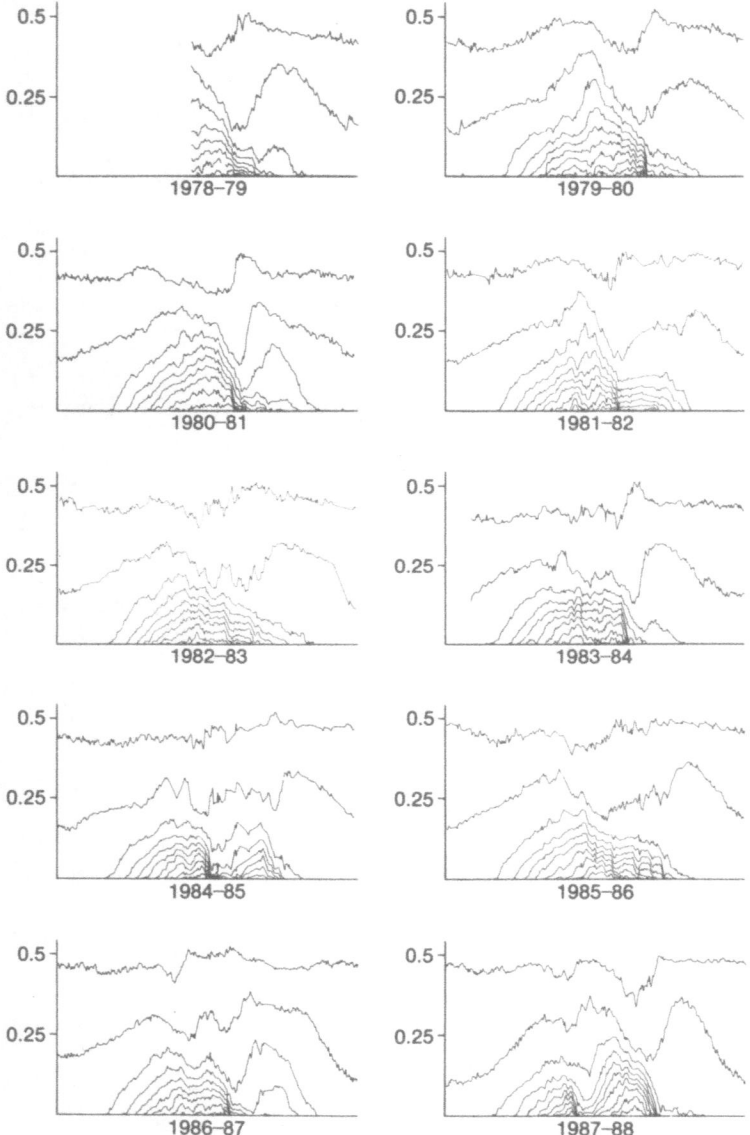

Fig. 7   As Fig. 6 but with each plot for one year from 1 July
to 31 June in the northern hemisphere.

in Fig. 7 display, to some extent, a double-peak structure. This
structure results from the following sequence of events: first, the
radiative increase in areas in autumn (associated with the build-up
of the westerly vortex); second, the rapid decrease in areas due to
non-conservative processes during strong warmings (radiation and
apparent loss of Q in unresolved scales of motion); and third,
increases in area due to strong radiative cooling during and after
the warming. As noted earlier, the evolution of areas during the
second stage of this sequence has been taken to characterize the
formation of a main vortex/surf zone in the stratosphere. It is
during this stage that radiation augments any irreversible dynamical
rearrangement of Q in breaking down the westerly vortex. Notice
that, in contrast to the southern hemisphere, the double-peak
structure is found in area curves for high values of Q. This is a
further indication that planetary-scale disturbances in the northern
hemisphere have significant non-conservative effects much closer to
the centre of the westerly vortex than they do in the southern
hemisphere.

In the northern hemisphere, the intensity and timing of
stratospheric warmings vary a great deal from year to year. Hence,
the timings of the above-mentioned changes in area vary, as do the
magnitudes of the maxima and minima of the double peak. Moreover,
the nonconservative effects of disturbances penetrate much further
into the westerly vortex in some years (e.g. 1987/88) than in others
(e.g. 1982/83).

6. CONCLUSIONS

We used data from a stratospheric sounding unit to outline some of
the salient features of the seasonal evolution in the extra-tropical
stratosphere of both hemispheres. Planetary-scale disturbances are
well-known to be an important element of the stratospheric
circulation during winter and spring. They lead to stratospheric
warmings whose evolution depends on the time of year and on which
hemisphere they occur in. In the southern hemisphere, the most
pronounced warming is the springtime final warming. The
stratospheric westerly vortex is distorted and shrinks, with weak
horizontal gradients of potential vorticity forming around it. In
the northern hemisphere, strong warmings take place before and after
mid-winter. Some early winter warmings are of the so-called Canadian
type, and are characterised by strong distortions of the circulation
(manifested in extreme buckling of isopleths of potential vorticity
on synoptic maps), but only modest rises in temperature at middle and
high latitudes. Though displaced from the pole, the cyclone remains
strong. By contrast, late winter warmings in the northern
hemisphere, some of which are "major" warmings, are associated with
dramatic rises in temperature and a breakdown of the vortex. During
this breakdown, the cyclone shrinks rapidly, and weak gradients of
potential vorticity form, on average, around it. This evolution

marks the formation of what has been called (by McIntyre and Palmer 1983, 1984) the main vortex/surf zone structure. After the warming, there is strong radiative cooling, and some recovery of the westerly flow. A dynamically active final warming may succeed a strong mid-winter warming if there is enough time for radiation to re-establish the stratospheric westerly vortex (O'Neill, Pope and Mechoso, paper in preparation). But after the major warming of 1981, winds remained weak at middle and high latitudes, and the transition to summertime easterlies was gradual.

We used a co-ordinate independent diagnostic - the area $A(Q)$ enclosed by an isopleth of potential vorticity, Q - to highlight some of the systematic changes in the distribution of potential vorticity that occur during the seasonal cycle of the extra-tropical stratosphere. In particular, the evolution of the curves $A(Q)$ for differing values of Q clearly revealed the formation of the main vortex/surf zone structure in the stratosphere. In the stratosphere of the southern hemisphere, such a structure forms when planetary-scale disturbances develop in spring, and usually not before. In the stratosphere of the northern hemisphere, by contrast, the main vortex/surf zone structure typically begins to form much earlier in the season - typically not long after mid-winter. The structure forms earlier in winter in the northern stratosphere than in the southern because planetary-scale disturbances are stronger and more persistent in the northern hemisphere. In the northern stratosphere, the formation of the main vortex/surf zone is accompanied by an increase in areas $A(Q)$ for low values of Q, giving rise to a double-peak structure in the time series of $A(Q)$. The first peak corresponds to the build up of the westerly vortex by radiative cooling in early winter; the second to increases in Q on an isentropic surface when strong radiative cooling in the post-warming period brings down high values of Q from above.

A co-ordinate independent equation for the radiative contribution to the time rate of change of $A(Q)$ was used as the basis for an outline of the interaction between radiation and dynamics in the stratosphere, with emphasis on systematic effects. Dynamically induced departures from radiative equilibrium, and the associated radiative field, could account, qualitatively at least, for some of the features noted in the evolution of the $A(Q)$. In particular, during the build up of the westerly vortex in the stratosphere in early winter, there is a competition between radiation, which acts to create a circumpolar westerly vortex of high-Q air, and vortex erosion and mixing of Q by disturbances, which act to break down the vortex by redistributing Q irreversibly. The vortex builds more slowly than it would in a hypothetical, axisymmetric stratosphere in which the seasonal cycle was driven solely by the march of solar radiative heating. By contrast, during the breakdown of the westerly vortex after mid-winter (during spring in the southern hemisphere) radiation augments vortex erosion. The vortex breaks down more rapidly than it would in a hypothetical, axisymmetric stratosphere (Pope, 1989).

A major issue was which of the known non-conservative processes in the stratosphere - radiation, quasi-horizontal mixing or the effects of gravity-wave drag - contribute to the formation of the main vortex/surf zone structure, and what are their relative contributions. McIntyre and Palmer (1983, 1984) argued that the structure forms in the stratosphere of the northern hemisphere by vortex erosion and quasi-horizontal mixing of potential vorticity during so-called planetary-wave breaking. We argued here, however, that the effects of dynamically induced departures from radiative equilibrium can also lead to the formation of a main vortex/surf zone-like structure in the average horizontal distribution of Q. We referred to the detailed quantitative work of Pope (1989), who demonstrated, by using data from a numerical model and from satellites, that radiation was a major contribution to the formation of the structure during the year (1980/81) discussed in this paper.

REFERENCES

Butchart, N. and Remsberg, E.E. (1986) The area of the stratospheric polar vortex as a diagnostic for tracer transport on an isentropic surface. J. Atmos. Sci., 43, 1319-1339.

Clough, S.A., Grahame, N.S. and O'Neill, A. (1985) Potential vorticity in the stratosphere derived using data from satellites. Q.J.R. Meteorol. Soc., 111, 335-358.

Fairlie, T.D.A., Fisher, M. and O'Neill, A. (1990) The development of narrow baroclinic zones and other small-scale structure in the stratosphere during simulated major warmings. Q.J.R. Meteorol. Soc., 116, 287-316.

Geller, M.A., Wu, M.-F. and Gellman, M.E. (1984) Troposphere-stratosphere (surface-55 km) monthly winter general circulation statistics for the northern hemisphere - interannual variations. J. Atmos. Sci., 41, 1726-1744.

Juckes, M.N. and O'Neill, A. (1988) Early winter in the northern stratosphere. Q.J.R. Meteorol. Soc., 114, 1111-1126.

Klenk, K.F., Bhartia, P.K., Hilsenrath, E. and Fleig, A.J. (1983) Standard ozone profiles from balloon and satellite datasets. J. Clim. Appl. Met., 22, 2012-2022.

Labitzke, K. (1977) Interannual variability of the winter stratosphere in the northern hemisphere. Mon. Wea. Rev., 105, 762-770.

Labitzke, K. (1982) On the interannual variability of the middle stratosphere during the northern winters. J. Met. Soc. Japan, 60, 124-139.

McIntyre, M.E. and Palmer, T.N. (1983) Breaking planetary waves in the stratosphere. Nature (London), 305, 593-600.

McIntyre, M.E. and Palmer, T.N. (1984) The "surf zone" in the stratosphere. J. Atmos. Terr. Phys. 46, 825-849.

Mechoso, C.R. O'Neill, A., Pope, V.D. and Farrara, J.D. (1988) A study of the stratospheric final warming of 1982 in the southern hemisphere. Q.J.R. Meteorol. Soc., 114, 1365-1384.

Pick, D.R. and Brownscombe, J.L. (1981) Early results based on the stratospheric channels of TOVS on the TIROS-N series of operational satellites. Int. Counc. Sci Unions, Comm. Space Res., Adv. Space Res., 1, No. 4, 247-260.

Pope, V.D. (1989) The interaction between radiation and dynamics in the observed and simulated stratosphere. PhD University of Reading, pp. 1-204.

Shine, K.P. (1987) The middle atmosphere in the absence of dynamical heat fluxes. Q.J.R. Meteorol. Soc., 113, 603-633.

THE FINAL WARMING OF THE STRATOSPHERE

CARLOS R. MECHOSO
Department of Atmospheric Sciences
University of California Los Angeles
Los Angeles, California 90024, USA

ABSTRACT. The dramatic changes during spring in the shape of the polar vortex in the stratosphere of the two hemispheres are illustrated using perspective plots of the three-dimensional structure of the isotach and potential vorticity fields. In the southern hemisphere, the vortex shape evolves from a cone, slightly expanding with height and nearly symmetric about the pole, to an inverted cone, distorted and displaced from the pole. At 10 mb, in particular, the flow reverses to easterlies some time after the temperature gradient at high latitudes reverses. In the northern hemisphere, there is much more variability in both the structure of the flow and the time of the breakdown of the westerly vortex. At 10 mb, in particular, the reversal of the flow may precede or follow that of the temperature gradient.

There are also important interhemispheric differences in the location and magnitude of the largest temperature increases over the polar regions during spring. Those in the southern hemisphere are in the lower stratosphere whereas those in the northern hemisphere are in the upper stratosphere, the former being almost twice as large as the latter. The values of minimum temperatures in the lower stratosphere suggest that in early spring conditions suitable for the formation of polar stratospheric clouds, thought to play a key role in ozone destruction, are the rule in the southern hemisphere and the exception in the northern hemisphere.

1. INTRODUCTION

The polar region in the stratosphere can become warmer than its surroundings more than once during winter and spring. The most spectacular of these "warmings" is exclusive to the northern hemisphere: every so often in mid-winter the westerly circulation is violently disrupted and the temperature in the polar region increases by tens of degrees in a matter of days. This paper deals with the last of these warmings, referred to as the "final warmings", after which the summer circulation becomes established.

The winter and summer stratospheric circulations have attracted a great deal of attention, and major progress in our understanding of the stratosphere has been achieved in attempting to unravel their distinct phenomena. The relative quietness of the summer circulation strongly contrasts with the sporadic bursts of planetary-wave activity in winter.

An insight into the reasons for these behaviors comes from the analyses of vertical propagation of planetary waves by Charney and Drazin (1961) and Matsuno (1970); wave-mean flow interaction by Andrews and McIntyre (1976); and the introduction of concepts of planetary-wave breaking and "surf-zones" by McIntyre and Palmer (1983).

A. O'Neill (ed.),
Dynamics, Transport and Photochemistry in the Middle Atmosphere of the Southern Hemisphere, 55–69.
© 1990 Kluwer Academic Publishers.

Comparatively, the spring and fall circulations have received less attention. Nevertheless, the intriguing spring warming is recognized in radiosonde observations over Antarctica during the early 1940s. Court (1942) asserts that the lack of exchanges between Antarctica and surroundings in winter allows the air there to cool down in isolation until it has lower temperatures than anywhere else on earth. He also reports the downward progress of the springtime warming over Little America, Antarctica. Godson (1963) points out that the final warming in 1961 over Byrd, Antarctica, occurred in a very gradual fashion at both 100 and 30 mb, with peak temperatures around mid-December. He also reports that other years – and stations – show a tendency toward a stepwise warming earlier in the season. Labitzke and van Loon (1972) report that winter temperatures in the lower stratosphere are lower and summer temperatures higher over the South Pole than over the North Pole. Hirota et al. (1983), based on an analysis of satellite data, state that eddy activity in the period from spring to summer is more vigorous in the southern hemisphere than in the northern hemisphere. Shiotani and Hirota (1985) find that for the stratosphere of the southern hemisphere in spring 1981 there is an enhancement in the activity of the flow component with zonal wavenumber two, followed by a downward shift in the zonal-mean jet core and an enhancement in the flow component with zonal wavenumber one. For the northern hemisphere, on the other hand, they find no systematic pattern in the evolution of the wave activity and jet core. Yamazaki and Mechoso (1985) investigate the final warming in the southern hemisphere during 1979 and describe a planetary-wave event that unfolds during the warming.

The interest in the final warming of the stratosphere increased dramatically after the discovery of the Antarctic "ozone hole" phenomenon (Farman et al., 1985). The current consensus is that ozone is depleted in reactions with radicals resulting from the destruction of man-made chlorofluorocarbons. The reactions would occur on the surface of particles forming polar stratospheric clouds (PSCs). PSCs require special meteorological conditions to develop. The air inside the polar vortex during winter and spring appears to satisfy those conditions. Consequently, the behavior of the vortex in the southern hemisphere during winter and its breakdown in spring have been the object of an unprecedented scrutiny, which has extended to the northern hemisphere.

Recent studies of the final warming in the southern hemisphere using satellite data include those of Farrara and Mechoso (1986), Newman (1986), Yamazaki (1987), Garcia and Solomon (1987), Newman and Randel (1988) and Mechoso et al. (1988). Farrara and Mechoso (1986) draw attention to the fact that the final warming in the southern hemisphere shows interannual variability in association with that in planetary-wave activity. Newman (1986) classifies the final warming in the southern hemisphere into two categories: 1) those in mid-October accompanied by a major wave event and 2) those in mid-November with no strong October wave event. Yamazaki (1987) points out that in 1982 the final warming in the southern hemisphere is more rapid and intense than in the northern hemisphere. Garcia and Solomon (1987) find that higher and lower October temperatures are apparently associated with the easterly and westerly phase of the tropical quasi-biennial oscillation (QBO). Mechoso et al. (1988) show that in the southern hemisphere stratosphere the warming starts to develop in - or near - the same longitudinal quadrant year after year, that eastward-travelling anticyclones develop over a preferred geographical region in September, and that a quasi-stationary anticyclone lies roughly between 90°E and 180°E in October. They suggest that these are three pieces of evidence indicative of a connection with surface features, particularly in the orography.

In this paper, we present a three-dimensional view of the vortex evolution during the final warming in both hemispheres. We look at the vortex structure, interannual variability, and interhemispheric differ-ences. The dataset used consists of daily geopotential height and tem-perature analyses compiled since October 1978 by the U.S. National Meteorological Center (NMC) and the U.K. Meteorological Office (UKMO) using retrievals from stratospheric sounding units (SSUs) and microwave sounding units (MSUs) on board the National Oceanic and Atmospheric Administration (NOAA) polar orbiting satellites.

We begin in Section 2 by showing the vortex evolution from early winter to early spring in terms of the isotach field. The changes in zonal-mean temperatures during spring are shown in Section 3. Our pre-sentation of the evolution of vortex structure during the final warming is given in Section 4. In Section 5 we discuss sudden warming events during the final warming. The timing of the final warming in both hemi-spheres is compared in Section 6. A summary and discussion of our current knowledge and future directions in the subject concludes the paper in Section 7.

2. THE VORTEX EVOLUTION FROM EARLY WINTER TO EARLY SPRING

The vortex evolution is presented in terms of the isotach field. Isotachs are shown for one level in the upper troposphere (300 mb) and four levels from the lower to the upper stratosphere (100, 50, 10, and 2 mb). Winds are computed by using the geostrophic relation. All procedures for the estimate of actual winds are open to criticism in regard to their accuracy (Elson, 1986; Boville, 1987). However, geostrophic winds are adequate for our purpose of outlining the major features in the circulation.

2.1. The Southern Hemisphere

We select the final warming in 1982 as an example. This year is charac-terized by intense dynamical events, but its broad features are largely common to other years in the dataset.

Monthly-mean isotachs for June and September are shown in Fig. 1. All contours and shading in the figures in this section correspond to winds with westerly zonal component. One of the interesting features in Fig. 1 is that zonal asymmetries in the isotachs in the upper troposphere and lower stratosphere appear connected in the vertical. In the upper troposphere, the subtropical jet spirals polewards from south of Australia to the coast of East Antarctica, where it ends at a local wind maximum. In the lower stratosphere, the local maxima appear as extensions of the subtropical jet and the mid-latitude jet south of Africa.

In June, winds in the stratosphere strengthen with increasing height to values larger than 110 ms^{-1} at 2 mb. In September, winds in the lower and middle stratosphere are stronger than in June, but those in the upper

JUNE 1982 ISOTACHS   SEPTEMBER 1982 ISOTACHS

Figure 1. Monthly-mean isotachs in the southern hemisphere for June and September 1982 at 300, 100, 50, 10 and 2 mb. Polar stereographic projections are plotted at an angle to give an impression of the three dimensional structure. Map and grid projections are shown at the top and bottom of each panel to fix the position of the flow features. The shading corresponds to wind speeds greater than the following values: 35 ms^{-1} at 300 mb, 40 ms^{-1} at 100 mb, 45 ms^{-1} at 50 mb, 60 ms^{-1} at 10 mb and 75 ms^{-1} at 2 mb. At each level, the contour outside the shaded region corresponds to 25 ms^{-1}. The contour within the shaded region corresponds to 75 ms^{-1} at 10 mb and 110 ms^{-1} at 2 mb for June; 55 ms^{-1} at 100 mb, 60 ms^{-1} at 50 mb and 90 ms^{-1} at both 10 and 2 mb for September.

stratosphere are weaker. There are values in excess of 60 ms^{-1} at 50 mb in the Indian Ocean sector, where winds are barely above 45 ms^{-1} in June. This reflects the on-going process of vortex decay, which starts at upper levels and spreads down in time as the season advances. Another view of this process is given in the next section. From the point of view of the isotachs, therefore, the stratospheric vortex in early winter is strong, broader with height, and fairly symmetric about the pole. The vortex in early spring is stronger and more symmetric in the lower stratosphere, but is weaker and more asymmetric in the upper stratosphere.

2.2. The Northern Hemisphere

We also take the final warming of 1982 as an example. This event is not highly representative, however, since final warmings in the northern

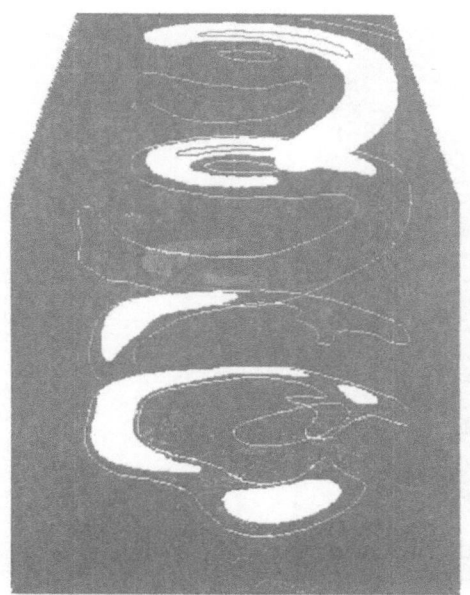

Figure 2. As in Fig. 1, except for the northern hemisphere in December 1981 and March 1982. In December, the shading corresponds to wind speeds greater than the following values: 35 ms^{-1} at 300, 100 and 50 mb, 60 ms^{-1} at 10 mb and 75 ms^{-1} at 2 mb. The contour within the shaded region corresponds to 75 ms^{-1} at 10 mb and 110 ms^{-1} at 2 mb. In March, the shading corresponds to wind speeds greater than 30 ms^{-1} at all levels.

hemisphere show much more interannual variability than in the southern hemisphere.

The monthly-mean isotachs for December 1981 and March 1982 are shown in Fig. 2. There are several similarities between panels for corresponding months in Figs. 1 and 2. Both hemispheres show an apparent connection between wind maxima in the upper troposphere and lower stratosphere. Further, winds in early winter (June and December for the southern and northern hemisphere, respectively) increase with increasing height to values in excess of 110 ms^{-1} at 2 mb. There are also notable differences. The stratospheric vortex in the northern hemisphere is more asymmetric about the pole than in the southern hemisphere, particularly in early spring. Also, the vortex is broader in the northern hemisphere than in the southern hemisphere.

To illustrate the interannual variability in the northern hemisphere, we show the monthly-mean isotachs for March 1985 (Fig. 3). In this case, the vortex is stronger in the upper stratosphere than in the middle stratosphere.

MARCH 1985 ISOTACHS

Figure 3. As in Figure 2, except for March 1985. The shading corresponds to wind speeds greater than 30 ms^{-1} at all levels.

3. THE CHANGES IN ZONAL-MEAN TEMPERATURES DURING SPRING

The spring warming is concisely represented by the difference between zonal-mean values at the end and at the beginning of the period. The corresponding fields averaged for the years 1979-1986 are shown in Fig. 4. In the northern hemisphere, largest temperature increases in the spring are over the polar region in the upper stratosphere and amount to approximately 25 K. In the southern hemisphere, largest temperature increases are also over the polar regions, but develop in the lower stratosphere and are almost twice as large (45 K) as those in the northern hemisphere. Corresponding to these temperature changes, easterly winds generally appear first in the upper stratosphere and then gradually spread down to lower levels (Mechoso *et al.*, 1989).

4. THE FINAL WARMING IN TERMS OF THE VORTEX STRUCTURE

We define the outside edge of the stratospheric vortex on a surface of constant potential temperature (isentropic surface) as the isoline of Ertel's potential vorticity where the gradients of this field change from

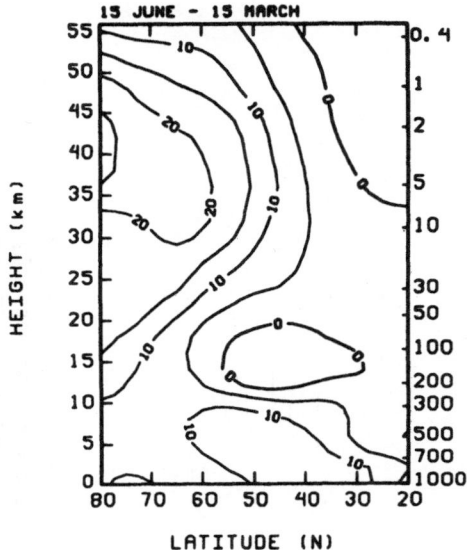
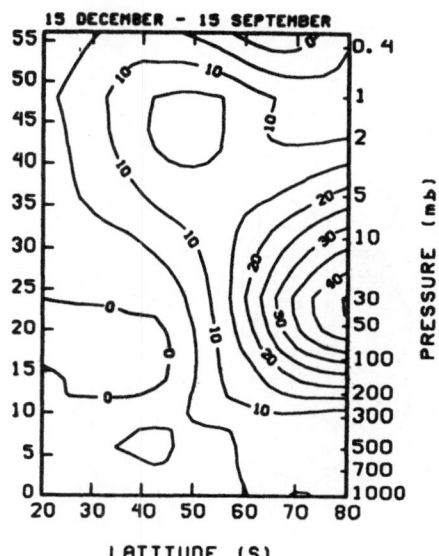

Figure 4. Average for the years 1979-1986 of the difference between zonal mean temperatures at the end and the beginning of spring for the northern hemisphere (left) and the southern hemisphere (right).

strong (inside) to weak (outside). In the region of strong gradients, Rossby-wave disturbances result in small transversal displacements of the air parcels. In the region of weak gradients, the displacements can be comparatively large and eventually lead to Rossby-wave breaking. The breaking is associated with mixing of the potential vorticity, which can become irreversible.

Perspective plots of the three-dimensional structure of the outside edge of the vortex can be depicted by plotting the corresponding locations for several isentropic surfaces distributed throughout the stratosphere. In this section we present such plots for selected days in the spring of each hemisphere. The isentropic surfaces range from 400 K to 1200 K.

4.1. The Southern Hemisphere

Perspective plots of the westerly vortex in the southern hemisphere for selected days in June, September, October and November 1982 are shown in Fig. 5. On 16 June, the vortex is almost symmetric around the pole and covers a large area that increases with height. On 11 September, the vortex has shrunk considerably throughout the depth of the stratosphere and is of cylindrical shape. These views are consistent with those of the vortex provided by the isotachs (see Fig. 1). On 21 October, the area covered by the vortex is greatly reduced in the upper stratosphere and the structure looks like an inverted cone distorted and displaced

62

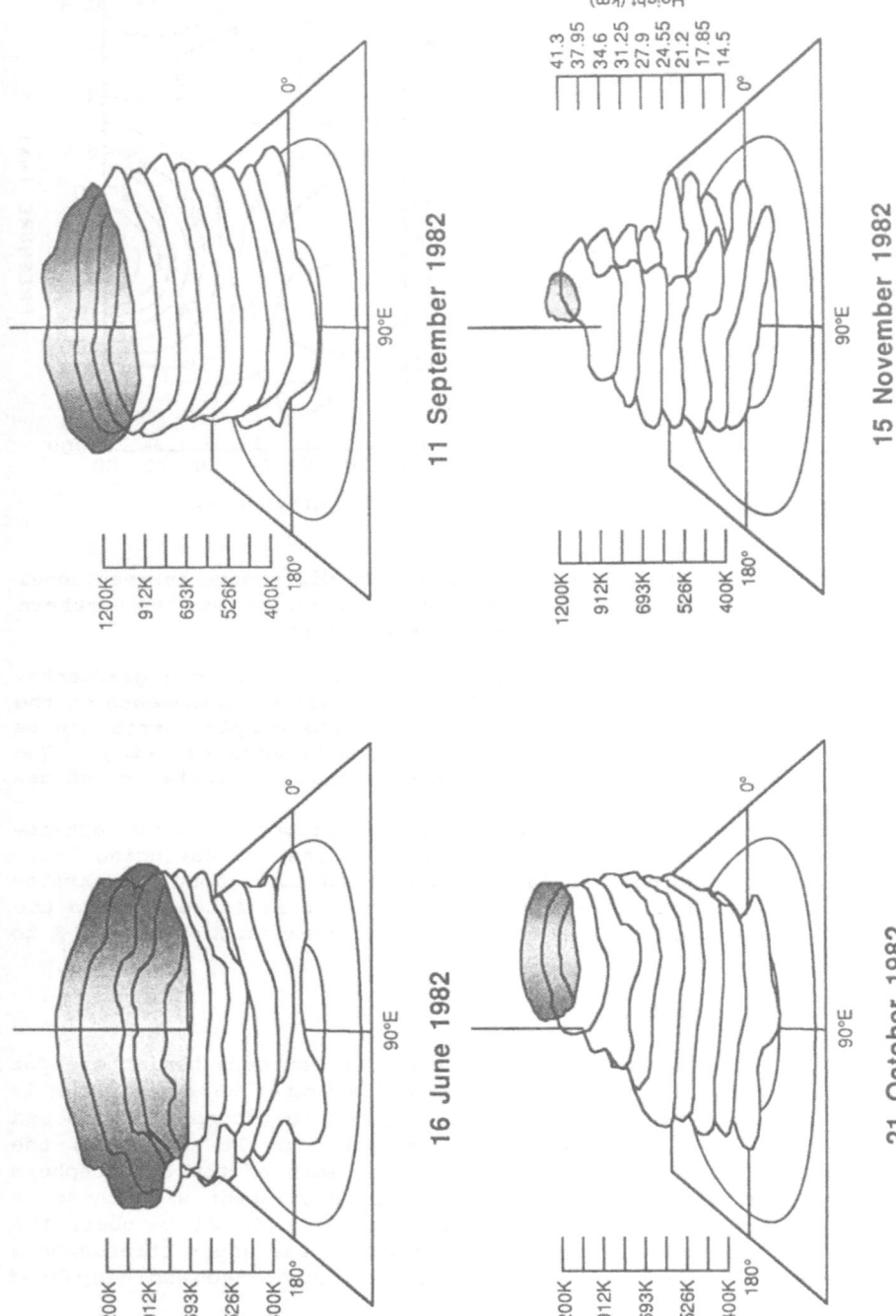

Figure 5. Perspectives of the three-dimensional structure of the westerly vortex in the southern hemisphere from the lower stratosphere (400K) to the upper stratosphere (1200K).

from the pole towards the eastern half of the hemisphere. On 15 November, the area covered by the vortex is further reduced, particularly in the upper stratosphere. The changes in vortex structure during spring, therefore, are much larger in the upper stratosphere than in the lower stratosphere.

There are a number of systematic features that unfold during the evolution described above. In September, synoptic plots show anti-cyclonic disturbances travelling eastward along the periphery of the vortex (Mechoso et al., 1988). These anticyclones correspond to an enhancement of the eastward travelling component of the flow with zonal wavenumber two (ETR-wave 2). This enhancement was reported several years back by Leovy and Webster (1976), but its precise originating mechanisms remain unclear. In October, there is an amplification of a quasi-stationary anticyclone that corresponds to an enhancement of the flow component with zonal wavenumber one (QS-wave 1). Mechoso et al. (1988) attribute the accompanying rapid shrinkage of the cyclone to the combined effects of the drawing away of its high potential vorticity and the dynamically induced radiative effects produced when the cyclone is displaced from the pole by the growing anticyclone.

4.2. The Northern Hemisphere

Perspective plots of the vortex structure on 24 January and 10 March 1982 are shown in Fig. 6. On 24 January, the vortex splits in the upper stratosphere, a behavior not found in the southern hemisphere during mid-winter. On 10 March, the vortex is displaced from the pole, has cylindrical shape in the lower and middle stratosphere, and expands in the upper stratosphere. In mid-April the vortex decays rapidly and by the end of the month easterlies prevail in the middle stratosphere (not shown). In 1982, therefore, there are some similarities in the evolution of the final warming in both hemispheres.

As indicated above, this similarity does not extend to all cases. Final warmings in the northern hemisphere can be schematically divided into those of the disturbed type and those of the quiescent type. Final

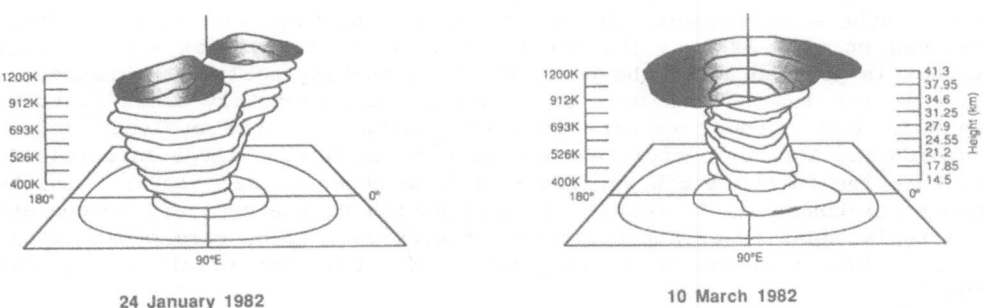

24 January 1982   10 March 1982

Figure 6. As in Figure 5, except for the northern hemisphere.

warmings of the disturbed type, such as those of 1982 and 1985, are somewhat similar to final warmings in the southern hemisphere in that a persistent anticyclone highly disturbs the westerly circulation.

Final warmings of the quiescent type, such as that in 1981, proceed
slowly in a stratosphere with weak zonal asymmetries. A major warming in
early February 1981 was accompanied by a weakening of the westerly vortex
and a rapid deceleration of zonal-mean winds throughout the stratosphere.
In the upper stratosphere, where the radiative damping time is short,
radiative cooling after the warming contributed to the re-establishment
of a westerly circulation during March. In the lower stratosphere, where
the damping time is longer, the cyclonic vortex and westerly winds remained weak. The strong disturbances that developed periodically in the
troposphere during March 1981 had little effect on the upper stratosphere,
where the circulation remained almost symmetric about the pole. The wind
reversal in the upper stratosphere was therefore mainly radiatively
driven. The final warming in 1981 for the northern hemisphere was even
more gradual than its counterpart in the southern hemisphere.

5. SUDDEN WARMINGS DURING THE FINAL WARMING

The spring decelerations of the stratospheric flow are punctuated by
strong events of planetary-wave disturbances. These events involve large
temperature increases in polar regions. In general, temperature increases
in the polar region during sudden warmings in spring are much larger than
those during similar events in the fall (Mechoso et al., 1989).

5.1. The Southern Hemisphere

In the southern hemisphere during spring of 1982 winds decelerate steadily throughout the stratosphere (Fig. 7). The largest and most rapid
changes are in the middle and upper stratosphere, notably in late October
when westerlies are replaced by easterlies during a strong warming pulse.
In the vertical at 60°S at this time, the westerly jet core descends
rapidly from the middle to the lower stratosphere, and thereafter it
descends more slowly. In the horizontal at 10 mb, the westerly jet moves
poleward throughout the period as is characteristic of final warmings in
the southern hemisphere (Hartmann, 1976; Shiotani and Hirota, 1985;
Mechoso et al., 1988). The jet fluctuates in strength as the westerly
vortex is jostled about the polar cap by planetary-scale disturbances.

In the lower and middle stratosphere, the planetary-scale disturbance in late October results in a displacement of the cyclonic vortex
towards the African sector, and the establishment of a large anticyclonic
circulation in the South Pacific sector (Mechoso et al., 1989). In the
upper stratosphere, it results in destruction of the cyclonic vortex and
its replacement by an anticyclonic circulation. Also, warm pools of air
take a polar position where they remain for the rest of the spring and
summer.

5.2. The Northern Hemisphere

In the northern hemisphere during spring of 1982, the westerly jet is in
high latitudes during March (Fig. 8). The jet decelerates strongly

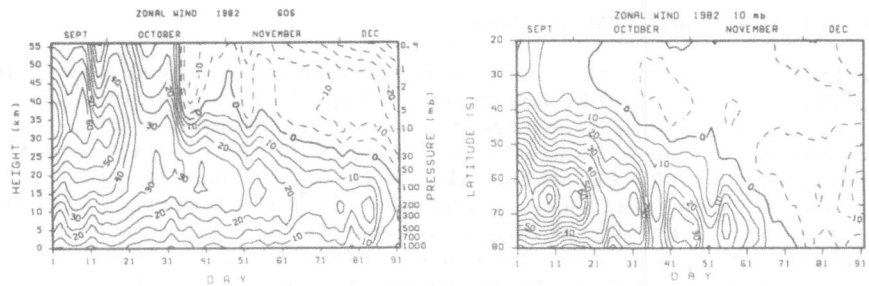

Figure 7. Variation of zonal-mean wind (ms^{-1}) during spring 1982 in the southern hemisphere. Height-time section at 60°S (left), latitude-time section at 10 mb (right).

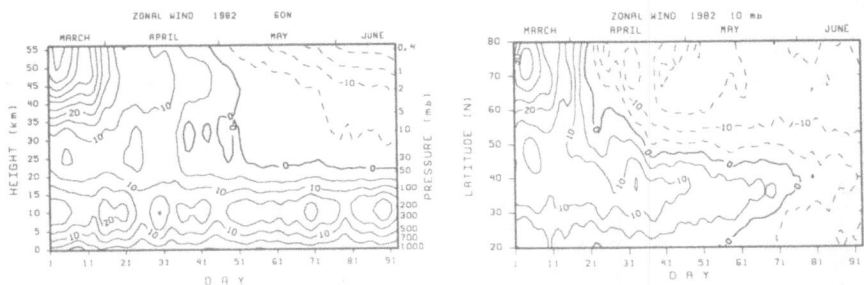

Figure 8. As in Figure 7, except for the northern hemisphere.

throughout the middle and upper stratosphere during the late part of the month and early April as a strong planetary scale disturbance develops. In the vertical at 60°N, the zero wind line descends from the upper stratosphere to about 50 mb in the lower stratosphere. In the horizontal at 10 mb, easterlies appear almost simultaneously at middle and high latitudes and spread more gradually into lower latitudes.

The planetary-scale disturbance in early April results in circulation changes that resemble those described above for the southern hemisphere in late October (Mechoso et al., 1989). In the lower and middle stratosphere, anticyclonic and cyclonic circulations are established over the North American and European sectors, respectively. In the upper stratosphere, an anticyclonic circulation is established over the pole. Vortices in the northern hemisphere are smaller than in the southern hemisphere and the flow is not so strongly dominated by its zonal wavenumber one component.

6. THE TIMING OF THE FINAL WARMING

The timings of the wind and temperature gradient reversals at 10 mb during the springs of 1979 to 1987 are shown in Fig. 9. In the northern hemisphere the timing of these reversals can vary by a month or more from one year to the next (see also Labitzke, 1982). For example, the zonal-mean wind reversal in 1981 was about six weeks later than it was in 1982. Notice that for some years (e.g., 1981 and 1982) the temperature gradient reversal precedes the wind reversal, whereas for other years (e.g., 1985) the converse applies. This difference reflects the different vertical structures found in the northern hemisphere during spring. From the thermal-wind relation, the former situation is consistent with zonal-mean easterlies first appearing in the upper stratosphere and then descending into the middle stratosphere; the latter with easterlies first appearing in the middle stratosphere when westerlies prevail aloft. There is less interannual variability in the timings of the reversal in the southern hemisphere. Consistent with the gradual downward progression of the zonal-mean jet, the wind reversal occurs some time after the temperature gradient reversal.

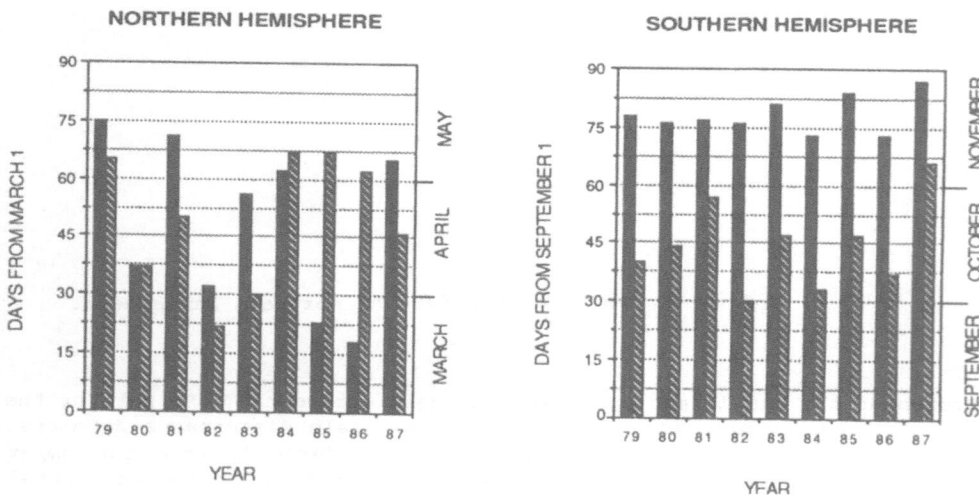

Figure 9. Time of the reversal, at 10 mb, of the zonal wind (filled bars) at 60° and of the temperature difference between 60° and 85° (hatched bars) for the northern hemisphere (left) and southern hemisphere (right).

7. SUMMARY

The final warming of the stratosphere has emerged as one of the richest and most challenging phenomena of this region of the atmosphere. It is host to large dynamical events and drastic changes in the radiative heating.

Final warmings in the southern hemisphere are characterized by their "predictability", evolving in a broadly similar way each year. The lower polar stratosphere warms up strongly (about 45 K), the vortex decays in the upper stratosphere first, ETR-wave 2 amplifies in September and QS-wave 1 amplifies in October at a preferred geographical location (the first quadrant). The temperature gradient in the upper stratosphere reverses first and, a few weeks later, easterlies prevail at all latitudes. The evolution can be punctuated by sudden warmings, and the reversal of the temperature gradient can occur in unison with one of them.

Final warmings in the northern hemisphere are much more variable in the structure of the flow at the time of the breakdown of the westerly vortex. In this hemisphere, final warmings can be schematically divided into disturbed and quiescent. Disturbed warmings are much like warmings in the southern hemisphere, with the vortex decaying at upper levels first and a quasi-stationary anticyclone developing over the Aleutian sector. During quiescent warmings, the flow is weak in the stratosphere and highly symmetric about the pole as Rossby waves do not propagate vertically through the weak potential vorticity gradients remaining after the strong mixing in winter.

The differences between the evolution of the stratospheric flow during final warmings in the northern and southern hemispheres extend to the timing of the final reversal to summer easterlies. In the northern hemisphere, there is large interannual variability in this timing. Further, the reversal of the flow may precede or follow that of the temperature gradient at high latitudes. In the southern hemisphere, on the other hand, there is less interannual variability in the timing, and the flow reversal follows that of the temperature gradient by a few weeks.

There is some indication that the evolution of the stratosphere during winter can influence the nature and timing of the final warming in spring. A major mid-winter warming may lead to an almost complete breakdown of the westerly flow in the stratosphere. It seems that if this happens towards the end of winter, particularly in February, strong gradients of potential vorticity may not be re-established radiatively in the lower and middle stratosphere. Then, tropospheric disturbances can no longer propagate upwards, and a dynamically induced final warming is precluded. The comparatively late major warmings of February 1979, 1981 and 1984 were all succeeded by late, quiescent wind reversals. However, in view of the complexity and variability of the atmospheric circulation, exceptions to such a rule are to be expected.

In Mechoso et al. (1990) we emphasize an important interhemispheric difference in the circulation in the lower stratosphere. This was revealed by inspection of the daily values of minimum brightness temperatures from channel 24 of the MSU during winter and spring for the period 1980 to 1988. These values represent layer-mean temperatures centered at about 90 mb, i.e. in the lower stratosphere. The range of these daily values shows that the largest interannual variability is in late winter in the northen hemisphere and in spring in the southern hemisphere. If 195 K can be taken as the threshold for the formation of polar stratospheric clouds (PSCs) for the conditions prevailing in the lower stratosphere, then late winter temperatures in the lower stratosphere of the southern hemisphere are always below the threshold for the formation of PSCs, while those in the northern hemisphere are rarely below that threshold.

All the results presented in this paper were obtained from analyses of geopotential height and temperature fields using a set of approximate relations to estimate the horizontal component of the wind. While the procedure is of questionable accuracy if the resulting winds are to be used in detailed studies of specific phenomena, we find it acceptable in order to achieve a broad picture of the stratospheric circulation. The time has come, however, to embark on the investigation of largely unresolved problems that will require detailed calculations, such as the instability of stratospheric flows, tracer transports during warmings, parcel trajectories, and several others of similar importance. One of the outstanding problems is the relative "permeability" of the walls of the stratospheric vortex and the validity of the "containment vessel" paradigm. These investigations will require new datasets including wind information. The forthcoming EOS (earth observing systems) program carries the promise of yielding some of this information.

Acknowledgements. The material presented here is part of a comprehensive study of the general circulation of the stratosphere being performed in collaboration between groups at the University of California Los Angeles and United Kingdom Meteorological Office. Members of the UCLA group are the author, J. D. Farrara, D. Pan and G. L. Manney. Members of the UKMO group are A. O'Neill, V. D. Pope and M. Fisher (presently at UCLA). The research at UCLA was supported by NSF under Grant ATM 8814892 and by NASA under Grant NAGW-1021. Additional support was provided by NATO under Grant 0391/87.

References

Boville, B. A., 1987: The validity of the geostrophic approximation in the winter stratosphere and troposphere. *J. Atmos. Sci.*, **44**, 443-457.
Charney, J. G., and P. G. Drazin, 1961: Propagation of planetary-scale disturbances from the lower into the upper atmosphere. *J. Geophys. Res.*, **66**, 83-109.
Court, A., 1942: Tropopause disappearance during the Antarctic winter. *Bull. Amer. Meteor. Soc.*, **23**, 220-238.
Elson, L. S., 1986: Ageostrophic motions in the stratosphere from satellite observations. *J. Atmos. Sci.*, **43**, 321-330.
Farman, J. C., B. G. Gardiner and J. D. Shanklin, 1985: Large losses of total ozone in Antarctica reveal seasonal ClO_x/NO_x interaction. *Nature*, **315**, 207-210.
Farrara, J. D. and C. R. Mechoso, 1986: An observational study of the final warming in the Southern Hemisphere stratosphere. *Geophys. Res. Lett.*, **13**, 1232-1235.

Garcia, R. R. and S. Solomon, 1987: A possible relationship between interannual variability in Antarctic ozone and the quasi-biennial oscillation. *Geophys. Res. Lett.*, **14**, 848-851.

Godson, W. L., 1963: A comparison of middle stratosphere behavior in the Arctic and Antarctic, with special reference to final warmings. *Meteor. Abhandl.*, **36**, 161-206.

Hartmann, D. L., 1976: The structure of the stratosphere in the Southern Hemisphere during late winter 1973 as observed by satellite. *J. Atmos. Sci.*, **33**, 1141-1154.

Hirota, I., T. Hirooka and M. Shiotani, 1983: Upper stratospheric circulations in the two hemispheres observed by satellites. *Q. J. Roy. Meteor. Soc.*, **109**, 443-454.

Juckes, M. N., and M. E. McIntyre, 1987: A high-resolution one-layer model of breaking planetary waves in the stratosphere. *Nature*, **328**, 590-596.

Labitzke, K., and H. van Loon, 1972: The stratosphere in the Southern Hemisphere. In *Meteorology of the Southern Hemisphere*. C. W. Newton (ed.), Meteor. Mono., **35**, Amer. Meteor. Soc., Boston, MA, 113-138.

Labitzke, K., 1982: On the interannual variability of the middle stratosphere during the northern winters. *J. Met. Soc. Jap.*, **60**, 124-139.

Leovy, C. B., and P. J. Webster, 1976: Stratospheric long waves: Comparison of thermal structure in the Northern and Southern Hemispheres. *J. Atmos. Sci.*, **33**, 1624-1638.

Matsuno, T., 1970: Vertical propagation of stationary planetary waves in the winter Northern Hemisphere. *J. Atmos. Sci.*, **27**, 871-883.

McIntyre, M. E., and T. N. Palmer, 1983: Breaking planetary waves in the stratosphere. *Nature*, **305**, 593-600.

McIntyre, M. E., and T. N. Palmer, 1984: The 'surf zone' in the stratosphere. *J. Atmos. Terr. Phys.*, **46**, 825-849.

Mechoso, C. R., A. O'Neill, V. D. Pope and J. D. Farrara, 1988: A study of the stratospheric final warming of 1982 in the Southern Hemisphere. *Q. J. Roy. Meteor. Soc.*, **114**, 1365-1384.

Mechoso, C. R., A. O'Neill, J. D. Farrara, V. D. Pope, and D. Pan, 1989: Interhemispheric comparison of the fall and spring circulations in the middle atmosphere. Third International Conf. on Southern Hemisphere Meteorology and Oceanography, November 13-17, 1989, Buenos Aires, Argentina. Amer. Meteor. Soc., 421-428.

Newman, P. A., 1986: The final warming and polar vortex disappearance during the Southern Hemisphere spring. *Geophys. Res. Lett.*, **13**, 1228-1231.

Newman, P. A., and W. J. Randel, 1988: Coherent ozone-dynamical changes during the Southern Hemisphere spring, 1979-1986. *J. Geophys. Res.*, **93**, 12,585-12,606.

Randel, W. J., 1988: The seasonal evolution of planetary waves in the Southern Hemisphere stratosphere and troposphere. *Q. J. Roy. Meteor. Soc.*, **114**, 1385-1409.

Shiotani, M., and I. Hirota, 1985: Planetary wave-mean flow interaction in the stratosphere: A comparison between the Northern and Southern Hemispheres. *Q. J. Roy. Meteor. Soc.*, **111**, 309-334.

Yamazaki, K., and C. R. Mechoso, 1985: Observations of the final warming in the stratosphere of the Southern Hemisphere during 1979. *J. Atmos. Sci.*, **42**, 1198-1205.

Yamazaki, K., 1987: Observations of the stratospheric final warmings in the two hemispheres. *J. Meteor. Soc. Jap.*, **65**, 51-65.

COMPARISON OF THE SOUTHERN HEMISPHERE SPRINGS OF 1988 AND 1987

Paul A. Newman, Mark R. Schoeberl, and Leslie R. Lait[*]
Code 616
NASA/Goddard Space Flight Center
Greenbelt, MD 20771

ABSTRACT. Differences between southern hemisphere (SH) springs of 1988 and 1987 in the stratosphere are discussed. The two years present a case study of opposite phases of the equatorial quasi-biennial oscillation (QBO) and the QBO's effect on SH spring stratospheric conditions. During 1988 (easterly QBO phase), mid-latitude temperatures were warmer than 1987 (westerly QBO phase) during July and August while polar temperatures were similar. During September, October, and November, 1988 polar temperatures were substantially higher than in 1987. Total ozone values reflected these thermal differences, with record low ozone values in October 1987 following the September ozone hole depletion phase, and higher total ozone values in October 1988. The large temperature differences result from a larger warming rate in 1988 than in 1987. Similarly, the higher total ozone amounts in October 1988 resulted from a weaker depletion during September 1988. The faster warming rate of temperature, and slower depletion rate for ozone in 1988 do not occur in a smooth linear fashion, but occur as a series of events, which result from strong planetary wave one events.

1. Introduction

The southern hemisphere (SH) springs of 1988 and 1987 were vastly different. Not only were 1988 late-Spring stratospheric temperatures much warmer than 1987, but 1988 total ozone values were substantially larger than 1987. The reasons for these differences involve an apparent quasi-biennial oscillation (QBO) of meteorological conditions. The purpose of this paper is to give a short description of the differences between 1988 and 1987 as an illustration of this effect, and to indicate the effect on the Antarctic ozone depletion.

The Antarctic ozone depletion results from a chemical mechanism (Journal of Geophysical Research, Special Issue on Antarctic Ozone, 1989).

---

[*] National Academy of Science/National Research Council Fellow

A. O'Neill (ed.),
Dynamics, Transport and Photochemistry in the Middle Atmosphere of the Southern Hemisphere, 71–89.
© 1990 Kluwer Academic Publishers.

The depletion begins in late-August and typically ends by early October. The depletion was quite small in 1979, but was very intense in 1987 (Schoeberl et al., 1989). The general reason for the large depletions during the last few years is caused by the increased concentrations of chlorine-containing species in the polar stratosphere in combination with the unique SH meteorological conditions (Anderson et al., 1989). The rate of depletion has a strong interannually varying component. For example, while the depletion was quite strong in 1987, it was much weaker in 1988. This spring total ozone interannual variability is related to the variability of the unique SH polar meteorological conditions. Years with relatively lower total ozone values are associated with both low levels of planetary wave activity, and lower temperatures (Newman and Randel, 1988).

Much of the interannual variability of the ozone hole and the SH meteorological conditions is synchronized with the QBO in tropical winds and temperature, as first pointed out by Garcia and Solomon (1986). A more comprehensive study of the Antarctic depletion - QBO connection was made by Lait et al. (1989). The latter study showed a clear out of phase relationship between the midlatitude and equatorial QBO for both total ozone and temperature (see also Oltmans and London, 1982). Lait et al. established that this QBO in ozone hole depletion rates could not be accounted for by direct equatorial QBO effects (i.e., the secondary circulation, see Plumb and Bell (1982) for a review of the QBO and the associated circulation). They further showed that the spring decline rate of total ozone showed a better correlation with the QBO than did the minimum ozone value, suggesting that the QBO was modulating the chemical loss rate.

Lait et al. suggested that the polar ozone QBO modulation was modulated by midlatitude eddy activity which was modulated by the equatorial QBO. This polar ozone QBO modulation could result from two mechanisms. The first mechanism is the direct modulation of the Polar Stratospheric Cloud (PSC) extent through the temperature of the Antarctic strato-sphere. Since PSC formation is a necessary step in the heterogeneous chemical processes which increase the number of chlorine containing radicals while denitrifying polar air, a reduction in PSC amounts would presumably reduce the magnitude of the chemical perturbation.

The second mechanism which would effectively reduce the magnitude of the ozone depletion is the mixing of midlatitude air into the polar region, across the polar vortex boundary. Midlatitude air is relatively rich in nitrogen-containing radicals. If this air is mixed with polar vortex air then the reactive radical chlorine monoxide (ClO) would be converted to the reservoir species chlorine nitrate, and the ozone depletion would be drastically reduced.

A third mechanism, which is related to the first mechanism, is the transport of ozone rich air into the depleted region. Planetary waves provide the primary driving mechanism of stratospheric processes. These wave events produce a poleward and downward secondary circulation. This

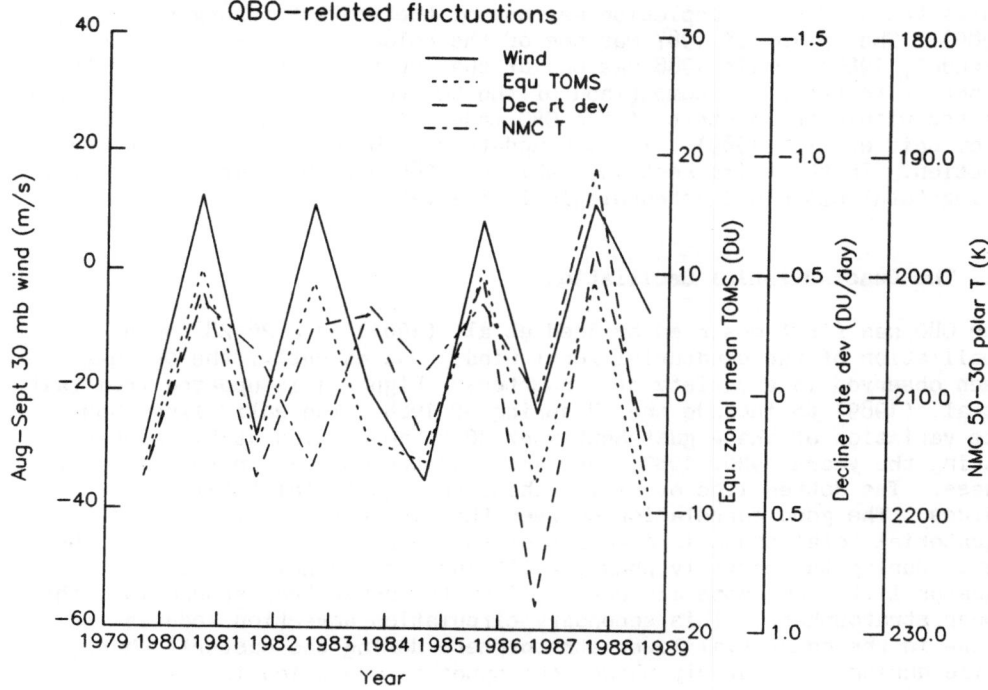

Figure 1. Zonal wind at 30 mb over Singapore, averaged over August-September for each year 1979-1988. Also shown are the equatorial zonal mean TOMS values and South polar NMC 50-30 mb layer mean temperatures averaged over the same months, as well as interannual fluctuations in the polar total ozone decline rates during late-August through September.

secondary circulation will have a number of effects among which are: an adiabatic increase of temperatures (see mechanism 1), a net transport of high ozone mixing ratios from mid-latitudes and above into the ozone depleted vortex, and a net deceleration of the polar night jet.

These mechanisms may not be independent, since temperature increases due to adiabatic heating (i.e. through descending air) will be accompanied by eddy activity increases. This will also increase the probability of mid-latitude entrainment (see WMO, 1986 for a discussion of the angular momentum balance). However, both the eddy forcing mechanism (and associated temperature effects) and the entrainment mechanism act in the same sense to decrease the magnitude of the ozone depletion. Thus any increase in eddy activity during the period of ozone hole formation should act to reduce the magnitude of the depletion.

The 1988 and 1987 spring Antarctic conditions represent two extreme cases which illustrate the processes discussed above. Whereas 1987 was in the easterly phase of the QBO, 1988 was in the westerly phase. The 1987 ozone hole was the deepest ever measured (Schoeberl et al., 1988) while the 1988 ozone depletion was one of the weakest (Krueger et al., 1989). The spring of 1987 was one of the coldest springs on record (Randel, 1988), while 1988 was one of the warmest. The purpose of this paper is to provide a comparison of the two years. This comparison will be set within the context of the QBO modulation which has been updated from Lait et al. (1989). The QBO update will be presented in the next section. In the third section, 1987 and 1988 will be compared directly using total ozone and meteorological observations.

2. The Quasi-Biennial Oscillation

The QBO was first observed by Reed et al. (1961) as a 26-28 month oscillation of the equatorial zonal wind. Subsequently, the QBO has been observed in a variety of parameters. Figure 1 is updated from Lait et al. (1989) to include the SH spring of 1988. The solid line shows the variation of the August-September 30 mb zonal wind at Singapore. During the years 1980, 1983, 1985 and 1987 the QBO was in the westerly phase. The dotted line of Fig. 1 shows the equatorial total ozone values. The good correlation between the Singapore zonal wind and equatorial total ozone is a result of the secondary circulation of the QBO. During the easterly phase, a QBO-induced ascending motion at the equator lifts low ozone air parcels from the upper troposphere into the lower stratosphere. This secondary circulation advection reduces the ozone in the equatorial lower stratosphere during the easterly phase, while during the westerly phase, the opposite advection increases equatorial ozone. This easterly phase ascending secondary circulation produces a descending branch in the sub-tropics, which acts to increase ozone in the sub-tropics. Hence, the QBO produces an opposite total ozone effect in the sub-tropics.

The SH "polar QBO" is only observed during the SH spring. Other seasons show substantially less coherence between polar ozone and the equatorial or mid-latitude QBO. The theoretical connection of the tropical QBO to the spring SH polar QBO has not been established. However, the apparent relationship is fairly robust. Figure 1 shows the National Meteorological Center (NMC) polar temperatures (dash-dot line, note the reversed plotting scale), and the decline rate deviation of TOMS total ozone (dashed line, note the reversed plotting scale). The decline rate deviation is that part of the August-September ozone loss rate that deviates from the year-to-year linear trend (determined over the years 1979 through 1989). Years with negative decline rate deviations have higher ozone losses, while those with positive decline rate deviations have relatively smaller ozone losses. Spring sub-polar lower strato- spheric heat flux values for planetary wave one (not shown here) are also highly correlated with polar total ozone and temperatures. Thus, 1987 was a year with a westerly QBO phase, cold temperatures, and a large ozone loss, and small values of planetary wave 1 heat flux, whereas 1988 had an easterly phase, warmer temperatures, a smaller ozone loss, and large values of planetary wave 1 heat flux. Note that while 1988 and 1979 have similar decline rate deviations, the ozone loss was much greater in 1988, since the overall year-to-year negative trend has been removed from the decline rates.

The correlation between all of the variables in Fig. 1 is obvious. The only exception to the correlation is the 1982 polar temperature, where the temperature is warmer during a westerly phase instead of colder. Nevertheless, all of the remaining polar temperatures show an excellent correlation with the QBO phase, as defined by the 30 mb Singapore winds.

The relationship between the decline rate deviation and the equatorial wind can be tested by hypothesizing that the two processes are independent. Assuming equal probabilities between equatorial QBO phases, and equal probabilities for positive and negative polar decline rate deviations, we can treat their correspondence as a series of bernoulli trials and use the binomial distribution to compute the probability that the 10 years of data will have the same sign between the decline rate and the equatorial QBO. An examination of Fig. 1 indicates that the polar QBO and the equatorial QBO signs are always the same, with the exception of 1983. The probability that 9 or more of these 10 years should have had identical signs is 1.1%. Since the probability is low, our independent data assumption is unlikely. Hence, the polar decline rate and the equatorial QBO are statistically related.

The springs of both 1987 and 1988 show a remarkable contrast, hence these two years are chosen as a case study of opposite QBO phases. While meteorological conditions during the springs of 1988 and 1987 are generally considered to be extremes, the extreme differences did not develop until after the decline of total ozone had ended in early October. During the depletion phase (i.e. August to early October) these two years were fairly typical in behavior. The temperatures of Fig. 1 show that the differences of 1988 and 1987 are not outside the range of differences which were observed prior to 1987.

3. 1988 - 1987 Comparison

In this comparison of the 1987 and 1988 SH springs, we will begin by looking at monthly, zonal mean statistics, and then follow with a more detailed analysis of daily zonal means. As we proceed, it will become apparent that the large differences between 1987 and 1988 primarily result from a small number of wave events that occurred on July 30, 1988 (Day 212), August 10, 1988 (Day 223), August 27, 1988 (day 240), September 23, 1988 (day 267) and October 28, 1988 (day 302). These events considerably warm the stratosphere and increase stratospheric ozone. During the following discussion, the reader should focus on these 1988 events, and the relative absence of similar events in 1987 as the primary cause of the 1987-1988 differences. This section is divided into three parts, beginning with total ozone data in part 3.1, temperatures in part 3.2, and finally eddy diagnostics in part 3.3.

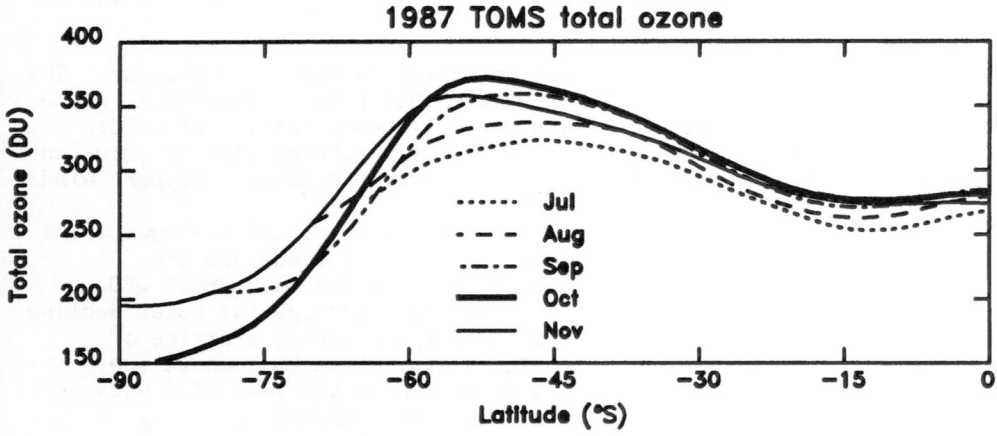

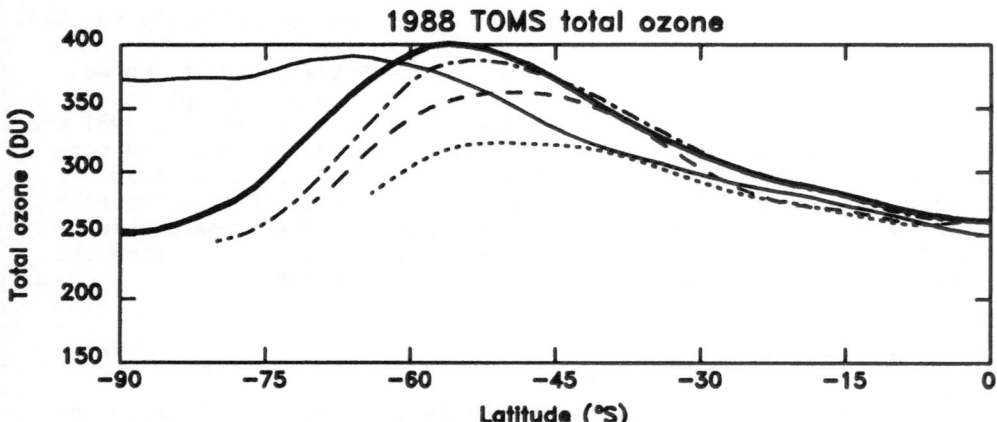

Figure 2. Monthly mean zonal mean total ozone for July (dots), August (dash), September (dash-dot), October (heavy solid), and November (solid) during both 1987 (top) and 1988 (bottom).

3.1. TOTAL OZONE

Figure 2 displays the monthly, zonal mean total ozone for the SH spring months during 1987 and 1988, top and bottom, respectively. The July patterns (dotted line) in both years show a broad maximum in the mid-latitudes of about 325 Dobson units (DU). In both years, this maximum intensifies and moves poleward. By August (dashed line), values in both years at 75S are near 275 DU, with a polar maximum near 350 DU. September (dash-dot line) begins to show dramatic differences between 1987 and 1988, with the mid-latitude maximum stronger and more poleward in 1988. October 1987 (thick solid line) shows polar values below 150

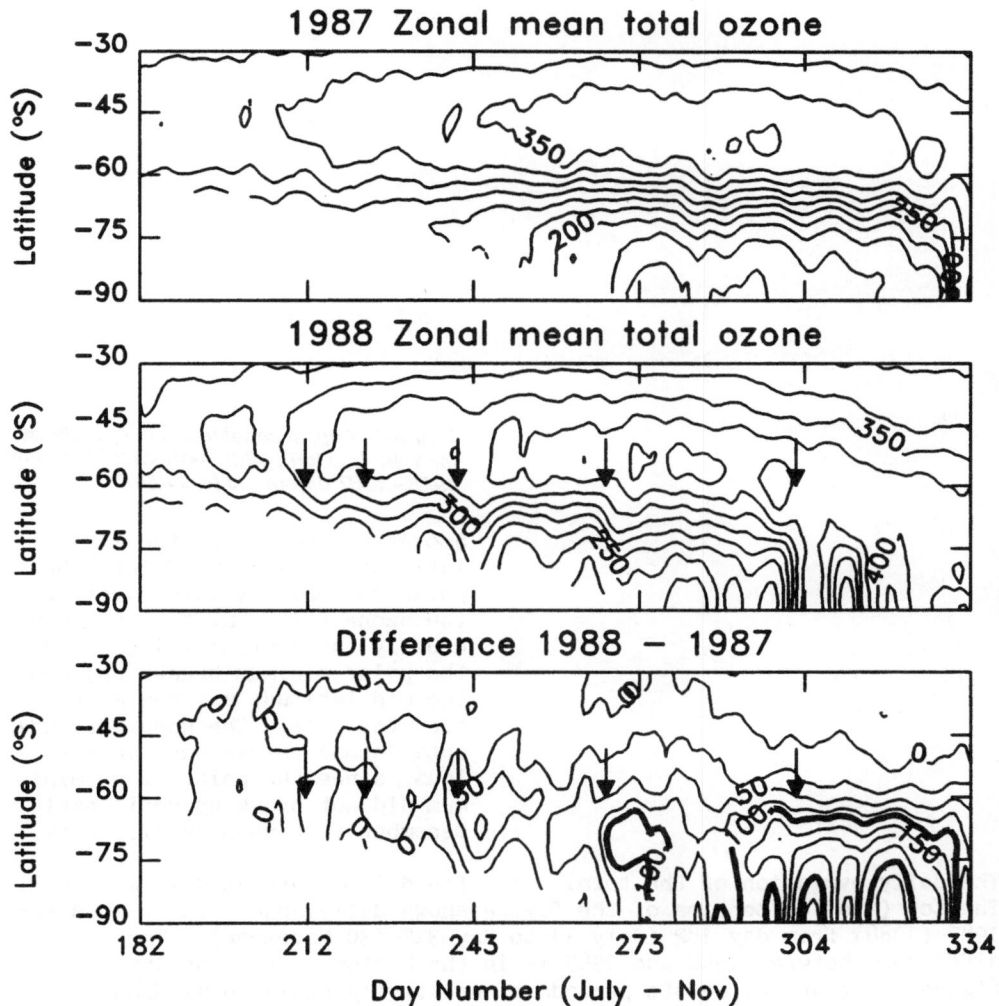

Figure 3. Daily zonal mean total ozone for July through November (days 182 to 335) during 1987 (top) and 1988 (middle). The difference between 1988 and 1987 is shown at the bottom. The contour increment is 25 Dobson units.

DU, while 1988 has a polar value nearer 250 DU, more than 100 DU difference. November 1988 (thin solid line) has nearly double the November 1987 total ozone value in the polar region, with a barely distinguishable mid-latitude maximum near 65S.

The poleward movement of the mid-latitude maximum seen in Fig. 2 is typical behavior for the SH spring. During September the ozone hole appears but by early-October this depletion has ceased, and polar ozone

78

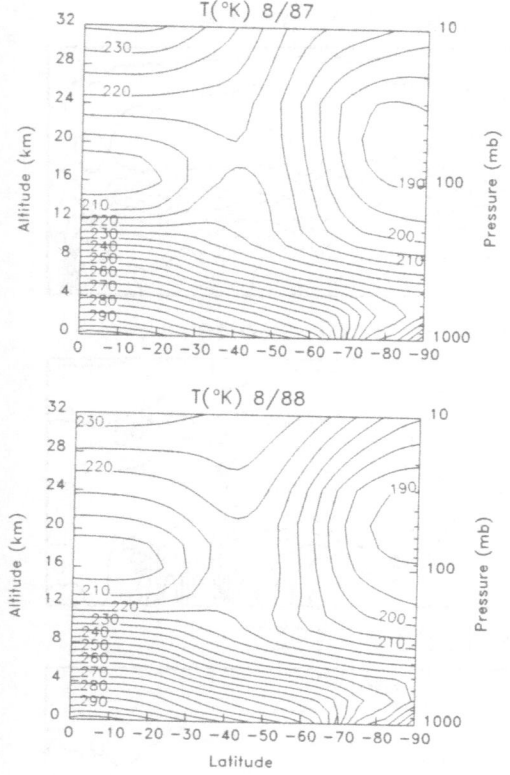

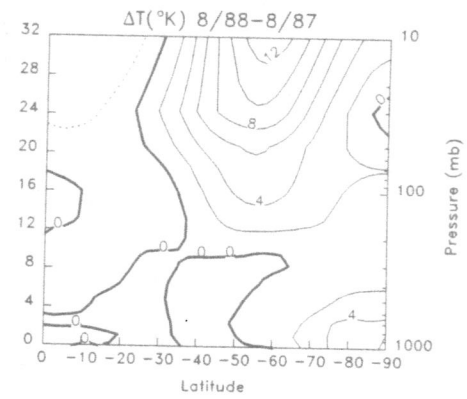

Figure 4. August zonal mean temperatures for 1987 (top left), 1988 (bottom left), and the 1988 - 1987 difference (top right).

amounts increased as the mid-latitude maximum moved into the polar region. By late-November the ozone values at the pole have usually substantially increased as the polar minimum breaks up, and the mid-latitude maximum moves over the pole. The November 1987 total ozone values were anomalous, since the polar ozone minimum did not break up until early December (Atkinson et al., 1989).

The daily evolution of the total ozone field is illustrated in Figure 3. The top (middle) portion of the figure shows daily zonal mean values for 1987 (1988) from day 182 (July 1) to day 335 (30 November). The difference between 1988 and 1987 is in the bottom portion of the figure. The arrows denote periods of relatively rapid total ozone increases which occurred on days 212, 223, 240, 267, and 302. These events are associated with large planetary wave events, as will be shown later in section 3.3. The winter differences (days 182 to 220) are generally small, but have increased to 25 DU in August from 40 to 60S. The differences increase in late August to 50 DU by day 244, and further develop to values in excess of 100 DU in late-September. Following the breakup of the polar minimum in early November 1988, the differences exceed 200 DU, and maximize at 250 DU in mid-November. These differences do not appear in a smooth linear fashion, but develop during the wave events (corresponding to July 30, August 10, August 27, September 23 and October 28, 1988).

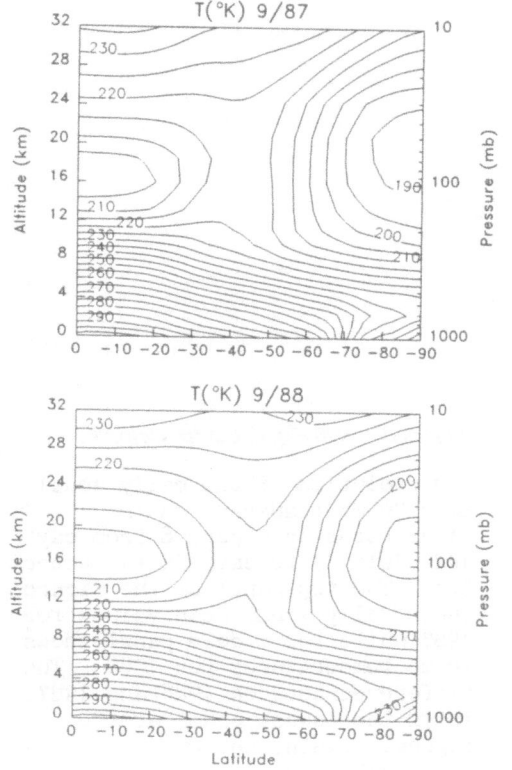

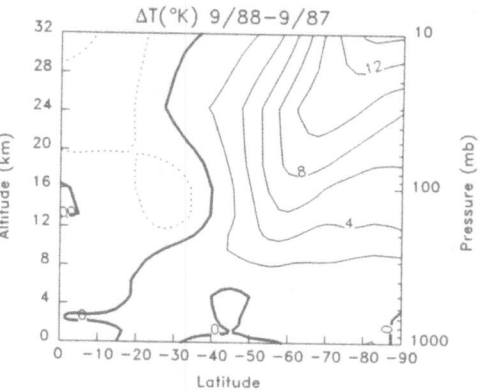

Figure 5. As in Fig. 4, but for September.

3.2 TEMPERATURES

The temperature and total ozone fields evolve similarly during the July-November period. July temperatures (not shown here) are close to the climatological average (Wu et al., 1987; Nagatani et al., 1988) in both 1987 and 1988 with values below 190 K in the 30-50 mb region over Antarctica. The July 1988 - 1987 difference is small (4 K) in the mid-latitude region of the stratosphere. Figure 4 illustrates the August zonal mean temperatures for 1987 (top left), 1988 (bottom left), and the 1988 - 1987 difference (top right). August 1988 lower stratospheric temperatures are slightly warmer in the mid-latitudes, more than 12 K warmer near 60N at 10 mb, and 2 K cooler in the equatorial region. September temperatures (Figure 5), show the 1988 temperatures to be warming at a faster rate than 1987. October temperatures (Figure 6) show 1988 to be substantially warmer in the polar region than in 1987 (18 K at 30 mb in the polar region). Finally, 1988 November temperatures (Figure 7) indicate a full reversal of the thermal gradient, while 1987 temperatures still show a cold pool of air over Antarctica. The November difference at 70 mb is 30 K at 90S.

As in the total ozone data, the 1988 temperatures warm at a more rapid rate than 1987. The warming and downward movement of the polar minimum is typical behavior. The evolution of 70 mb temperatures shows a good correlation with total ozone. The large build up of total ozone shown in Fig. 1 is also evident in 70 mb temperatures, where 90S temperatures have increased from 209 K in October to 235 K in November. This is the same period over which total ozone values increase by 125 DU.

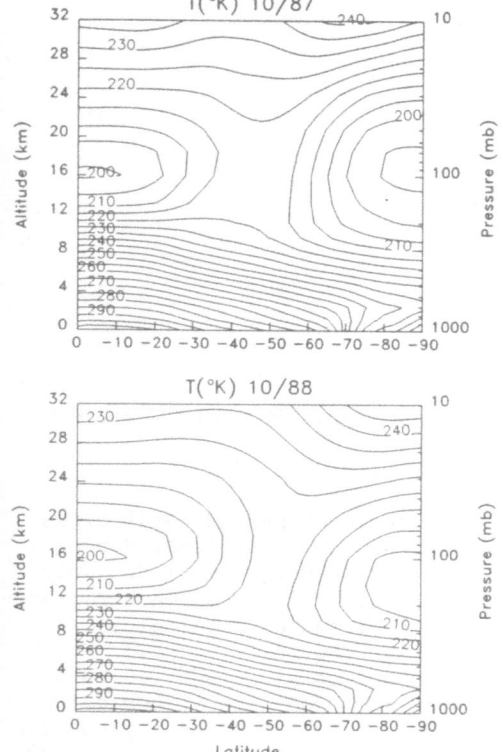

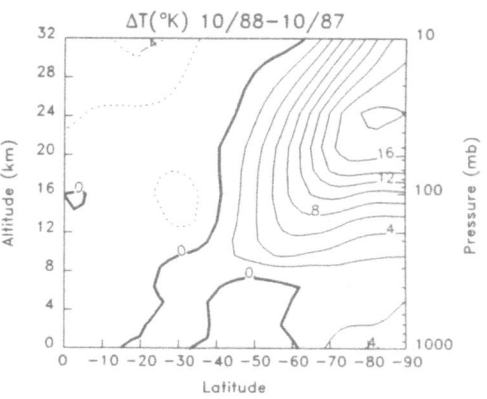

Figure 6. As in Fig. 4, but for October.

The evolution of the daily zonal mean 70 mb temperatures is illustrated in Figure 8 from day 182 (July 1) to day 335 (November 30). The top panel of the figure shows daily zonal mean values for 1987 while the middle panel shows 70 mb temperatures for 1988. The difference between 1988 and 1987 is in the lowest panel of the figure. Again, the wave events are denoted by arrows. The differences during winter (days 182 to 220) are generally small (less than 3 K), but have increased to 6 K by mid-August (day 230) near 60S. The differences increase in late August to 9 K on day 244, and further develop to values in excess of 12 K in late-September. Following the breakup of the polar minimum in early November 1988, the differences exceed 30 K, and maximize at 36 K in mid-November. As for total ozone, the 70 mb differences do not appear in a smooth linear fashion, but develop as a series of events (days 21?, 223, 240, 267, and 302, corresponding to July 30, August 10, August 27, September 23 and October 28, 1988). Comparison to the total ozone data of Figure 3 indicates the 70 mb temperatures evolve identically.

Figure 9 illustrates the movement of the mid-latitude temperature maximum using the temperature gradient. The thin solid contours indicate positive gradients, while the dashed contours indicate negative gradients. The thick solid line (zero gradient) indicates the position of the mid-latitude temperature maximum. Again, the five wave events are denoted by arrows. Note that in 1987 (upper panel) the temperature field is fairly stable, whereas in 1988 (lower panel) the temperature gradient is stronger and more variable in time. Clearly, the important

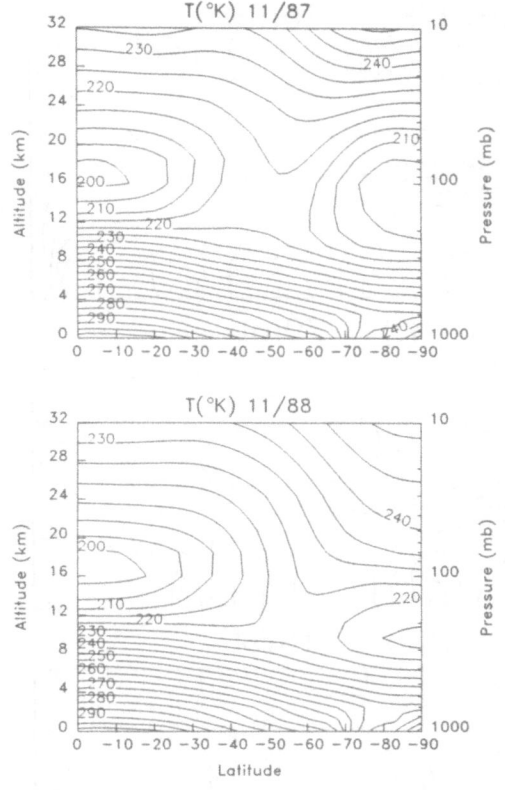

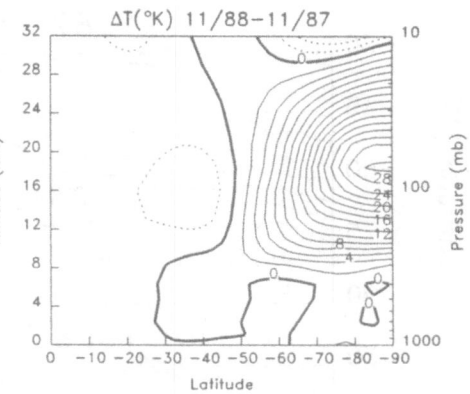

Figure 7. As in Fig. 4., but for November.

events in 1988 occur on August 27 (day 240), September 23 (day 267), and October 28 (day 302). These events are demarked by rapid changes of the thermal gradient, and the poleward movement of the thermal maximum (particularly during the breakup on October 28). The first two events on days 212 and 223 produce virtually no change at the 70 mb pressure level. Nevertheless, at lower pressures (higher altitudes), these two events lead to substantial variations of the meridional gradient.

Although zonal mean temperatures provide an important component to the understanding of the differences between 1987 and 1988, the more important question for polar ozone chemistry is the areal extent of polar stratospheric clouds (PSC's), and the quantity of air which is exposed to these PSC's. Heterogeneous reactions occur on the surfaces of PSC's, and these reactions have a double effect: 1) they remove the nitrogen species which inhibit the catalytic destruction of ozone by chlorine-containing radicals, and 2) they release the chlorine-containing radicals from the chlorine reservoir species. Since PSC's only form in regions of extremely cold temperatures, the temperatures can be used as a surrogate for the potential presence of PSC's. Figure 10 displays a contour plot of the SH fractional areal extent of cold temperatures with 1987 on top, and 1988 on the bottom. Each contour represents the fraction of the SH which is covered by temperatures indicated on the left hand scale. For example, on 30 June 1988 (day 182) 3% of the SH is covered by temperatures below 188.4 K, while 9% is below 196 K. The lower thick solid line denotes the frost point (type

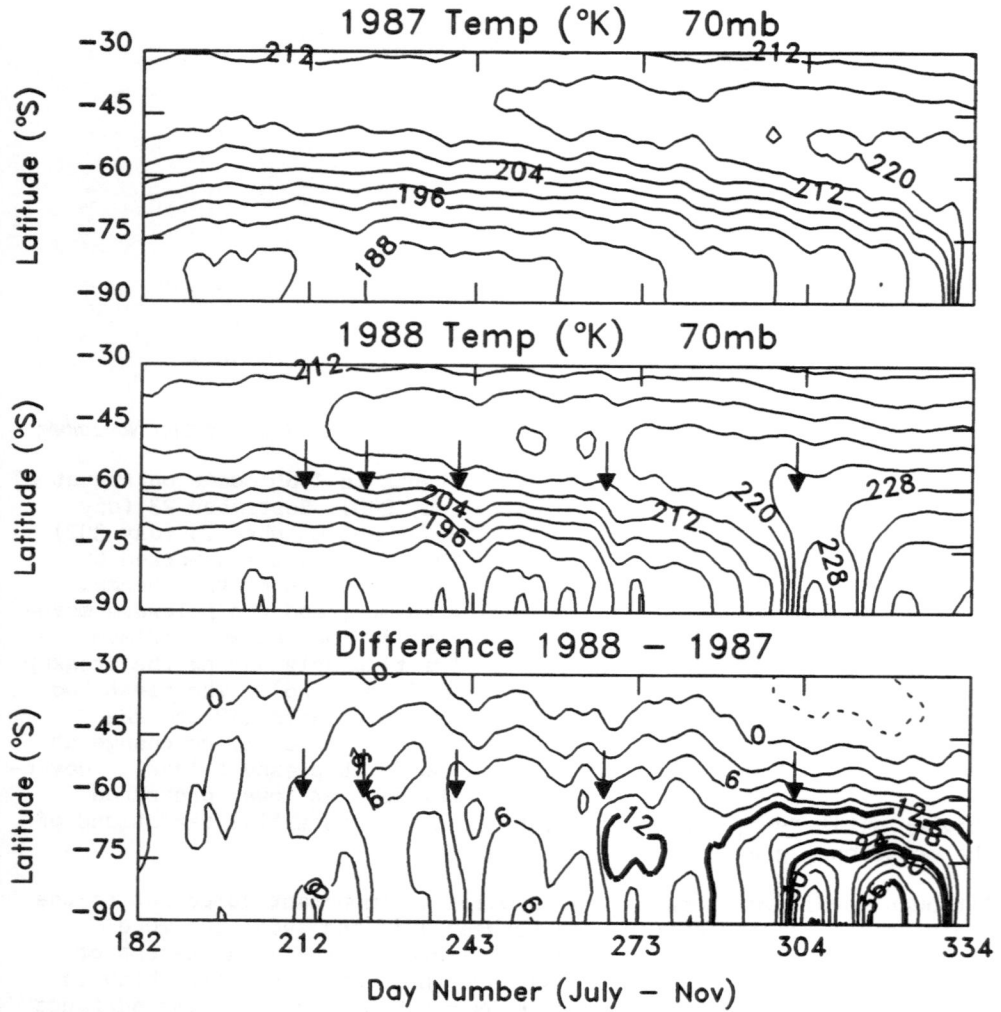

Figure 8. Daily zonal mean 70 mb temperatures for July through November (days 182 to 335) during 1987 (top, 4 K contour increment) and 1988 (middle, 4 K contour increment). The difference between 1988 and 1987 is shown at the bottom (3 K contour increment).

II PSC) for an assumed water vapor of 5 ppmv, while the upper thick solid line is the nitric acid tri-hydrate PSC (type I PSC) formation temperature (assuming 5 ppmv of water and 10 ppbv of nitric acid, Hanson and Mauersberger, 1988). The lower dashed line indicates the minimum temperature observed in the polar region (i.e. the zero area contour). Again, the five wave events are denoted by arrows.

During July the extent of cold temperatures was larger in 1987 than in 1988 by about 1/3, with marginally colder extremes. This situation

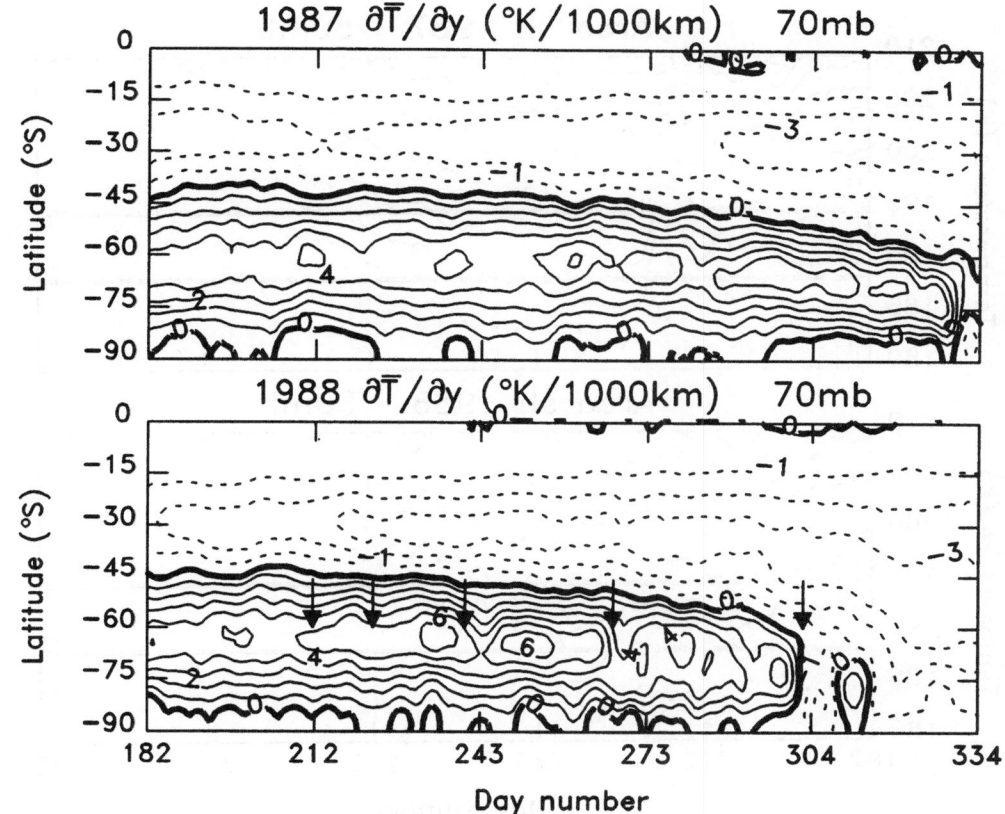

Figure 9. Daily zonal mean 70 mb meridional temperature gradient for July through November during 1987 (top), and 1988 (bottom). Units are in K/1000km.

remained relatively static until late August, when 1988 temperatures rapidly warmed and exceeded the type II PSC limit at the end of the month. The two years continue to diverge during the large warming episode in late September 1988. By the end of October 1988, cold temperatures had virtually disappeared, while in 1987 they continued to persist for a month. Using the 195 K thick line to judge the disappearance of type I PSC's, the type I's disappear at the start of October in 1988, while they disappear in late October in 1987. The rapid diminishment of cold temperature extent occurs in association with these warming/ozone enhancement events. The main exception is the event on day 223 (August 10). As with the thermal gradient, the effect of this wave event is substantially greater at lower pressures (higher altitudes).

3.3. EDDY DIAGNOSTICS

As mentioned earlier, the large episodic increases of temperature and

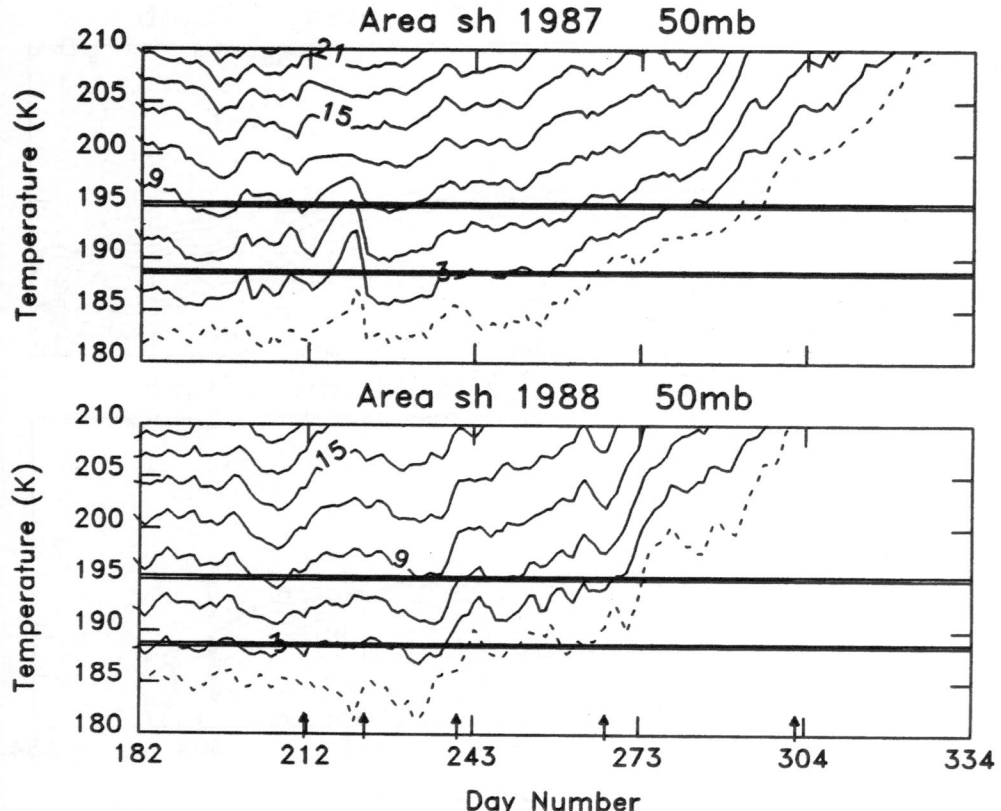

Figure 10. *The solid lines represent the percent of the southern hemisphere covered by a temperature indicated on the left hand side for July through November during 1987 (top) and 1988 (bottom). The dashed line is the minimum polar temperature. The two thick solid lines on each graph represent approximate PSC formation temperatures for type I (upper line, 195 K), and type II (lower line, 188.5 K). The arrows at the bottom of the lower figure denote strong 1988 wave events.*

total ozone in 1988 are accompanied by rather large wave events. Figure 11 displays the eddy variance at 50 mb during the spring for 1987 (top) and 1988 (bottom). The dashed line is the 50 gpm contour, the thin solid lines are the 100 gpm contours, and the thick solid line is the 400 gpm contour. Again, the wave events are denoted by arrows. The 1988 data show a remarkable contrast to 1987. The variance during 1987 is generally below 300 gpm over the entire period, with the exception of a mid-October event (day 283), and a late-November event (day 332). The 1988 variance clearly shows the strong wave pulses with persistent values above 300 gpm from mid-September (day 255) to mid-November (day 314). This eddy variance is generally dominated by wave number one. The eddy variance is also in good agreement with the eddy heat flux at 50 mb. Figure 12 displays the daily 50 mb eddy heat flux averaged from 40 to 75S for 1987 (top) and 1988 (bottom). Again, the wave events are

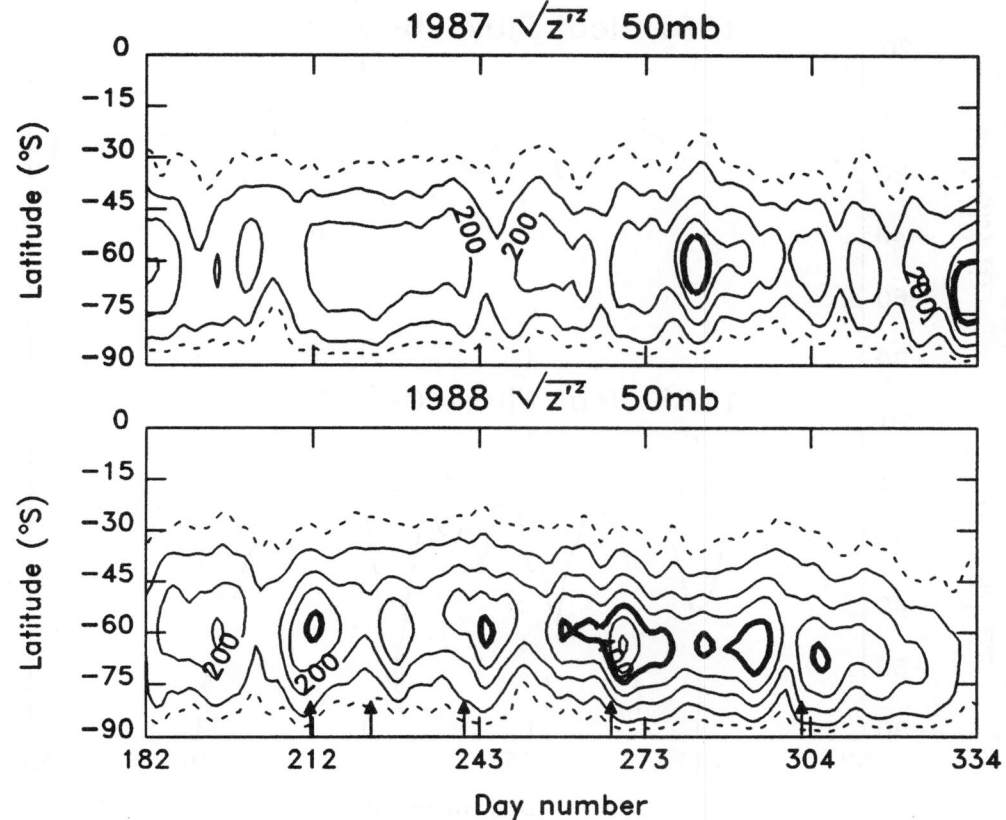

Figure 11. Daily zonal mean 50 mb eddy variance values for July through November during 1987 (top) and 1988 (bottom). Unit are geopotential meters.

denoted by arrows. In conjunction with the warming/ozone enhancement events are strong pulses of eddy heat flux. The 1987 heat flux is considerably less than the 1988 values, with many fewer peaks in the heat flux.

4. Interpretation

As indicated in the introduction, the increased levels of total ozone in 1988 with respect to 1987 could result from at least three sources. The first source results from increased temperatures, which lead to an early disappearance of PSC's and the consequent shut-down of heterogeneous reactions. The second source is the mixing of mid-latitude air which contain high amounts of nitrogen species into the nitrogen depleted polar air, effectively cutting off catalytic destruction of ozone by chlorine-containing radicals. The last source is an increase of eddy activity, leading to increased transport circulation which brings high

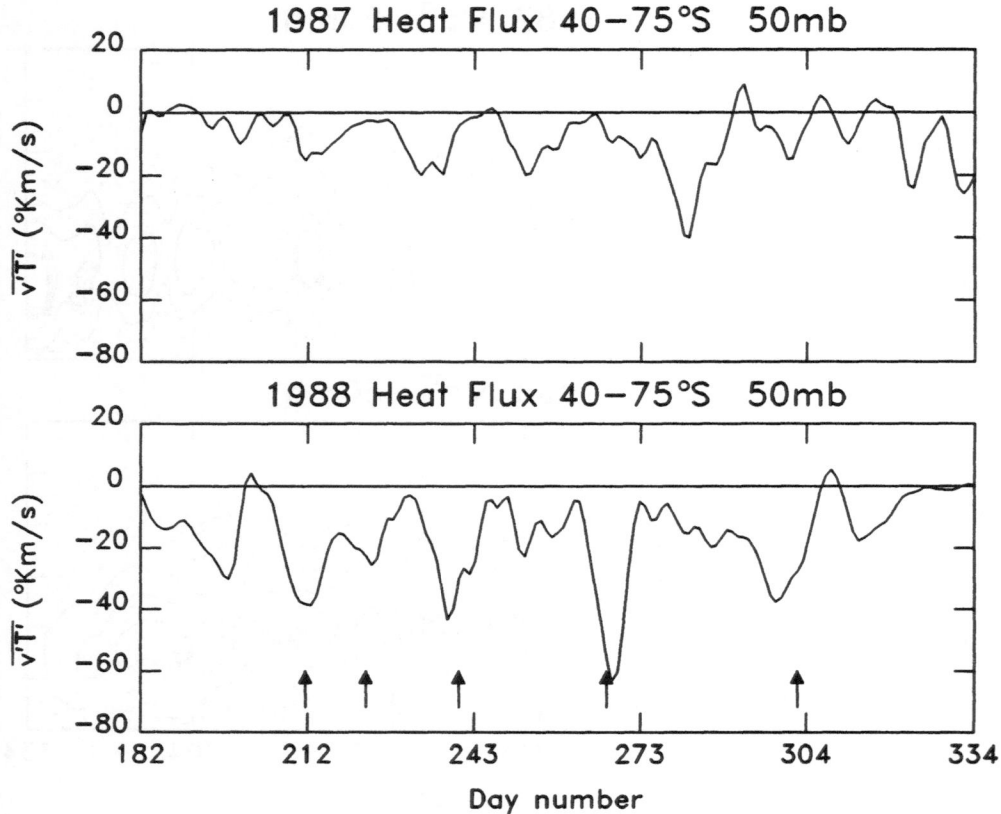

Figure 12. Daily zonal mean and latitudinal mean (40-75S) 50 mb eddy heat flux for July through November during 1987 (top) and 1988 (bottom). Units are K m/s.

ozone air from the mid- stratosphere to the ozone depleted lower stratosphere, and warms the lower stratosphere at the same time. The first and third mechanisms are not independent, but the third source includes the effect of the increased transport of ozone rich air from above.

The evidence for the third mechanism (increased eddy activity) arises from the good correspondence of the total ozone and the temperature field. The effect of a wave propagating into the stratosphere (as evidenced by the strong heat flux during the wave events of 1988) is to decelerate the zonal mean flow. This deceleration will be balanced by a meridional transport at the upper levels, which will result in a downward circulation at the mid-to-high latitudes, and a rising motion in the tropics. The rising motion in the tropics will cause an adiabatic cooling, while adiabatic heating will occur in the mid-to-high latitudes (as occurred during the 1988 wave events). This same circulation pattern will act to increase ozone by bringing down ozone from the photochemical production region.

The direct evidence for the first mechanism (PSC modulation) is less sound, since we only infer the presence of PSC's from the temperature analyses. Nevertheless, this mechanism is in general agreement with the third mechanism, since the wave events warm the lower stratosphere above the temperatures necessary for PSC formation. If PSC's are necessary for ozone destruction, then the ozone destruction would have ceased at an earlier date. However, since total ozone values increase on the time scale of only a few days, it seems reasonable that the depleted region could only be filled by transport, not by photochemical replacement.

The second mechanism calls for increased transport across the vortex boundary in order to bring higher amounts of nitrogen-containing species from the mid-latitudes into the polar vortex. This mechanism will reduce ozone destruction by converting chlorine-containing radicals into chlorine nitrate. While the increased levels of wave driving of the stratosphere would indicate stronger transport in 1988 than in 1987, the evidence for mid-latitude air motion into the vortex is inconclusive. During the wave events of August and September, Ertel's potential vorticity (a stratospheric tracer, see Clough et al., 1985) does not show a strong entrainment of air into the vortex, necessary to shut off the depletion. However, smaller-scale mixing events which are unresolved in the NMC data may be playing a role in moving air across the boundary.

5. Summary and Conclusions

The SH springs of 1987 and 1988 were substantially different. Temperatures during mid to late-winter were generally similar between the two years, but as the spring evolved, the stratosphere warmed at a much greater rate in 1988 than in 1987. By November, the differences were up to 30 K in the polar lower stratosphere. The reason for the higher rate of warming in 1988 is the presence of a series of planetary wave events. These wave events increased both temperatures and total ozone values (see section 4 for a discussion of total ozone depletion modulation mechanisms). Thus, the primary reason for the QBO modulation of the SH polar stratosphere appears to be the presence of planetary scale wave events. While 1988 had a series of five strong events, 1987 had only two events in mid-October and late-November.

The reason for the increased level of wave activity in 1988 seems to be the planetary wave forcing, not differences in propagation characteristics of the stratosphere. Index of refraction calculations for planetary waves show small differences during July. In addition, planetary wave amplitudes in the upper troposphere also show rather weaker amplitudes in 1987 than in 1988. Note that, while the differences of propagation characteristics are large by October 1988, it is the first three wave events which lead to the modulation of the ozone destruction rates. However, the effect of small differences in propagation characteristics of the stratosphere may enhance and feedback on the effects of only slightly larger amplitudes.

6. References

Anderson, J. G., Brune, W. H., Lloyd, S. A., Toohey, D. W., Sander, S. P., Starr, W. L., Loewenstein, M., Podolske, J. R. (1989) "Kinetics of O_3 destruction by ClO and BrO within the Antarctic vortex: an analysis based on in situ ER-2 data", in press, J. Geophys. Res..

Atkinson, R. J., Mathews, W. A., Newman, P. A., and Plumb, R. A. (1989) "Evidence of the mid-latitude impact of Antarctic ozone depletion", Nature, 340, 290-294.

Clough, S. A., Grahame, N. S., and O'Neill, A. (1985) "Potential vorticity in the stratosphere derived using data from satellites", Q. J. R. Met. Soc., 111, 335-358.

Garcia, R. R. and Solomon, S. (1987) "A possible relationship between interannual variability in Antarctic ozone and the quasi-biennial oscillation", submitted to Geophys. Res. Lett.

Hanson, D., and Mauersberger, K. (1988) "Laboratory studies of the nitric aric trihydrate: implications for the south polar stratosphere", Geophys. Res. Lett., 15, 855-858.

Krueger, A. J., Stolarski, R. S., and Schoeberl M. R. (1989) "Formation of the 1988 Antarctic ozone hole", Geophys. Res. Lett., 16, 381-384.

Lait, L. R., Schoeberl, M. R., and Newman, P. A. (1989) "Quasi-biennial modulation of the Antarctic ozone depletion", J. Geophys. Res., in press.

Nagatani, R. M., Miller, A. J., Johnson, K. W., and Gelman, M. E. (1988) "An eight-year climatology of meteorological and SBUV ozone data", NOAA Technical Report NWS 40, pp. 125.

Newman, P. A. and Randel, W. J. (1988) "Coherent ozone-dynamical changes during the southern hemisphere spring, 1979-1986", J. Geophys. Res., 93, 12585-12606.

Oltmans, S. J. and London, J. (1982) "The quasi-biennial oscillation in atmospheric ozone", J. Geophys. Res., 87, 8981-8989.

Plumb, R. A. and Bell, R. C. (1982) "A model of the quasi-biennial oscillation on an equatorial beta-plane", Q. J. R. MET. SOC., 108, 335-352.

Randel, W. J. (1988) "The anomalous circulation in the southern hemisphere stratosphere during spring 1987", Geophys. Res. Lett., 15, 911-914.

Reed, R. J., Campbell, W. J., Rasumssen, L. A., and Rogers, D. G. (1961) "Evidence of a downward-propagating, annual wind reversal in the equatorial stratosphere", J. Geophys. Res., 66, 813-818.

Schoeberl, M. R., Stolarski, R. S., and Krueger, A. J. (1989) "The 1988 Antarctic ozone depletion: comparison with previous year depletions", Geophys. Res. Lett., 16, 377-380.

Stolarski, R. S., Krueger, A. J., Schoeberl, M. R., McPeters, R. D., Newman, P. A., and Alpert, J. C. (1986) "Nimbus-7 SBUV/TOMS measurements of the springtime Antarctic ozone decrease", Nature, 322, 808-811.

Wu, M.-F., Geller, M. A., Nash, E. R., and Gelman, M. E. (1987) "Global atmospheric circulation statistics - Four year averages", ASA Technical Memorandum 100690.

A COMPARISON OF THE DYNAMIC LIFE CYCLES OF TROPOSPHERIC MEDIUM-SCALE WAVES AND STRATOSPHERIC PLANETARY WAVES

William J. Randel
National Center for Atmospheric Research
P. O. Box 3000
Boulder, CO 80303

ABSTRACT. This paper presents a comparison of the dynamic life cycles exhibited by a) baroclinically-forced medium-scale waves in the Southern Hemisphere (SH) troposphere, and b) tropospherically-forced, vertically-propagating planetary waves in the SH winter stratosphere. Cross correlation analyses of nine years of daily data reveal the characteristic EP flux signatures and wave-induced mean flow tendencies during wave growth, maturity and decay. Wave source and sink regions are shown, and strong midlatitude Rossby wave radiation towards low latitudes is observed in both instances. Tropospheric waves are shown to influence the zonal mean flow deep into the lower stratosphere (up to 30 mb). Wave-induced mean flow accelerations in the troposphere are to a large degree reversible between wave growth and decay, whereas the stratospheric waves exert a predominant one-way influence only (zonal flow deceleration). The midlatitude stratospheric waves are also shown to have a delayed, secondary influence in the subtropical upper stratosphere; this may be the statistical signature of planetary wave reflection (or overreflection) from a low latitude critical region.

1. Introduction

The life cycles of baroclinic waves in the troposphere have been studied extensively in the last decade, beginning with the pioneering modeling study by Simmons and Hoskins (1978). The dynamic life cycles of the waves in that idealized experiment were documented further in Edmon et al. (1980) and Hoskins (1983), based on zonal mean diagnostics. Observationally, Randel and Stanford (1985) documented a particularly clear baroclinic wave life cycle in the Southern Hemisphere (SH) troposphere during December 1979. The zonal wave number scale of these waves is observed to be $\sim k = 4\text{-}7$, and they are termed 'medium-scale waves.' Medium scale waves often dominate circulation patterns in the SH troposphere (Salby, 1982; Hamilton, 1983).

Randel (1990) has studied the dynamical signatures of tropospheric wave life cycles statistically, by application of cross-correlation analyses to seven years of

91

A. O'Neill (ed.),
Dynamics, Transport and Photochemistry in the Middle Atmosphere of the Southern Hemisphere, 91–109.

operational data from the European Center for Medium Range Weather Forecasts (ECMWF). The advantage of using a statistical technique is that an immense amount of data may be studied concisely, and only coherent wave behavior which occurs repeatedly (and are thus in some sense 'real') are highlighted. Case studies of individual events are also valuable, and compliment statistical analyses. The statistical methodology is capable of revealing wave structure with high detail: for example, Randel (1990) documents Ferrel cell fluctuations associated with tropospheric baroclinic waves. Cross-correlation analyses are used here, and tropospheric results based on National Meteorological Center (NMC) data are presented.

Vertically propagating stratospheric planetary waves also exhibit clear life cycle characteristics, as illustrated in the cross-correlations and composited case study in Randel et al. (1987). The focus of this paper is to present a statistical analysis of stratospheric planetary wave life cycles, and make a comparison with the evolution observed of tropospheric waves. These comparisons reveal similarities in some aspects, such as the strong meridional propagation of Rossby wave activity from middle towards low latitudes during wave maturity and decay. There are also important differences; wave-induced mean flow changes in the stratosphere exhibit much less reversibility between wave growth and decay than that found in the troposphere. The correlations suggest some process which effectively damps stratospheric wave activity at low latitudes, with spatial and temporal signatures consistent with 'planetary wave breaking.' Additionally, the results show that a consistent feature of the stratospheric life cycles is a delayed reversal of the EP flux in the low latitude upper stratosphere. The nature of this remote, coherent behavior is not yet explained; one possibility is that it is the statistical signature of planetary wave reflection from a low latitude critical layer.

2. Data and Analyses

The data analyzed here are daily zonal mean wind, temperature and wave flux quantities derived from geopotential height grids over 1000-1 mb. Geopotential data over 1000-100 mb are NMC operational analyses, while data from 70-1 mb are Climate Analysis Center (CAC) operational products. Data are analyzed over 1979-1987. Horizontal winds are derived using linearized momentum equation balances. Additionally, daily analyses of zonal mean variables over 1000-100 mb from ECMWF are used; these are discussed in detail in Randel (1990).

Zonal mean diagnostics are based on the transformed Eulerian mean formalism. The zonal mean zonal momentum equation is:

$$\frac{\partial \overline{u}}{\partial t} = \hat{f}\overline{v}^* + D_F. \qquad (1)$$

Here $\hat{f} = 2\Omega\sin\phi - \frac{1}{a\,\cos\phi}\frac{\partial}{\partial\phi}(\cos\phi\overline{u})$, $\overline{v}^*$ is the meridional residual circulation, and

D_F is the quasi-geostrophic Eliassen-Palm flux divergence:

$$D_F = (\rho_s a \, \cos\phi)^{-1} \left[\frac{1}{a \, \cos\phi} \frac{\partial}{\partial \phi}(F_\phi) + \frac{\partial}{\partial z}(F_z) \right] \qquad (2a)$$

$$(F_\phi, F_z) = \rho_s a \, \cos\phi \left[-\overline{u'v'}, \frac{R}{H}\hat{f}\frac{\overline{v'T'}}{N^2} \right]. \qquad (2b)$$

Notation is standard, and may be referenced in Dunkerton et al. (1981).

The quasi-geostrophic wave activity equation is also used here, in the form

$$\frac{\partial}{\partial t} A + D_F = -\alpha A, \qquad (3)$$

where the $-\alpha A$ term is a linear damping factor. A is defined as

$$A = \frac{1}{2} \ \overline{q'^2}/\overline{q}_y,$$

where q' is the quasi-geostrophic wave potential vorticity, and $\overline{q}_y$ the zonal mean potential vorticity gradient. Equation (3) is tested below by correlating $(\frac{\partial}{\partial t}+\alpha)$ A and D_F time series; for these calculations the time series are smoothed in both latitude and time with a 1-2-1 running filter.

Time lag cross-correlation analyses are used to find spatially coherent structure in the meridional plane and delineate the time evolution. The methodology used here is to choose one time series as a reference for midlatitude wave amplitude variations, and then use this time series as reference for correlating all other zonal mean variables throughout the meridional plane. The correlations here use wave kinetic energy at 300 mb, averaged over 40-50°S, as reference time series for the tropospheric waves, while the stratospheric calculations are referenced with respect to wave kinetic energy at 10 mb, averaged over 40-60°S. Results are not sensitive to the use of kinetic energy reference time series; identical results are obtained for geopotential variance or wave enstrophy. Correlation statistics are calculated from 90-day time series centered on each season; the tropospheric correlations are averaged over the four seasons and both hemispheres (because there is little variability with season or hemisphere), while stratospheric correlations shown here were calculated over August-October (the time of maximum stratospheric wave activity in the SH; qualitatively similar patterns are found for the NH winter). The SH stratospheric results do change somewhat with season; early and late winter correlations are contrasted in Randel (1988). 5% significance levels for the tropospheric correlations are near 0.06, and near 0.16 for the stratospheric patterns. Further details of these calculations may be found in Randel (1990).

94

3. Results

a. Tropospheric life cycles

Figure 1 shows EP flux correlation diagrams for the tropospheric wave life cycle. The vectors in Fig. 1 represent correlations (with respect to the 300 mb, 40-50° wave kinetic energy) at each position with the separate components of the EP flux vector; the horizontal component measures correlation with F_ϕ, likewise the vertical component with F_z. The contours denote correlations with the EP flux divergence D_F (Eq. 2a). Time lags in Fig. 1 are with respect to the wave energy time series, hence lags -2 and -1 day correspond to upper tropospheric, midlatitude wave growth, lag 0 to maturity, and lags +1 to +2 to wave decay. Important features revealed in Fig. 1 include:

1. Strong correlations with F_z (poleward heat flux) in the lower troposphere during wave growth and maturity; the heat flux correlations increase in the upper troposphere and lower stratosphere with time lag, consistent with vertical propagation of wave activity.

2. There is strong convergence of EP flux in midlatitudes (negative contours in Fig. 1) at lags -2 and -1 days (denoting local wave growth), and these correlations extend into the lower stratosphere (up to 30 mb) as the wave amplifies.

3. As the wave matures the upper tropospheric F_y (poleward momentum flux) correlations become strong, denoting propagation of wave activity to low latitudes. Coincident with this equatorward propagation is the loss of midlatitude wave activity (positive midlatitude D_F correlations at lags +1 to +2 days) and increase in low latitudes (negative correlations).

Note that these correlations represent deviations from the time mean patterns. As shown in Randel (1990), the actual daily variability in D_F is similar to that in Fig. 1 (with sign reversal in midlatitudes), whereas the time mean EP vectors are only modulated by these daily variations (the downward pointing arrows at lag +2 days in Fig. 1 indicate a reduction in poleward heat flux rather than absolute equatorward flux). The patterns in Fig. 1 are in good agreement with the baroclinic grow-barotropic decay life cycles modeled in Simmons and Hoskins (1978). Note the high degree of reversibility in the midlatitude correlations between wave growth and decay in Fig. 1.

Figure 2 presents correlation diagrams for zonal mean zonal wind acceleration $\frac{\partial \bar{u}}{\partial t}$ (top) and zonal mean temperature tendency $\frac{\partial \bar{T}}{\partial t}$ (bottom) for time lags -2, 0 and +2 days. During wave growth (lag -2 days) there is strong zonal wind deceleration in midlatitudes (negative correlations), flanked to both north and south by acceleration regions. The correlations are in phase vertically, extend deep into the lower stratosphere, and exhibit a meridional half wavelength of 15-20° latitude. These patterns are completely reversed at lag +2 days (wave decay), with midlatitude acceleration flanked meridionally by deceleration regions. There is good

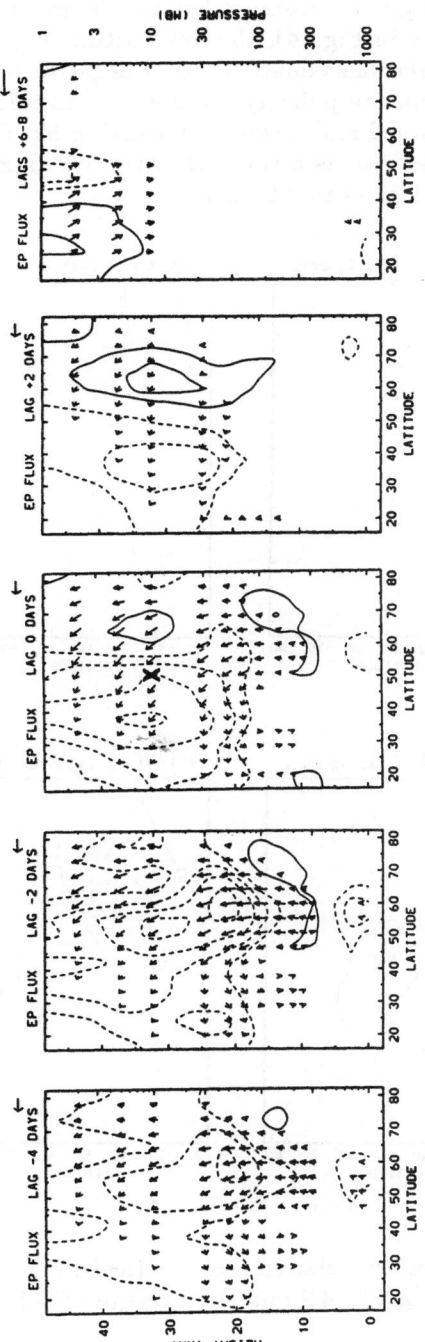

Figure 4. EP flux correlation diagrams (as in Fig. 1) for stratospheric planetary wave life cycles (note reference position at 10 mb in lag 0 plot). Time lags of -4, -2, 0, +2 and the average of +6 to +8 days. Arrows in the last panel are doubled for clarity. Contours of ± 0.1, 0.2, ..., with 5% significance levels near 0.16. Note the vertical axis is 1000-1 mb, versus 1000-10 mb in Fig. 1.

3. Approximately one week following the maximum kinetic energy at 10 mb in midlatitudes (lags +6 to +8 days in Fig. 4), the low latitude upper stratosphere exhibits positive D_F correlations equatorward of negative patterns (a north-south dipole pattern of opposite polarity to that seen in midlatitudes at lag +2 days). There are poleward and downward pointing EP flux correlations in this region at this time; this is a reversal of the EP flux patterns seen in midlatitudes throughout lags -4 to +2 days.

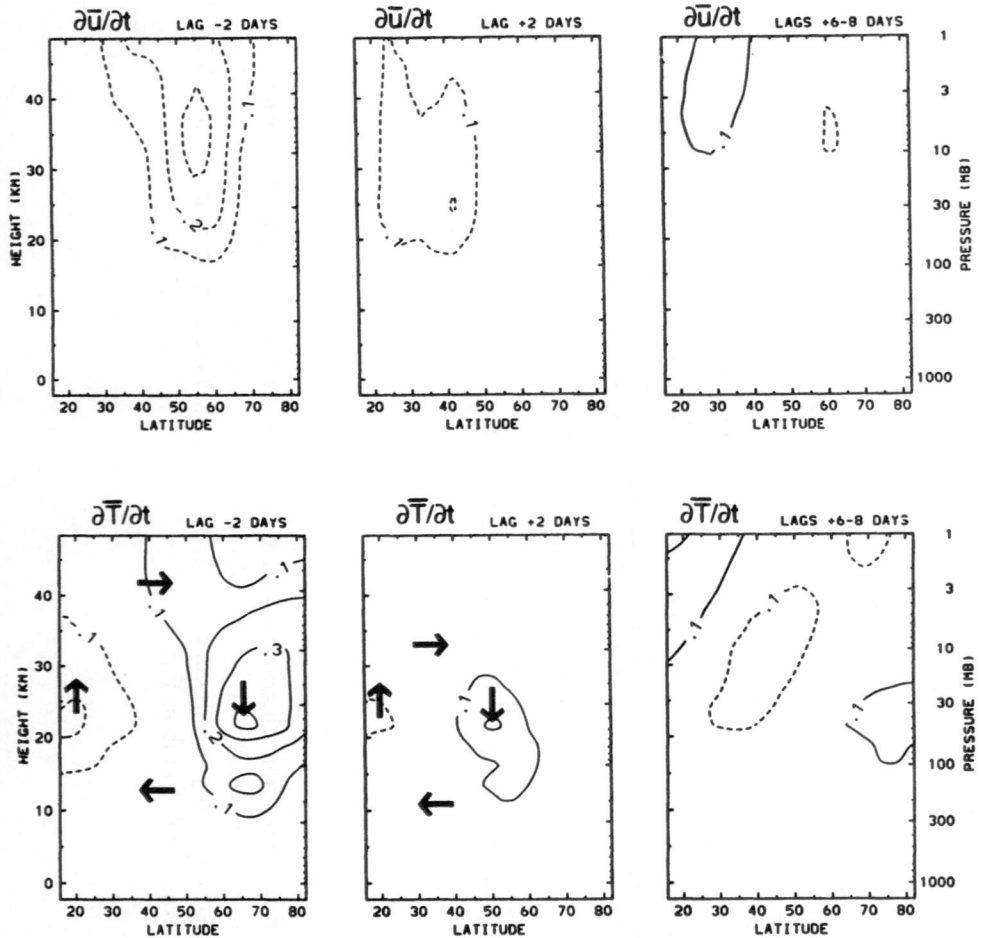

Figure 5. $\frac{\partial \bar{u}}{\partial t}$ (top) and $\frac{\partial \bar{T}}{\partial t}$ (bottom) correlation sections (as in Fig. 2) for stratospheric life cycles, for time lags -2, +2 and the average of +6 to +8 days. Contours of $\pm$ 0.1, 0.2, ...

Figure 5 shows meridional sections of $\frac{\partial \bar{u}}{\partial T}$ and $\frac{\partial T}{\partial t}$ correlations for the stratospheric life cycles, at lags -2, +2 and +6 to +8 days. There is strong middle to
high latitude $\bar{u}$ deceleration over 100-1 mb (and above) at lag -2 days, weakening
and moving to lower latitudes at lag +2 days. The lag -2 day $\frac{\partial T}{\partial t}$ patterns show
strong polar warming and low latitude cooling centered near 100-10 mb, and there
are presumably patterns of opposite polarity above 1 mb (e.g., Garcia, 1987) not
seen in these data. The $\bar{u}$ deceleration regions approximately follow the equatorward movement of the negative D_F correlations seen in Fig. 4. This is shown
more clearly in the 10 mb latitude-time lag sections for $\frac{\partial \bar{u}}{\partial t}$ and D_F correlations in
Figs. 6a-b. At 1 mb (Figs. 6c-d), there is less evidence of meridional movement of
the correlation patterns. Randel (1988) shows that these $\frac{\partial \bar{u}}{\partial t}$ correlations change
seasonally in such a way that early winter wind changes are at lower latitude than
those in late winter-spring, i.e., the waves are more confined to high latitudes in
spring.

The lag +6 to +8 day $\frac{\partial \bar{u}}{\partial t}$ correlations in Fig. 5 show subtropical acceleration
in upper levels in the same location as the positive D_F contours at lags +6 to +8
days in Fig. 4. These similar spatial patterns are also seen in the 1 mb sections in
Figs. 6c-d, where the time lag of approximately 1 week with respect to midlatitude
events is emphasized. NH correlations (not shown) exhibit qualitatively similar
patterns, with a somewhat longer time delay (of order $\sim$ 10 days).

The delayed low latitude patterns seen in both wave and mean flow statistics
is an intriguing result which deserves further comment. Figure 7 shows time
series of selected variables for one year (1981) used for the correlation statistics.
The middle stratospheric reference kinetic energy time series is shown, along with
100 mb heat flux; these series demonstrate the episodic nature of planetary wave
events. Also shown are time series for low latitude, upper stratospheric $\overline{u'v'}$,
D_F and $\frac{\partial \bar{u}}{\partial t}$. Arrows in Fig. 7 denote episodes of poleward F_ϕ (positive $\overline{u'v'}$)
and positive D_F (and $\frac{\partial \bar{u}}{\partial t}$) which lag by approximately one week the midlatitude
kinetic energy maxima. Inspection of individual time series from other years shows
some events with positive D_F and poleward F_ϕ in these regions (as in Fig. 7),
but more often there is only a modulation (not a reversal) of the climatological
values here. A similar low latitude EP flux vector and D_F pattern is shown in
Hitchman et al. (1987), their Fig. 6, with a similar time lag relationship with
midlatitude Rossby wave events. They suggest such patterns may result from
inertial instability in the tropical mesosphere, forced as a result of meridionally
propagating midlatitude Rossby waves. However, the patterns shown here are
substantially removed from the equator, and thus not likely due to that mechanism.
The location and timing of these reversed F_ϕ patterns, in conjunction with weak
zonal winds in low latitudes, is suggestive of a reflection of planetary wave activity
near a critical line (where wave phase velocity equals zonal wind speed). These
patterns may be the statistical signature of such planetary wave reflection. In the
case where additionally $\bar{q}_y < 0$, the waves may 'overreflect' and extract energy
from the mean flow (Dunkerton, 1987). Figure 8 shows $\bar{u}$ and $\bar{q}_y$ during 18-24
August, 1981 (c.f. Fig. 7), and indeed there is a region of $\bar{q}_y < 0$ in the upper
stratosphere on the equatorward flank of the jet.

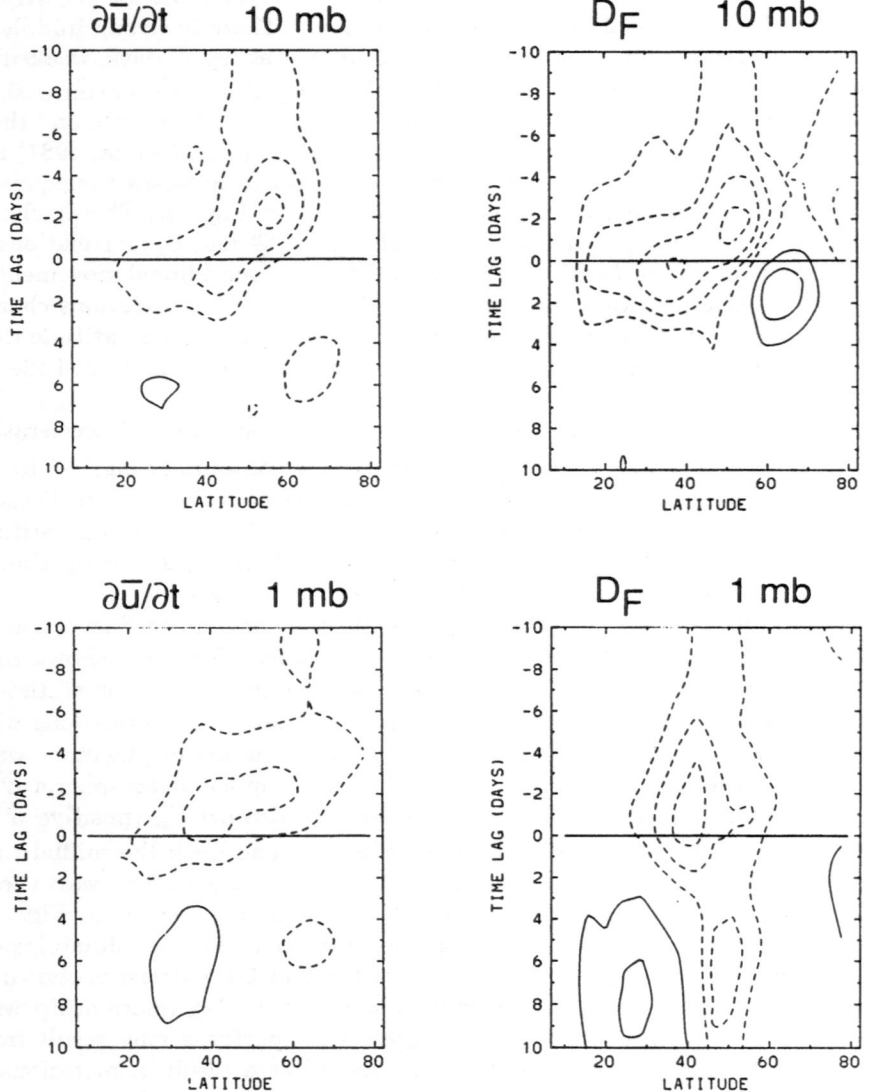

Figure 6. Latitude-time lag correlation sections for stratospheric life cycles. Shown are $\frac{\partial \bar{u}}{\partial t}$ (left) and D_F (right) patterns at 10 mb (top) and 1 mb (bottom). Contours of $\pm 0.1, 0.2, \ldots$. Note the overall spatial and temporal agreement in the $\frac{\partial \bar{u}}{\partial t}$ and D_F patterns at both levels.

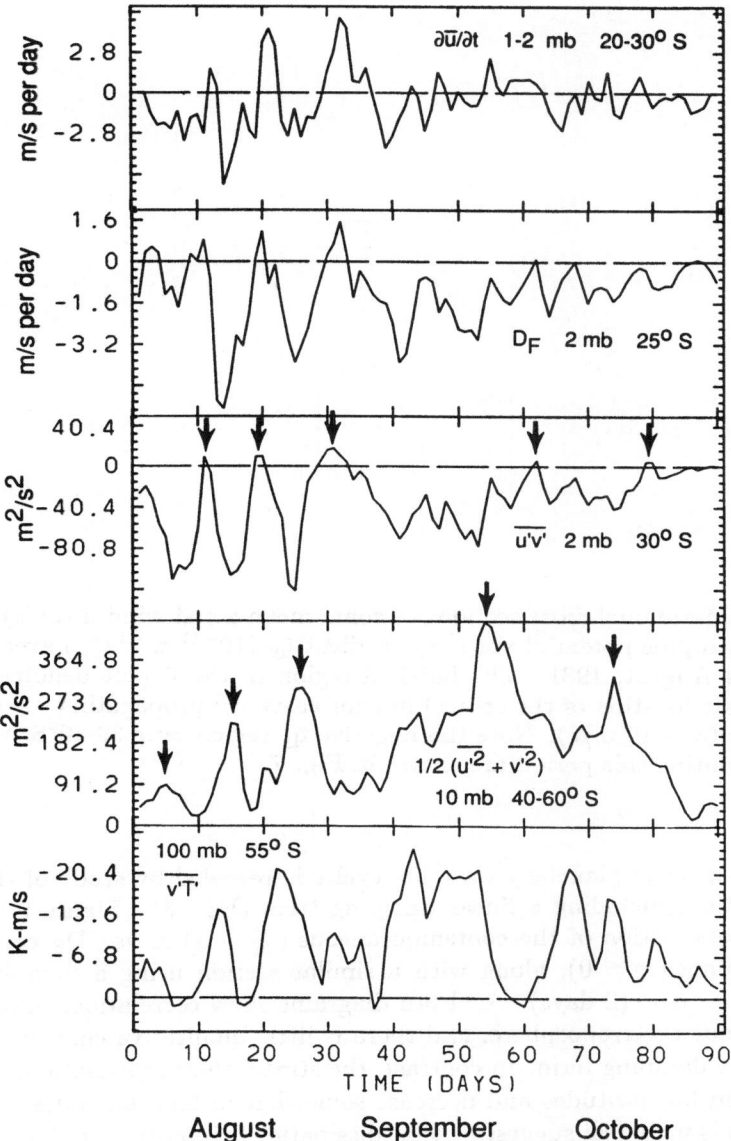

Figure 7. Time series of planetary wave variations during August-October, 1981. The variables and locations are listed; note the kinetic energy time series (second from bottom) which is used as reference for all the correlation sections. Arrows denote midlatitude wave events and the time-lagged, positive $\overline{u'v'}$ in the subtropical upper stratosphere.

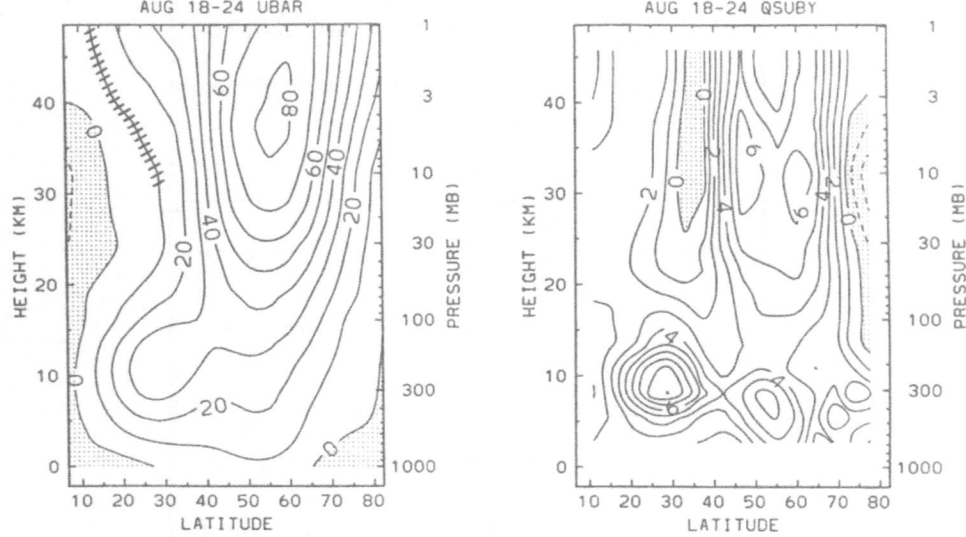

Figure 8. Meridional cross sections of zonal mean zonal wind $\bar{u}$ (m/s) and quasi-geostrophic potential vorticity gradient $\bar{q}_y$ (10^{-11} $m^{-1}s^{-1}$) averaged over 18-24 August, 1981. The hatched region in the $\bar{u}$ plot denotes the approximate location of the critical line for eastward propagating waves in SH winter (c $\sim$ 10 m/s). Note the negative $\bar{q}_y$ region over 30-40°S. Wave statistics during this period are shown in Fig. 7.

A further aspect of planetary wave life cycles is revealed by study of the wave activity equation, including a linear damping term (Eq. 3). Figure 9 shows a meridional cross-section of the contemporaneous ($\frac{\partial}{\partial t} + \alpha$) A vs. D_F correlation with no damping ($\alpha = 0$), along with a similar section using a damping time scale of 2 days ($\alpha = (2 \text{ days})^{-1}$). Both diagrams show correlations of order 0.2 - 0.4 throughout the troposphere, and there is little qualitative change with the addition of the damping term. In contrast, the stratospheric correlations increase dramatically in low latitudes and decrease somewhat in high latitudes when the damping term is included, suggesting that dissipation is important in low (but not high) latitudes.

Figure 10 shows plots of the 20°-70° average ($\frac{\partial}{\partial t} + \alpha$) A vs. D_F correlation as a function of α at 400, 50 and 2 mb; a similar plot from NH data is also included. These plots show a dramatic increase in 2 mb correlation as more damping is added, up to 1-2 day time scales; this is the reason 2 days was chosen for Fig. 9 (note that patterns similar to Fig. 9 are found for NH data). Correlations at 50

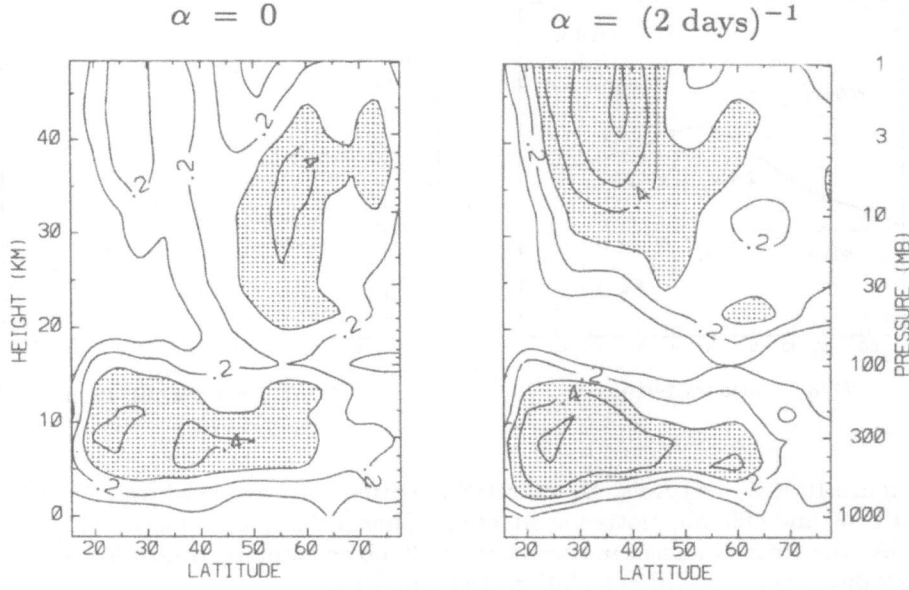

Figure 9. Meridional cross sections of the contemporaneous correlation between $(\frac{\partial}{\partial t} + \alpha)$ A and D_F (Eqn. 3) for $\alpha = 0$ (no damping) and $\alpha = (2\ \text{days})^{-1}$ (2 day damping time scale). Contours of 0.1, 0.2, ..., and correlations above 0.3 are stippled. Note the strong increase in correlation in the low latitude upper stratosphere when the damping term is included.

and 400 mb show near constant values as the damping is increased up to 1-2 day time scales, deteriorating substantially for stronger values. Overall, the correlations in Figs. 9-10 suggest some effective wave damping in the low latitude upper stratosphere, with an effective time scale of order 2 days. This location and time scale is in good agreement with stratospheric surf zone – planetary wave breaking ideas (McIntyre and Palmer, 1984), because wave breaking should occur preferentially in low latitudes (Baldwin and Holton, 1988), with a time scale consistent with advective processes ($\sim$ a few days). Note that radiative damping times in this region are of order $\sim$ 10-20 days (Kiehl and Solomon, 1986, their Fig. 5), and hence radiative processes are probably not the source of this damping.

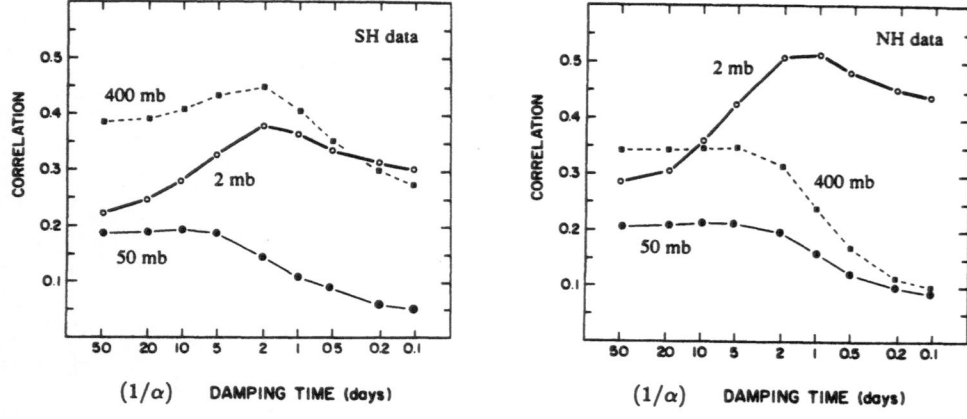

Figure 10. ($\frac{\partial}{\partial t} + \alpha$) A vs. D_F correlation coefficients, averaged over 20-70°, at 2, 50 and 400 mb, plotted as function of linear damping time scale ($1/\alpha$). Note the strong increase in correlation at 2 mb as damping time approaches 1-2 days. Data for SH correlations (left) and NH correlations (right).

4. Comparison of Life Cycles: Similarities and Differences

The above correlation analyses have yielded statistical pictures of the wave and mean flow patterns which repeatedly occur in association with wave amplification/decay in the troposphere and stratosphere. Similarities are pointed out to underscore common dynamical processes, while differences are highlighted to focus on distinct physical situations.

a. Similarities

Both life cycles exhibit strong correlations with F_z (poleward eddy heat flux) below the region of and prior to wave growth. For tropospheric waves, this is lower tropospheric heat flux coincident with *in situ* baroclinic wave growth, while for stratospheric waves this corresponds to propagation from the upper troposphere (as noted above, *in situ* baroclinic generation of the planetary waves in the upper troposphere is prohibited because $\bar{q}_y > 0$).

An unmistakable similarity in these life cycles is the propagation of midlatitude wave activity towards low latitudes as the waves mature and decay. This is seen clearly in the strong equatorward F_y and north-south dipole patterns in D_F for the mature waves in Figs. 1 and 4. The preferential refraction of Rossby waves towards low latitudes results from the spherical geometry (see the discussion in Palmer, 1982).

b. Differences

Perhaps the most striking difference between the life cycles is the degree of reversibility seen in the tropospheric cycle (the lag +2 days patterns in Figs. 1-2 are all nearly opposite those at lag -2 days; note especially the mean flow tendencies), while there is little reversibility seen in the stratosphere. The midlatitude mean flow decelerates as the tropospheric waves grow, then re-accelerates as the waves radiate to low latitudes. Conversely, the stratosphere waves primarily decelerate the zonal flow during wave growth and maturity (Figs. 5-6); there is little evidence of rapid zonal acceleration as the waves decay. This is consistent with planetary waves which propagate into the stratosphere and dissipate (either mechanically or radiatively), as opposed to tropospheric waves which propagate strongly out of midlatitudes before they decay. Wave activity correlations also point to the importance of some damping process in the stratosphere, and show that it occurs preferentially in the low latitude upper stratosphere with a time scale of order 2 days. This location and time scale is consistent with a 'stratospheric surf zone – planetary wave breaking' hypothesis (McIntyre and Palmer, 1984).

The stratospheric correlations also show intriguing low latitude, upper stratospheric reversed EP flux and D_F patterns, which follow by approximately one week midlatitude, middle stratospheric wave amplitude maxima. The reality of these correlations are supported by similar spatial and temporal signatures in D_F and $\frac{\partial \bar{u}}{\partial t}$ (Figs. 6c-d), and in their similar appearance in NH data (not shown). The poleward F_ϕ and positive D_F correlations here are usually associated with a modulation ('turning off') of the climatological patterns in this region, although occasionally actual reversals occur. The location and timing of the reversed F_ϕ patterns suggest that they may be the statistical signature of planetary wave reflection from the low latitude critical region. The positive $\frac{\partial \bar{u}}{\partial t}$ accelerations at this time could result from radiative re-acceleration following wave events, or from direct wave-induced effects ('overreflection,' e.g., Dunkerton, 1987). Although details of these patterns await explanation, their occurrence in the present correlation analyses demonstrate that they are repeatable aspects of the planetary wave life cycle.

Acknowledgments

The author thanks Byron Boville for discussions and comments, and Tim Dunkerton and Michael McIntyre for comments on the text. Marilena Stone expertly prepared the manuscript. This work has been partly supported under NASA Grant W-16215. NCAR is sponsored by the National Science Foundation.

References

Baldwin, M. P. and J. R. Holton, 1988: Climatology of the stratospheric polar vortex and planetary wave breaking. *J. Atmos. Sci.*, *45*, 1123-1142.

Dunkerton, T. J., 1987: Resonant excitation of hemispheric barotropic instability in the winter mesosphere. *J. Atmos. Sci.*, *44*, 2237-2251.

Dunkerton, T. J., C.-P.F. Hsu and M. E. McIntyre, 1981: Some Eulerian and Lagrangian diagnostics for a model stratospheric warming. *J. Atmos. Sci.*, *38*, 819-843.

Edmon, H. J., Jr., B. J. Hoskins and M. E. McIntyre, 1980: Eliassen-Palm cross sections for the troposphere. *J. Atmos. Sci.*, *37*, 2600-2616; also corrigendum, *J. Atmos. Sci.*, *38*, 1115 (1981).

Garcia, R. R., 1987: On the mean meridional circulation of the middle atmosphere. *J. Atmos. Sci.*, *44*, 3599-3609.

Hamilton, K., 1983: Aspects of wave behavior in the mid and upper troposphere of the southern hemisphere. *Atmos. Ocean, 21*, 40-54.

Held, I. M. and P. J. Phillips, 1987: Linear and nonlinear barotropic decay on the sphere. *J. Atmos. Sci.*, *44*, 200-207.

Hitchman, M. H., C. B. Leovy, J. C. Gille and P. L. Bailey, 1987: Quasi-stationary, zonally asymmetric circulations in the equatorial middle atmosphere. *J. Atmos. Sci.*, *44*, 2219-2236.

Hoskins, B. J., 1983: Modelling of the transient eddies and their feedback on the mean flow. Large-Scale Dynamical Processes in the Atmosphere, B. J. Hoskins and R. P. Pearce, Eds. Academic Press, 397 pp.

Kiehl, J. T. and S. Solomon, 1986: On the radiative balance of the stratosphere. *J. Atmos. Sci.*, *43*, 1525-1534.

McIntyre, M. E. and T. N. Palmer, 1984: The 'surf zone' in the stratosphere. *J. Atmos. Terr. Phys.*, *46*, 825-849.

Palmer, T. N., 1982: Properties of the Eliassen-Palm flux for planetary scale motions. *J. Atmos. Sci.*, *39*, 992-997.

Randel, W. J., 1988: The seasonal evolution of planetary waves in the southern hemisphere stratosphere and troposhere. *Quart. J. Roy. Meteor. Soc.,* *114*, 1385-1409.

Randel, W. J., 1990: Coherent wave-zonal mean flow interactions in the troposphere. *J. Atmos. Sci.,* *47*, 439-456.

Randel, W. J. and J. L. Stanford, 1985: The observed life cycle of a baroclinic instability. *J. Atmos. Sci.,* *42*, 1364-1373.

Randel, W. J., D. E. Stevens and J. L. Stanford, 1987: A study of planetary waves in the southern winter troposphere and stratosphere. Part II: Life cycles. *J. Atmos. Sci.,* *44*, 936-949.

Salby, M. L., 1982: A ubiquitous wave number-5 anomaly in the southern hemisphere during FGGE. *Mon. Wea. Rev.,* *110*, 1712-1720.

Simmons, A. J. and B. J. Hoskins, 1978: The life cycles of some nonlinear baroclinic waves. *J. Atmos. Sci.,* *35*, 414-432.

TRAVELING PLANETARY WAVES IN THE MIDDLE ATMOSPHERE

ISAMU HIROTA
Department of Geophysics
Kyoto University
Sakyoku, Kyoto 606
Japan

ABSTRACT. Recent progress in the study of free traveling planetary waves in the middle atmosphere is briefly reviewed, by paying special attention to interhemispheric differeces in connection with the seasonal variation of mean flows. Observational evidence is presented for various westward traveling modes (normal mode Rossby waves) and eastward traveling waves in the southern hemisphere.

1. INTRODUCTION

The study of planetary-scale wave motions in the atmosphere has a long history. Among various wave modes, large-scale traveling waves in extratropical latitudes have been well known as Rossby waves of zonal wavenumber 1 and 2 with characteristic periods ranging from a few days to a few weeks.   Early observational and theoretical studies on the traveling Rossby wave in the atmosphere are comprehensively reviewed by Madden (1979) and Salby (1984).

In the last decade, progress in the satellite observation made it possible to expand the scope of investigations of this topics, and further observational evidence has been presented for planetary waves in the middle atmosphere on a global basis.

2. NORMAL MODE ROSSBY WAVES

It is now well recognized that westward traveling Rossby waves are the manifestation of a solution to Laplace's tidal equation which has a global structure on a rotating sphere. Therefore they are often called as "normal mode Rossby waves".

Each mode is represented by an ordered pair of intejers (s, n-s), where s is the zonal wavenumber and n is an index of meridional structure. For the geopotential height, the meridional structure is symmetric with respect to the equator when n-s is odd, and antisymmetric when it is odd. The wave period is given as an eigenvalue of

A. O'Neill (ed.),
Dynamics, Transport and Photochemistry in the Middle Atmosphere of the Southern Hemisphere, 111–116.
© 1990 Kluwer Academic Publishers.

each mode. However, the mode structure, as well as period, is influenced to some extent by the distribution of backgroud flows that differ
between the northern and southern hemispheres.

With the aid of TIROS-N/NOAA satellite observations, Hirota and
Hirooka (1984) and Hirooka and Hirota (1985) showed the global
existence of various types of normal mode Rossby waves in the stratosphere. Fig. 1 shows a typical example of the first symmetric modes of
zonal wavenumbers 1 and 2, i.e., (1, 1) mode and (2, 1) mode, which are
called as the 5-day wave and 4-day wave, respectively.

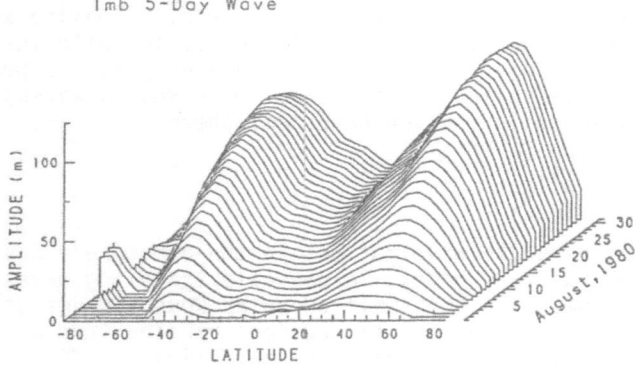

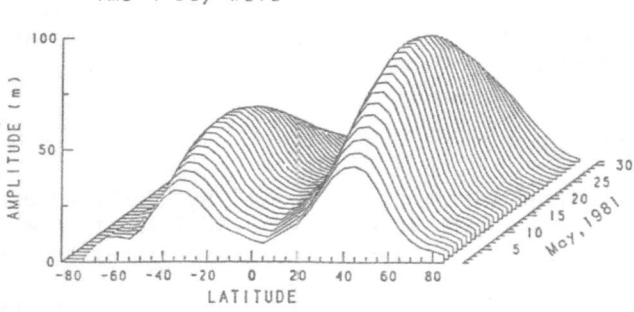

Figure 1. Three-dimensional plot of the 5-day wave (above) and 4-day
wave (below) amplitude versus latitude and time.

Quite recently, Hirooka and Hirota (1989) presented further evidence of normal modes with higher meridional numbers, and summarized the results of eight modes (s = 1, 2; n-s = 1, 2, 3, 4) for the 6 years of 1980 - 1985 as an "Appearance Calendar" (Fig. 2).

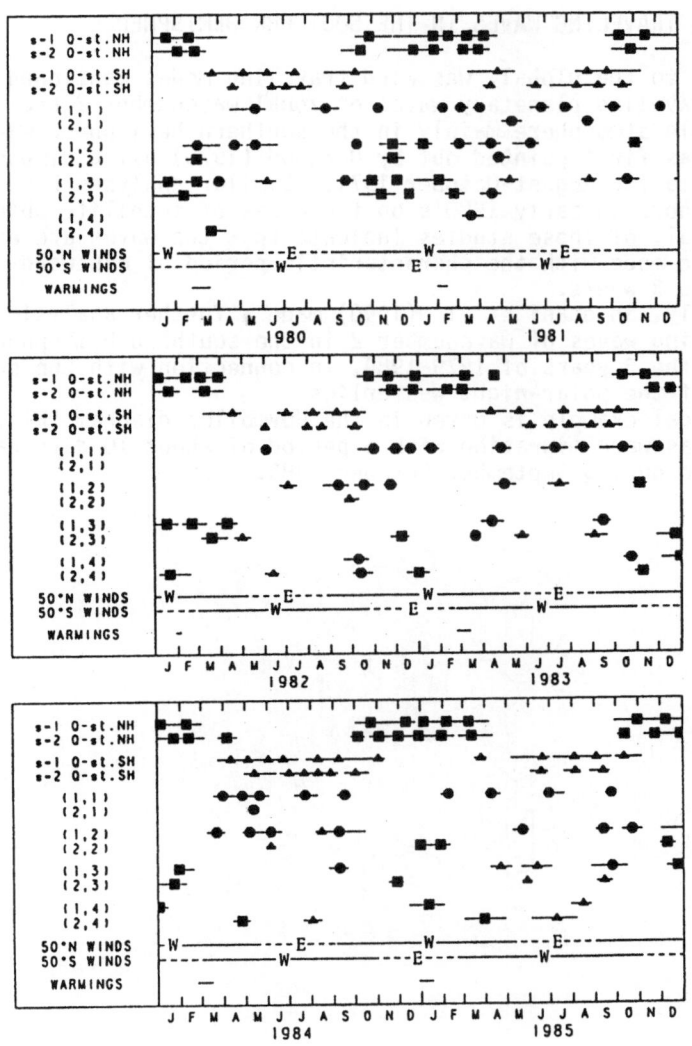

Figure 2.   Appearance calendar of normal mode Rossby waves at 1 mb. See Hirooka and Hirota (1989) in detail.

It is found from this calendar that each mode appears irregularly throughout the year and that the year-to-year variation is fairly large. This irregurality suggests that the mechanism for the excitation of free oscillations in the atmosphere is highly complicated, so that further studies are needed for the connection of these traveling waves between the lower and middle atmosphere.

3. EASTWARD TRAVELING WAVES IN THE SOUTHERN HEMISPHERE

In addition to the global, westward traveling modes mentioned above, eastward traveling planetary waves of zonal wavenumber 2 are dominant in the middle atmosphere mainly in the southern hemisphere winter.

This was first pointed out by Harwood (1975) using Nimbus-4 SCR radiance data for August-October 1971. Similar analyses have been made by many authors in early 1980's on the basis of satellite observations, and almost all of these studies indicate that the waves are essentially a barotropic mode with the characteristic period of eastward migration ranging 1 to 3 weeks.

Recently, Shiotani et al. (1990) made a further analysis of east-ward traveling waves of wavenumber 2 in the southern hemisphere strato-sphere for the 6 years of 1979-1984, in connection with the seasonal variation of the polar-night westerlies.

A typical example is given in the Hovmöller diagram (Fig. 3), in which the eastward migration with a period of about 10 days can be clearly seen during September-October 1983.

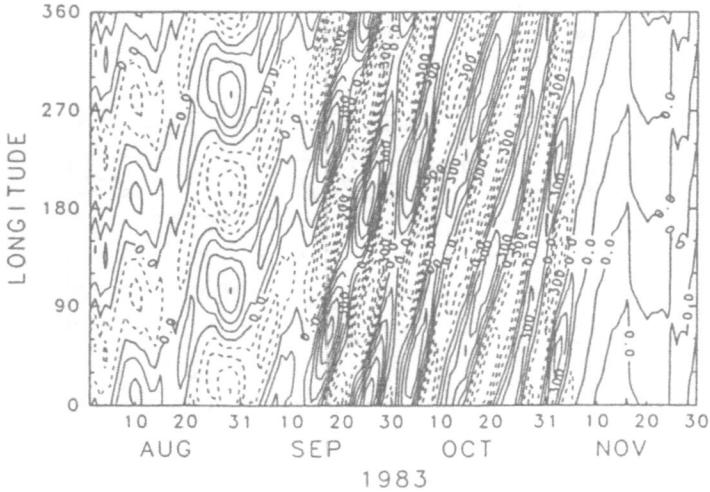

Figure 3. Time-longitude section of the geopotential height of zonal wavenumber 2 at 10 mb and 60°S.

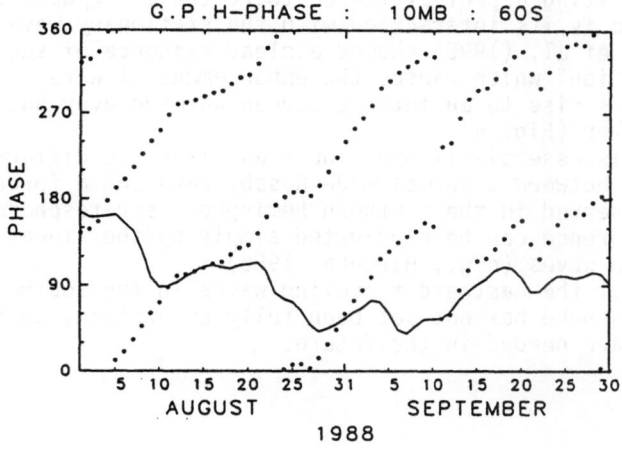

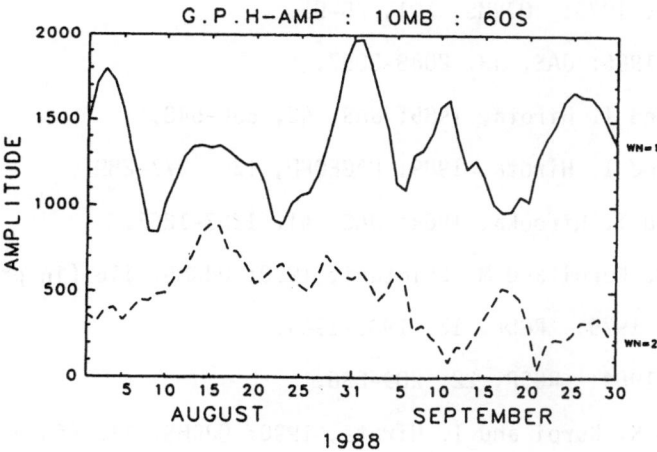

Figure 4.   Time change of phase angle of ridge (upper) and
amplitude (lower) of wavenumber 1 and 2 at 10 mb along 60°S.

Another interesting aspect of the eastward traveling wave 2 in the southern hemisphere is its interaction with the stationary wave of wave number 1. Hirota et al. (1990) showed a clear evidence of such a "wave-wave interaction" which causes the enhancement of wave 1 amplitude and hence gives rise to an intense sudden warming over Antarctica during the mid-winter (Fig. 4).

This process is essentially non-linear and is quite differnt from the "interference" between a normal mode Rossby wave and a forced stationary wave observed in the northern hemisphere stratosphere, because the interference can be elucidated simply by the linear superposition of the two waves (e.g., Hirooka, 1986).

The dynamics of the eastward traveling waves in the southern hemisphere middle atmosphere has not yet been fully understood, so that much more studies are needed in the future.

REFERENCES

Harwood, R.S., 1975: QJRMS, 101, 75-91.

Hirooka, T., 1986: JAS, 43, 2088-2097.

Hirooka, T. and I. Hirota, 1985: JAS, 42, 536-548.

Hirooka, T. and I. Hirota, 1989: PAGEOPH, 130, 277-289.

Hirota, I. and T. Hirooka, 1984: JAS, 41, 1253-1267.

Hirota, I., K. Kuroi and M. Shiotani, 1990: QJRMS, 116 (in press).

Madden, R.A., 1979: RGSP, 17, 1935-1949.

Salby, M.L., 1984: RGSP, 22, 209-236.

Shiotani, M., K. Kuroi and I. Hirota, 1990: QJRMS, 116 (in press).

DYNAMICAL PROPERTIES OF THE ANTARCTIC CIRCUMPOLAR VORTEX INFERRED FROM AIRCRAFT OBSERVATIONS

D. L. HARTMANN
Department of Atmospheric Sciences
University of Washington
Seattle, Washington 98195 USA

ABSTRACT. The amount of dynamical mixing and transport during winter and spring plays a key role in the seasonal evolution of total ozone in the middle atmosphere of the Southern Hemisphere. Too much mixing and heat transport will shut off the ozone-destroying chemistry. A small amount of transport can actually enhance the amount of ozone destroyed photochemically in the lower stratosphere. Aircraft observations taken during the Airborne Antarctic Ozone Experiment in August and September 1987 allow the estimation of the amount of mixing and transport of ozone during the time that the ozone hole developed in that year. Estimates based on conservative tracer data indicate that ozone was transported into the region where the ozone hole developed, thus requiring a photochemical sink for ozone that is at least as large as the observed rate of decline of ozone. By assuming a mean radiative cooling of 0.2 K day^{-1}, it is estimated that the net transport of ozone was relatively small, only 20% ± 10% of the observed trend and of the opposite sign. An independent estimate of the lateral mixing at the edge of the region of chemical depletion yields an effective meridional diffusion coefficient of 6 x 10^3 m^{2}s^{-1}, a rather small value that is consistent with weak transport. These estimates apply only to the period 23 August to 22 September 1987 when the measurements were taken.

1. Introduction

The amount of dynamical mixing of heat and mass between polar and middle latitudes, is important in determining the ozone concentration in the stratosphere. Mixing and net transport play a particularly interesting role in the lower stratosphere of the Southern Hemisphere. In this region the dynamical mixing is weak enough that the wintertime temperatures drop below the level required for polar stratospheric clouds to form. The persistence and extent of polar stratospheric clouds are modulated by variations in the temperature associated with dynamical activity. The presence of polar stratospheric clouds has a profound influence on the ozone chemistry of the lower stratosphere. Through condensation of polar stratospheric clouds (psc's), water and nitric acid are removed from the gas phase where they would otherwise paricipate in the chemistry that retains chlorine in inactive reservoir species in the lower stratosphere. If the cloud particles get large enough to precipitate, then the formation of psc's can lead to the removal of water and nitrate from the polar lower stratosphere. In addition to their influence on gas-phase chemistry, cloud particles in the lower stratosphere provide surfaces where heterogeneous chemical reactions can take place. Some of these reactions can lead to the rapid conversion of chlorine from reservoir species to species that can efficiently remove ozone (Molina and Molina, 1987).

Data from the Airborne Antarctic Ozone Experiment show a sharp boundary where the chemical composition of the lower stratosphere changed abruptly with latitude during September of 1987. Poleward of this boundary, which was inside the polar vortex, the chlorine monoxide is greatly elevated (Brune, et al. 1989) and the water and total odd nitrogen are reduced relative to the

117

A. O'Neill (ed.),
Dynamics, Transport and Photochemistry in the Middle Atmosphere of the Southern Hemisphere, 117–134.
© 1990 *Kluwer Academic Publishers.*

surrounding air farther from the pole (Kelly, et al., 1989, Fahey, et al. 1989, Proffitt, et al., 1989a). This boundary corresponds to temperatures colder than about 195K. These chemical differences were apparent early in the experiment, which began in the third week of August, but initially without a corresponding signature in the ozone mixing ratio. By the second week of September, however, a sharp drop in ozone developed across this boundary, indicating that ozone was being destroyed inside the region of altered chemistry, which we will call the chemically perturbed region (CPR) (Anderson, et al., 1989a, Proffitt, et al., 1989b, Starr and Vedder, 1989).

Dynamical isolation is important for the development of the springtime ozone depletion in the lower stratosphere above Antarctica. Dynamical isolation implies weak eddy mixing of heat and momentum and a weak mean diabatic circulation. With a weak diabatic circulation the polar stratospheric temperatures can remain near their radiative equilibrium values, which are very low during winter. Cold temperatures (near 195K) are necessary for the formation of polar stratospheric clouds. If the polar vortex is not adequately dynamically isolated, the temperatures will be too high for psc's to form. In additon, dynamical mixing will bring in water and nitrate from lower latitudes, thereby destroying the special chemical conditions that may have developed over the course of the winter.

The lower stratosphere in the Southern Hemisphere is characterized by a much more stable, symmetric vortex than the Northern Hemisphere. The difference is believed to be explained by the greater zonal symmetry of the topography and sea-surface-temperature of the Southern Hemisphere. This greater symmetry in the lower boundary conditions implies weaker forcing of stationary planetary waves. In the Northern Hemisphere the forcing of stationary waves is large enough to completely mix the lower stratosphere in midwinter or at the latest in early spring. In the Southern Hemisphere the Polar vortex retains its dynamical identity and cold core temperatures until late spring. The dynamical differences between the hemispheres are reflected in significant differences in the climatological annual and latitudinal variations of total ozone in the two hemispheres (Dütsch, 1974, Bowman and Krueger, 1985). Total ozone in the Northern Hemisphere reaches its maximum at the pole in early spring (about April 1). In the Southern Hemisphere the maximum ozone column occurs well away from the pole at about 60S for most of the year. The maximum at the south pole does not occur until late spring (mid November) which is about two months later than the corresponding season in the Northern Hemisphere.

While the qualitative differences between the stratospheric circulations of the two hemispheres are well known, quantitative estimates of the amount of mixing in the two hemispheres are rare and uncertain. It is important to have quantitative estimates of the rate of mixing and its relationship to mean conditions so that we may evaluate the relative roles of chemistry and dynamics in the Antarctic Ozone Hole, and determine the potential for a similar springtime depletion to develop in the Northern Hemisphere.

During the month of September 1987 a reduction of the total ozone column over Antarctica of about 50 percent occurred. Profiles showed that at altitudes between about 14 and 18 km the ozone destruction was near 100 percent. These ozone depletions were larger than those in the preceding and following years, which were more dynamically active. It seems therefore that the lowest column ozone amounts are achieved in the least dynamically active years. This is consistent with the idea that dynamical isolation and associated cold temperatures are important for the development of low ozone values (e.g. McIntyre, 1989), and also with the fact that transport brings ozone to the polar lower stratosphere (e.g. Tung and Yang, 1988).

There are two limiting cases on the effect of transport processes on the amount of ozone destroyed in the spring depletion season. Too much transport of heat and constituents will warm and mix the polar lower stratosphere sufficiently to shut down the fast chlorine destruction entirely. This apparently happens every year at the time of the final spring warming. Alternatively, if no

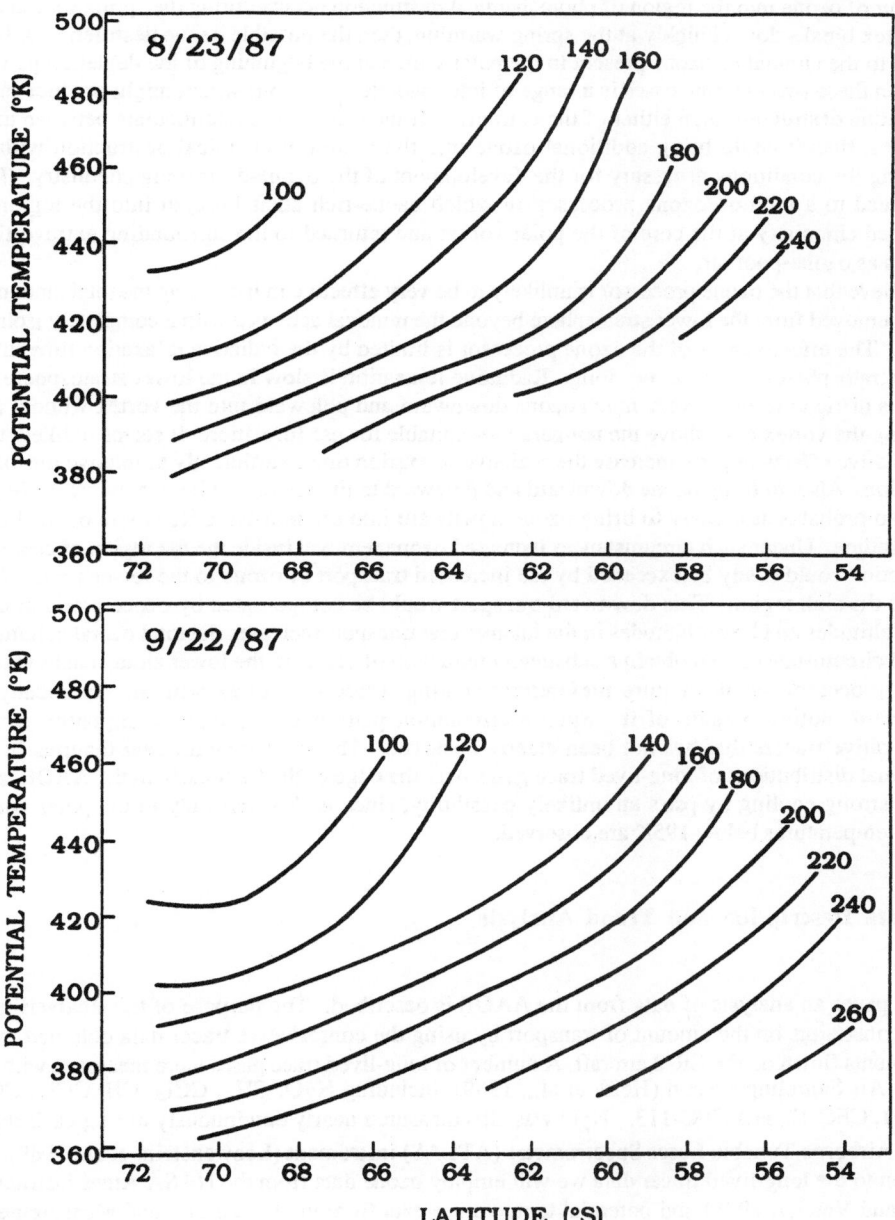

Figure 1. Cross sections of N_2O concentration in potential temperature versus latitude coordinates for a) 23 August and b) 22 September 1987, based on data from the ATLAS and MMS instruments. Units are parts per billion by volume(ppbv), the contour interval is 20 ppbv. (From Hartmann, et al., 1989b)

transport of ozone into the region of photochemical destruction occurs during the spring season and the vortex breaks down quickly at the spring warming, then the possible ozone destruction will be limited to the amount of ozone present in the polar vortex at the beginning of the depletion period. Between these two extreme cases is a range of intermediate conditions which might produce more total ozone destruction than either of the extremes. If the transport is intermediate between these extremes, then it could bring additional ozone into the region of chemical destruction without upsetting the conditions necessary for the development of the ozone-destroying chemistry. This could lead to a kind of 'ozone processor' in which ozone-rich air is brought into the region of perturbed chemistry at the core of the polar vortex and returned to the surrounding extravortical annulus as ozone-poor air.

I believe that the ozone processor is unlikely to be very effective in increasing the total amount of ozone removed from the lower stratosphere beyond the removal achieved with a completely isolated vortex. The effectiveness of the ozone processor is limited by the radiative relaxation time in the lower stratosphere, which is very long. Radiative relaxation is slow in the lower stratosphere, so that it is difficult to move very much ozone downward and poleward into the vortex without also warming the vortex core above the temperatures suitable for psc formation. It seems unlikely that the radiative effects of psc's increase the radiative relaxation times sufficiently to remove this basic limitation. Also, to bring ozone downward and poleward to the region of photochemical depletion, it is also probably necessary to bring ozone downward into the mid-latitudes where ozone has a long lifetime. Under such circumstances increased ozone removal inside the the region of chemical destruction could easily be exceeded by the increased transport of ozone to the lower stratosphere outside the sink region. This downward transport would be compensated by ozone production at higher altitudes and lower latitudes in the manner that transport increases the total ozone column in normal circumstances. To obtain a substantial reduction of ozone in the lower stratosphere via the 'ozone processor' would require preferential sinking inside the vortex without accompanying downward motion outside of it. Such a circulation pattern would leave a signature in the conservative tracers that has not been clearly observed. The lack of significant features in the latitudinal distributions of long-lived trace gases near the edge of the CPR early in the AAOE also makes strong cooling by psc's an unlikely possibility, since psc's occur only in the polar vortex where temperatures below 195K are observed.

2. Data Description and Trend Analysis

In this paper an analysis of data from the AAOE is described. The purpose of this analysis is to place constraints on the amount of transport by using the conservative tracer data obtained from instruments flown on the ER-2 aircraft. A number of long-lived trace gases were measured with the Whole Air Sampling system (Heidt et al.,, 1989), including N_2O, CH_4, CCl_4, CH_3CCl_3, CO, CFC-11, CFC-12, and CFC-113. N_2O was also measured nearly continuously during each flight by the Airborne Tunable Laser Spectrometer (ATLAS) instrument (Loewenstein, et al., 1989). In addition to the long-lived tracer data we will employ ozone data from the NASA Ames instrument (Starr and Vedder, 1988) and potential temperature data from the Meteorological Measurements System aboard the ER-2(Chan, et al., 1989).

Eleven ER-2 sorties originated at Punta Arenas, Chile and flew south along the Palmer Peninsula (Tuck, et al., 1989). The flight tracks for these varied, but most flew north-south roughly along surfaces of constant potential temperature. A dive and climb was executed near the maximum poleward extent of the flight track at 72S. These flight data allow estimates of three important quantities: the gradients of trace constituents with respect to latitude on a fixed potential temperature surface, the gradient with respect to potential temperature at a fixed latitude, and the time tendency at a fixed potential temperature and latitude. These quantities can in turn be used to make estimates of the net effect of transport on the ozone tendency. In the analysis that follows we assume that the zonal-mean behavior can be adequately characterized with a time-average at the longitude of the ER-2 flight track, and that this time average can be estimated with the data from the eleven ER-2 flights. The time-mean flow in the Southern Hemisphere stratosphere is nearly zonally symmetric on average and was so during the period in question.

With the continuous measurements from the ATLAS and MMS instruments on the ER-2 it is possible to draw cross-sections in the latitude-potential temperature plane. This is most appropriate for those flights when the ER-2 flew south on one potential temperature(hereafter theta) surface, performed a vertical profile near 72S and then returned on a different theta surface. Figure 1 shows cross sections of nitrous oxide for flights taken on 23 August and 22 September. The isolines slope downward toward the pole. This slope was established prior to the flight period, probably by intense radiative cooling of the polar stratosphere during the preceding autumn and early winter. While the meteorological conditions at 75W on 23 August and 22 September are different, it is nonetheless true that the basic structure of the N_2O distribution did not change dramatically over the period, while that of ozone did. During this period then the zonal mean distribution of conservative tracers was nearly in a steady equilibrium.

If we select the data from the dive and climb at the southernmost extent of the flight tracks, we can develop a time series of the ozone and nitrous oxide on theta surfaces, or alternatively of the ozone and potential temperature on N_2O surfaces. Analyses of temporal trends in the long lived trace species measured by the ER-2 are presented in Hartmann, et al (1989b). Figure 2 shows the evolution of the vertical profiles of potential temperature and ozone in N_2O coordinates. Potential temperature clearly shows unsystematic temporal variability associated with the meteorological changes, but no clear evidence of temporal trends over the one-month period of the experiment. The ozone cross section, however, show a very clear secular trend that is easily distinguishable from the meteorological variability. Since we can assume that N_2O is conserved at these altitudes, N_2O surfaces should correspond as closely as anything to material surfaces. The striking departure of the N_2O surfaces from the ozone surfaces is an indication of a photochemical sink.

We expect from arguments given in the next section that the dominant terms in the transport of trace substances will be vertical advection by the mean circulation and meridional eddy mixing roughly along theta surfaces. Since N_2O decreases toward the pole along theta surfaces, we expect that lateral eddy mixing will supply N_2O to the region near 70S. Since the N_2O concentration is steady on theta surfaces, it must be that a sufficient sinking motion exists near 70S to counterbalance any positive N_2O tendency provided by lateral mixing. We can therefore state, on the basis of the N_2O spatial structure and time tendency, that the mean vertical motion cannot be upward during the period of the rapid ozone decline. Lateral mixing cannot by itself be the cause of the ozone decline, since this would also require an N_2O increase in the same region as the ozone decline. Moreover, if the lateral mixing were approximately isentropic, as would be expected, then lateral mixing would lead to an increase in ozone over the pole, rather than the observed decline, because the ozone mixing ratio decreases toward the pole on theta surfaces..

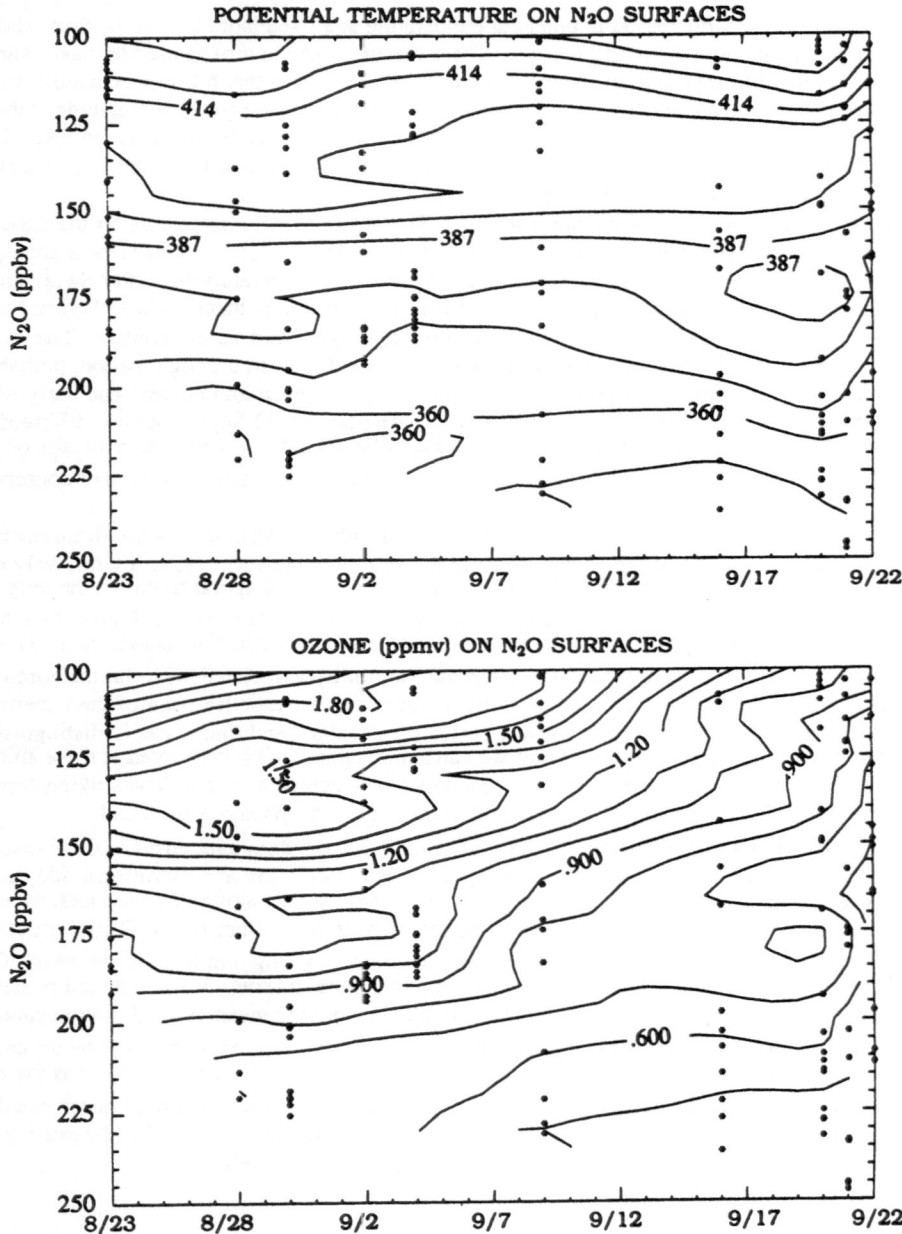

Figure 2. Contour plots of a) potential temperature (K) and b) ozone (ppmv) as functions of N_2O and time. taken from the region between 67 and 72S where the ER-2 performed vertical profiling. Asterisks denote locations of 50-second averaged data used to make the plot. (From Hartmann, et al., 1989b)

3. Formulation of Net Transport Analysis

3.1 ZONAL MEAN FORMULATION

To make estimates of the net transport from the tracer behavior we require a simple theoretical framework. Suppose that we write the conservation equation for some quantity q in potential temperature coordinates (e.g. Andrews, et al., 1987). After zonally averaging and using the continuity equation we obtain.

$$\frac{\partial \bar{q}}{\partial t} + \bar{v} \cos\phi \, \frac{\partial \bar{q}}{\partial y} + \dot{\bar{\theta}} \, \frac{\partial \bar{q}}{\partial \theta} = \bar{s} - \frac{\partial}{\partial y}\overline{(v'q')} - \frac{\partial}{\partial z}\overline{(\dot{\theta}'q')} + \overline{q' \frac{d}{dt}\left(\ln\left(\frac{\partial p}{\partial \theta} \right) \right)} \tag{1}$$

If the final two terms in (1) are neglected compared to the others, then we obtain an approximation for the transport terms which includes only the mean advection and the meridional eddy transport along isentropes. The meridional part of the mean advection is normally also negligible compared to these two terms. We thus obtain the following simplified equation for the time tendency of the zonal mean.

$$\frac{\partial \bar{q}}{\partial t} + \dot{\bar{\theta}} \, \frac{\partial \bar{q}}{\partial \theta} = \bar{s} - \frac{\partial}{\partial y}\overline{(v'q')} \tag{2}$$

If we average this over some range of sine of latitude,

$$\hat{()} = A_p^{-1} \int_{y_1}^{y_2} () \, dy \quad ; \quad A_p = \int_{y_1}^{y_2} dy \tag{3}$$

then we can obtain an equation for the time tendency over an arbitrary latitude band $y_1 - y_2$.

$$\frac{\partial \widehat{\bar{q}}}{\partial t} + \dot{\bar{\theta}} \, \frac{\partial \widehat{\bar{q}}}{\partial \theta} = \hat{\bar{s}} - \overline{(v'q')}\bigg|_{y_1}^{y_2} \cdot A_p^{-1} \tag{4}$$

If we employ a diffusive parameterization for the eddy flux, then we may write (4) as,

$$\frac{\partial \widehat{\bar{q}}}{\partial t} + \dot{\bar{\theta}} \, \frac{\partial \widehat{\bar{q}}}{\partial \theta} = \hat{\bar{s}} + \left(D_{yy}\frac{\partial \bar{q}}{\partial y} \right)\bigg|_{y_1}^{y_2} \cdot A_p^{-1} \tag{5}$$

If the meridional distribution on a potential temperature surface is approximately linear over the latitude belt of interest, where D_{yy} is changing rapidly, then (5) can be written approximately as

$$\frac{\partial \widehat{\bar{q}}}{\partial t} + \dot{\bar{\theta}} \, \frac{\partial \widehat{\bar{q}}}{\partial \theta} + \hat{v}_D \frac{\partial \bar{q}}{\partial y} = \hat{\bar{s}} \tag{6}$$

Where $\hat{v}_D = \left(D_{yy}|_{y_1} - D_{yy}|_{y_2} \right) \cdot A_p^{-1}$

and represents an effective meridional diffusive transport velocity. The form of (6) would apply without the assumption that the parcel dispersion characterized by D_{yy} varies more rapidly than the meridional gradient of q, if one of the boundaries of the latitudinal zone of averaging is taken in a region where either of these quantities, and therefore the eddy transport, is small. Such a region must of course exist near the pole inside the vortex, so that (6) must apply if the region of averaging is a polar cap. (6) will be used to interpret the meridional, vertical and temporal structure in the tracer data obtained from the ER-2.

3.2 ESTIMATES OF NET FLUXES

According to the analysis presented in the previous section, the long-lived tracer mixing ratios remained relatively constant in our reference grid box of 67-73S, 420-428K over the course of the experiment. We also know that these constituents are almost perfectly conservative at this location during early spring. Under these conditions our simplified conservation equation (6) reduces to the particularly simple form,

$$\hat{v}\frac{\partial \overline{q}}{\partial y} + \dot{\theta}\,\frac{\partial \overline{q}}{\partial \theta} = 0 \tag{7}$$

This near balance between vertical advection by the mean circulation and meridional mixing by waves was examined in a simple model by Tung(1982) and Holton(1986), tested in a numerical simulation model by Mahlman, et al.(1986) and Plumb and Mahlman (1987), and appears to be an excellent first approximation to an explanation for the mean tracer slopes in the stratosphere.

If (7) holds, it must be that the vertical and meridional characteristic velocities bear a simple relation to eachother that can be inferred from the tracer gradients. Since we have observations of a number of conservative tracers with very different gradients, we can test for consistency by calculating the relationship between the meridional and vertical gradients. If (7) applies then the gradients should have a constant ratio to eachother that is equal to minus the ratio of the velocities.

$$\gamma = \frac{\hat{v}}{\dot{\theta}} = -\frac{\partial \overline{q}}{\partial \theta} \Big/ \frac{\partial \overline{q}}{\partial y} = -\left[\left(\frac{d\theta}{dy}\right)_q\right]^{-1} \tag{8}$$

The condition for equilibrium of a conserved tracer expressed in (7) and (8) requires that the slope of constant mixing ratio surfaces be parallel to the slope of the parcel trajectories defined by the net transport velocity. These slopes should be the same for all conserved tracers. That this is so can be seen in Figure 3, which is a scatterplot of the gradient with respect to potential temperature versus that with respect to sine latitude. It can been seen that the constituents show a broad range of vertical and meridional gradients, but fall very nearly on a line. The slope of this line determines the ratio of the two transport velocities. The ratio γ is shown in Table 1 for the seven long-lived tracers. The mean value of γ for all of these constituents is 0.0032 ± 0.0002 K^{-1} (note that we have taken latitude to be positive in the Southern Hemisphere and left the radius of the earth out of the meridional derivative).

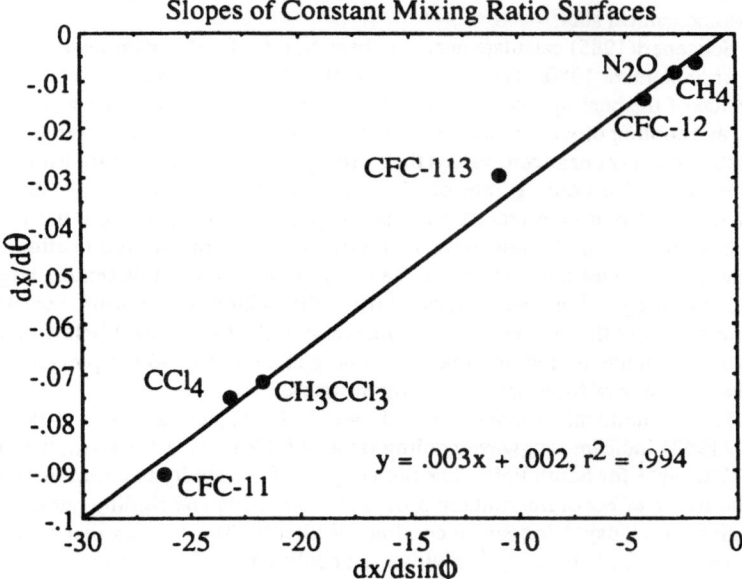

Figure 3. Scatter plot of tracer gradients with respect to sine latitude and potential temperature. The regression line is shown. See Table 1 for data. (From Hartmann, et al., 1989b)

Since the ratio of the two components of the transport velocity vector is known from the slopes of the conservative tracers, we may now write the conservation equation for any constituent with sufficiently slow sources as,

$$\frac{\partial \bar{q}}{\partial t} + \dot{\theta} \left\{ \gamma \frac{\partial \bar{q}}{\partial y} + \frac{\partial \bar{q}}{\partial \theta} \right\} = \bar{s} \tag{9}$$

The value of the quantity in brackets in (9) is approximately zero for the conservative tracers, but is not zero for ozone. The quantity in brackets can be calculated for ozone, however, since we have determined γ from the conservative tracers and know the gradients of ozone from the ER-2 observations. Table 1 shows the gradients of the normalized constituents with potential temperature and sine latitude and the value of γ for each. The mean value of γ for the seven long-lived trace gases of $\gamma^* = -0.0032$ K^{-1} is used to calculate tendency contributions per unit of mean vertical velocity, which are shown for lateral diffusion in the fourth column and for net lateral diffusion plus vertical transport in the fifth column of numbers in Table 1.

We do not know the strength of the photochemical sink of ozone or the strength of the meridional mass circulation, so that (9) still has two unknowns, the mean vertical velocity and the source strength. We estimate the net effect of transport by assuming a diabatic heating rate, and compare its magnitude with the observed trend. In this way we can gauge the likely importance of transport, compared to the photochemical sink.

Rosenfield and Schoeberl(1985) calculate near zero heating rates for the polar lower stratosphere for the conditions of September 1980. The net heating in the 100-50mb layer was positive with the largest absolute value of the heating rate equal to 0.06 K day^{-1}. September is a transition month between the longwave cooling of winter and the springtime solar heating of the lower stratosphere, so that it is reasonable to expect near zero values then. Using data from the AAOE, Rosenfield and Schoeberl(1988) calculated a cooling rate of about 0.2 K day^{-1} for the region near 70S in September of 1987. The radiative effect of polar stratospheric clouds(psc's) are not taken into account in these calculations, and could have a significant effect on the net heating. Shi, et al.(1985) argue that psc's produce a net heating in the lower stratosphere at this season, largely as a result of infrared warming. The psc's warm during the winter and spring because their temperatures are much colder than the emission temperature of the troposphere below them. This conclusion is dependent on the assumption that intervening cloud layers do not prevent the warm temperatures of the troposphere from being 'seen' by the psc's.

Radiometersonde measurements of longwave cooling rates for September taken in 1959-62 and reported by White(1963) indicate longwave cooling rates of 0.07 Kday^{-1} for Byrd, 0.34 Kday^{-1} for Hallett and 0.79 Kday^{-1} for South Pole. The high degree of variability between these stations may result from the strong effect of intermittent psc's on local heating and cooling rates. A straight average of these gives 0.4 Kday^{-1} longwave cooling. White(1963) estimates that solar heating during September will average 0.14 Kday^{-1}, giving a net cooling rate of -0.26 Kday^{-1}.

Hoffman, et al.(1988) estimate that aerosol layers during September of 1986 moved downward at a rate of about 1 kilometer per month. If this is equated with the Lagrangian mean vertical velocity in this region, then it corresponds to a cooling rate of about -0.3 Kday^{-1}. We take -0.2 Kday^{-1} as a best guess of the radiative cooling rate that was at work in September of 1987, though we recognize that this estimate is uncertain. It is on the low side of most of the estimates we have quoted, but consistent with the recent estimate of Rosenfield and Schoeberl(1988) for the time and place of interest

From (9) we can calculate the time tendency of a given tracer that would result from a mass circulation of a given strength, if the source term is set to zero. The strength of the circulation can be characterized by the diabatic vertical velocity, and the associated meridional mixing inferred from the tracer slopes. In the last column of Table 1 we present the e-folding time that transport acting alone would generate, if the mean value of $\gamma^* = -0.0032$ is used and the transport circulation has a magnitude corresponding to a cooling rate of -0.20 K/day. The e-folding time is just the inverse of the time tendency of the normalized tracer concentration. We have chosen γ^* to make the net transport of long-lived tracers small, because the observed tendencies are small and the sources are small, so that the e-folding transport time scales for these constituents are very long, hundreds of days.

For ozone the transport time scale also exceeds 100 days if a mean rate of potential temperature decrease of 0.4 K/day is assumed. Both sinking motion and lateral diffusion supply ozone to the polar region at this time of year, in contrast to the long lived tracers where these two fluxes cancel eachother. The gradients are sufficiently weak, however, that the transport is still relatively inefficient. To a certain extent it was to be expected that the net transport of ozone would be weak, once it was determined that the net transport of the conservative tracers was about zero, because ozone itself is nearly conservative at these altitudes, except within the chemically perturbed region. Indeed, without the most poleward point that is inside the CPR, regression on the remaining data points yields a latitudinal gradient of only -1.85(compared to -3.27 in Table 1) and the transport

Table 1: Data for transport calculation. Constituent mixing ratios have been normalized by the mean value given in the table which is time average for the samples falling in the latitude, potential temperature box; $67S<\phi<73S$, $420K<\theta<428K$. The time scale is the inverse of the normalized tendency that net transport would give if the diabatic circulation corresponded to a potential temperature change of 0.4 K/day. ND indicates that the quantity shown is nondimensional. (From Hartmann, et al., 1989b)

x	Mean (ppv)	$\partial\chi/\partial\theta$ (K⁻¹)	$\partial\chi/\partial\sin\phi$ ND	γ (K⁻¹)	$\gamma^* \partial\chi/\partial\sin\phi$ (K⁻¹)	$\gamma^* \partial\chi/\partial\sin\phi + \partial\chi/\partial\theta$ (K⁻¹)	Time Scale (days)
CH_4	0.99 [1]	-.006	-1.79	-.00334	.0058	-.00025	-10016.
N_2O	157.0 [2]	-.008	-2.77	-.00288	.0089	.00088	2840.
CFC-12	134.8 [3]	-.014	-4.19	-.00334	.0134	-.00059	-4222.
CFC-11	17.8 [3]	-.091	-26.26	-.00347	.084	-.00697	-358.
CH_3CCl_3	4.99 [3]	-.072	-21.72	-.00331	.0695	-.0025	-1001.
CCl_4	11.3 [3]	-.075	-23.2	-.00323	.0742	-.00076	-3289.
CFC-113	9.55 [3]	-.030	-10.9	-.00275	.0349	.00488	512.
Ozone	1.43 [1]	.008	-3.27	+.00245	.0105	.01846	135.
ClO	308.5 [3]	.018	+9.94	-.00181	-.0318	-.0138	-181.
H_2O	2.48 [1]	.002	-2.46	+.00081	.0079	.00987	253.

[1] Mixing ratio is in ppmv.
[2] Mixing ratio is in ppbv.
[3] Mixing ratio is in pptv.

time scale for ozone would be increased to 190 days. In the following section we look more carefully at the implications of the sharp ozone gradient at the edge of the CPR.

The gradients of ClO are stronger than those of ozone, so that the individual contributions from vertical and meridional transport are large. The effects of poleward and downward motion on the concentration of ClO tend to cancel eachother, however, giving a rather long time scale for transport changes to ClO also. The flux of ClO through the region could be significant, nonetheless, with high ClO brought in from above and exported to lower latitudes by quasi-isentropic mixing.

We estimate the time tendency of ozone by linear regression of the normalized ozone mixing ratio in our reference grid box in the y,θ plane versus time. This gives about -0.025 day^{-1} as the trend, or a time scale of about 40 days. We know, however, that the ozone mixing ratio only began to fall after the first of September and then dropped to 50% of its initial value by 22 September. This gives an e-folding time scale of about 32 days or a normalized trend of -0.031 day^{-1}. Since we know that the mean motion is downward, then the source strength (or rather the sink) must equal or exceed the tendency. If we assume a cooling rate of -0.2 K/day (a potential temperature decrease of 0.4 k/day), and use the mean value of γ from the long-lived tracer data of -0.0032, then the transport to into the region is about 20% of the time tendency determined from regression over the entire period.

We therefore estimate that ozone transport into the vortex increases the requirement for a photochemical sink by about 20% $\pm$ 10%, which is within the range that could be explained with current photochemical theory, and which does not allow chemistry to remove much more ozone than is initially present inside the region where the perturbed chemistry develops. This estimate of the ozone flux is supported by an independent estimate based on the meridional scale of the ozone discontinuity at the edge of the CPR developed in the next section.

4. Limits on Meridional Diffusion

In this section we attempt to place limits on the magnitude of the meridional mixing by using the observed meridional gradient of ozone at the boundary of the chemically perturbed region. Tracer data constrain the relationship between meridional mixing and the diabatic mean circulation, but do not constrain the magnitude of the transport circulation. The fact that the tracer distributions are relatively steady over the period of the experiment strongly suggests that the transports are weak. It seems unlikely that strong meridional mixing and diabatic circulation would exist and exactly cancel eachother in the generally transient environment of the real stratosphere. Radiative transfer calculations also suggest that the diabatic circulation is relatively weak, since radiative heating rates are small. Here we use the observed evolution with time of the photochemically-driven ozone gradient to place upper limits on the meridional mixing on isentropic surfaces. A simple model of the ozone distribution near the edge of the chemically perturbed region is used to infer rather weak values for the effective meridional diffusivity.

Tracer data suggest that the transport characteristics remained relatively steady over the period from late August to late September and that trends of conserved tracers were small. At the same time ozone decreased dramatically within the region where chlorine monoxide was elevated, the so-called chemically perturbed region. Outside of this region the ozone concentration remained relatively constant on a theta surface, much like the conserved tracers. We then suppose that ozone is relatively conservative outside of the chemically perturbed region and destroyed inside it by chemistry during the period of the experiment. Observations of ozone from the ER-2 suggest that the ozone was relatively constant with latitude along the 425K theta surface in late August(Starr and Vedder,1989; Proffitt, et al., 1989b). In early September a discernable drop in ozone concentration began to appear at the latitude of the chlorine monoxide increase. By the latter half of September, each flight was recording an approximately 50% drop in ozone concentration across a very narrow range of latitudes.

The model we propose to describe the development of the ozone gradient near the edge of the depletion region is a simple one-dimensional diffusion equation with a source term. Excluded from the model are the effects of the mean meridional circulation. Since the ozone mixing ratio has only a weak gradient along the 425K potential temperature surfaces at the beginning of the experiment, it seems safe to assume that the meridional circulation is not playing an important role in producing the gradient near the CPR boundary. It might be argued that mean subsidence could enhance the gradient if strong downward motion were occurring outside the boundary of the CPR, and thereby increasing ozone concentrations there, but was not operating inside the CPR. This by itself would imply a very sharp decrease in meridional mixing across the CPR boundary because of the known relationship between irreversible eddy mixing and the diabatic circulation(e.g. Mahlman, et al., 1984). Also, if a sharp feature in the effect of the mean meridional circulation on ozone gradients existed, we would expect to see it also in the conservative tracer data, which we do not.

We will assume for the purposes of this discussion that the first order balance at work in determing the meridional ozone gradient near the CPR boundary is between the photochemical sink term and lateral diffusion, with mean vertical advection playing a secondary role in this region. The model equation we will use is thus,

$$\frac{\partial \overline{q}}{\partial t} = D_{yy}\frac{\partial^2 \overline{q}}{\partial y^2} + \overline{S}_q \tag{10}$$

The initial conditions appropriate to the situation at hand are an intial q distribution that is independent of y. If a catalytic cycle like that proposed by Molina and Molina(1987) is accepted, then the ozone is destroyed at a rate proportional to the chlorine monoxide concentration squared and, for ozone concentrations that are not too small, independent of the ozone concentration itself. Since the chlorine monoxide concentration has a steplike increase at the boundary of the chemically perturbed region, it is reasonable to assume that the photochemical sink of ozone also has a step structure. The source is switched on at the initial time and has a discontinuity at the origin, which is identified with the latitude of the CPR boundary.

We are interested primarily in the dependence of the width of the transition on the diffusivity and the source strength. To make a solution simpler without any fundamental change in the problem, we assume that the source term has the following form,

$$\overline{S}_q = - \text{sgn}(y)\, Q \text{ for } t > 0. \tag{11}$$

where sgn(y) is the sign function. In this case the solution to (10) is,

$$\overline{q}(y,t) = - \text{sgn}(y)\, Q\, t \left[1 - \exp\left\{ \frac{-|y|}{(D_{yy}t)^{1/2}} \right\} \right]. \tag{12}$$

The solution is antisymmetric about y=0. The meridional scale of the variation of q near the origin is independent of the source strength, but depends only on the diffusion coefficient and time. The meridional diffusion coefficient is related to the length scale in the following way,

$$D_{yy} = L^2/t. \tag{13}$$

Where t is the elapsed time since the source term was switched on.

We test the applicability of the model for the discontinuity at the edge of the CPR and evaluate the appropriate value of the length scale L by fitting the solution to the composite of the ozone concentration relative to the CPR boundary derived in Hartmann, et al.(1989a). A composite latitudinal profile was obtained for each of the 425K and 450K surfaces by averaging the ozone mixing ratios from all the flights into latitude bins relative to the position where the ClO concentration rose above 130 pptv. Figure 4 shows the composite observed structure and the fits of the solution structure to the data at 450K and 425K for four degrees of latitude on either side of the ClO boundary. This analytic function fits the observed data very well, particularly equatorward of the ClO boundary. The largest value of the length scale that could be inferred from the composites is about 1 degree of latitude or about 100 kilometers. This is consistent with inspection of latitude profiles from individual flights.

The appropriate time scale to use can be inferred from the coevolution of the ClO and insolation in the CPR. According to calculations reported by Anderson, et al.(1989b), the destruction of ozone by chlorine chemistry began about the beginning of September in 1987, when the combination of ClO concentration and solar exposure passed the necessary threshold. As mentioned previously, this is also about the time when ozone discontinuities with latitude first began to be observed in association with the ClO edge. The ClO sink of ozone had thus been operating for about three weeks when the last ER-2 flight was taken on 22 September. The effective meridional diffusivity that is consistent with the observed ozone gradient is therefore calculated to be,

$$D_{yy} = \frac{L^2}{t} = \frac{\left(10^5 \, m\right)^2}{21 \cdot 86400 \, s} = 6 \times 10^3 \, m^2 \, s^{-1} \tag{14}$$

This is indeed a small diffusivity compared to the values typically used in two-dimensional models. It seems, therefore, that the meridional mixing at the edge of the CPR boundary was very weak during September of 1987. This is perhaps an upper limit on the effective diffusivity, since any mixing that occurs will also mix ClO-rich air into lower latitudes and blur the edge of the sink. In this calculation we have assumed that the sink has a perfectly sharp edge. If we had allowed for a less sharp edge to the photochemical sink, the inferred diffusivity would have been even smaller.

We can estimate the rate at which ozone is fluxed into the polar cap by diffusion across the CPR boundary by integrating the diffusive flux over a polar cap. Using a length scale of 100 km, a time of 21 days and a mean latitude for the CPR boundary of 65S, we obtain an e-folding time scale for ozone increase within the CPR resulting from diffusion across the boundary of 125 days. This is in reasonable agreement with the estimate given in Table 1 and with our estimate that the magnitude the effect of ozone transport is about 20% of the observed trend.

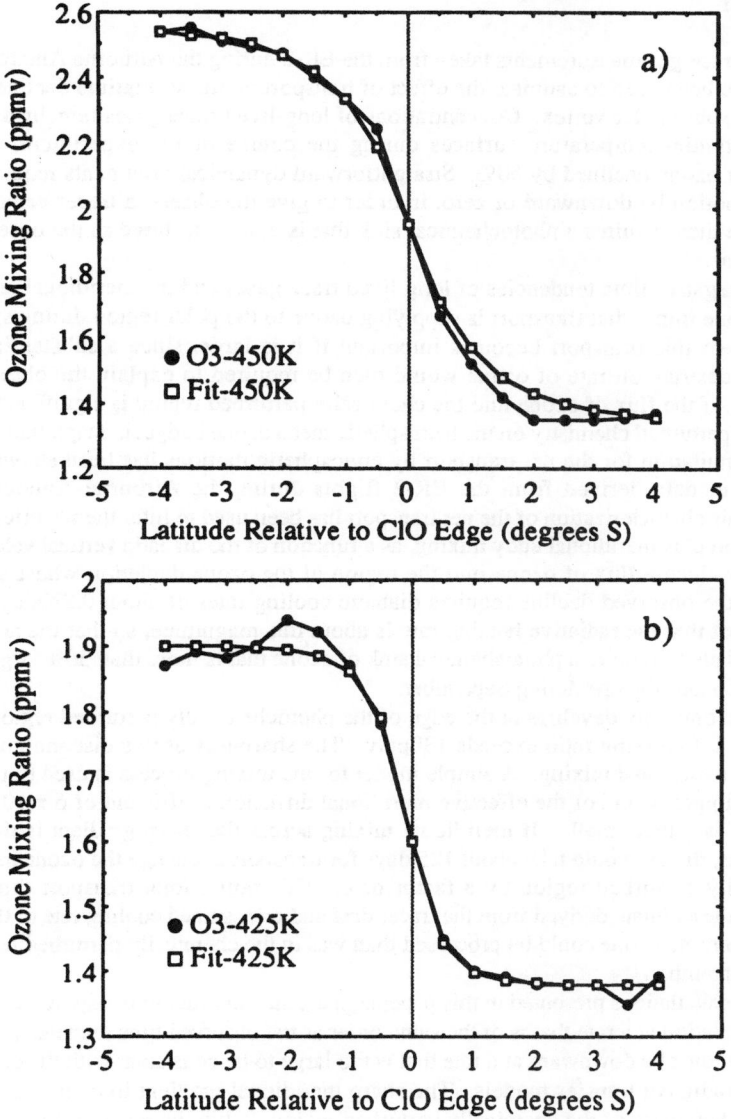

Figure 4. Ozone mixing ratio averaged for all the flights in latitude bins relative to the latitude where ClO first exceeds 130 pptv. Closed circles indicate data, open squares indicate the fit of the analytical solutions to the data. See text for explanation. Latitude is measured positive toward the South Pole. a) at 450K potential temperature surface, b) at 425K surface. (From Hartmann, et al., 1989b)

5. Discussion

The long-lived trace gas measurements taken from the ER-2 during the Airborne Antarctic Ozone Experiment have been used to estimate the effect of transport on the springtime decline in ozone observed in the south polar vortex. Concentrations of long-lived trace gases remained relatively constant on potential temperature surfaces during the course of the experiment, while the concentration of ozone declined by 50%. Straightforward dynamical arguments require that the mean vertical motion be downward or zero, in order to give the observed tracer behavior. The budget of ozone then requires a photochemical sink that is at least as large as the observed time tendency of ozone.

The zero or negative time tendencies of long-lived trace gases and the meridional and vertical gradients of ozone imply that transport is supplying ozone to the polar region during springtime. The magnitude of this transport becomes important if it is large, since a substantially larger photochemical destruction rate of ozone would then be required to explain the observed time tendency. Also, if the flux of ozone into the chemically perturbed region is significant, then the influence of the perturbed chemistry on the hemispheric mean ozone budget is magnified.

A simple formulation for the net transport by atmospheric motions has been shown to apply approximately to data derived from the ER-2 flights during the Airborne Antarctic Ozone Experiment. This characterization of the net transport has been used to infer the net effect of mean vertical advection plus meridional eddy mixing, as a function of the diabatic vertical velocity. It is found that to produce a flux of ozone into the region of the ozone depletion whose magnitude equals 20% of the observed decline requires diabatic cooling rates of about 0.2K/day. Current estimates suggest that the radiative heating rate is about this magnitude, so that the atmospheric transport is unlikely to require a photochemical sink of ozone that is more than 30% larger than the observed rate of ozone decline during September.

A marked discontinuity develops at the edge of the photochemically perturbed region over the pole, where the ClO mixing ratio exceeds 130pptv. The sharpness of this discontinuity implies relatively weak meridional mixing. A simple model for the mixing process is used to produce an estimate of the upper bound of the effective meridional diffusion coefficient of 6×10^3 m^2 s^{-1}. This diffusivity is rather small. If meridional mixing across the sharp gradient is the limiting transport process, then it would take about 125 days for transport to change the ozone mixing ratio in the chemically perturbed region by a factor of e. This rather long transport time scale is consistent with the estimate derived from the tracer data and an assumed cooling rate of 0.2Kday^{-1}. Only about 20% more ozone could be processed than was in the chemically perturbed region at the beginning of September.

The data and calculations presented in this paper suggest that, in order to transport ozone into the region of the depletion at a rate that is of the same order as the observed time tendency, the diabatic vertical velocity must be downward at a rate that is too large to be consistent with the cooling rates expected from radiative transfer models. The sharp meridional gradient in ozone concentration observed at the boundary of the chemically perturbed region within the polar vortex suggests that the lateral mixing is weak in this region. Since lateral mixing of potential temperature and diabatic sinking are interdependent, it is expected from the tracer data that the mean diabatic sinking within the vortex would in fact be rather small. Potential vorticity calculations from ER-2 data and from meteorological analysis(Hartmann, et al; 1988) show strong gradients on potential temperature surfaces in the region just outside of the chemically perturbed polar core of the vortex. This again suggests that the lateral mixing in this region is weak.

ACKNOWLEDGEMENTS:

This work was supported by the National Aeronautics and Space Administration under grants and contracts supporting the Airborne Antarctic Ozone Experiment. This paper contains material from Hartmann, et al.(1989b). I would like to acknowledge the collaboration of my coauthors on that paper and all of the participants in the Airborne Antarctic Ozone Experiment.

REFERENCES

Anderson, J.G., W.H. Brune and M.H. Proffitt (1989a) Ozone destruction by chlorine radicals within the antarctic vortex: The spatial and temporal evolution of ClO-O_3 anticorrelatin based on in situ ER-2 data, J. Geophys. Res., 94, in press.

Anderson, J.G., W.H. Brune, S.A. Lloyd, W.L. Starr, M. Loewenstein, and J.R. Podolske, (1989b) Kinetics of O_3 destruction by ClO and BrO within the Antarctic vortex: An analysis based on ER-2 data, J. Geophys. Res., 94, in press.

Andrews, D. G., J. R. Holton and C. B. Leovy (1987) Middle Atmosphere Dynamics, Academic Press, 489pp.

Borman, K. P., and A. Krueger (1985) A global climatology of total ozone from the Nimbus 7 Total Ozone Mapping Spectrometer, J. Geophys. Res. 90, 7967-7976.

Brune, W.H., J.G. Anderson and K.R. Chan (1989) In situ observations of ClO in the Antarctic: ER-2 aircraft results from 54S to 72S, J. Geophys. Res., 94, in press.

Chan, K.R., S.G. Scott, T.P. Bui, S.W. Bowen and J. Day (1989) Temperature and horizontal wind measurements on the ER-2 aircraft during the 1987 Airborne Antarctic Ozone Experiment, J. Geophys. Res., 94, in press.

Dütsch, H.U.(1974) The ozone distribution in the atmosphere, Can. J. Chem., 52, 1491-1504.

Fahey, D.W., K.K.Kelly, G.V. Ferry, L.R. Poole, J.C. Wilson, D.M. Murphy, M. Loewenstein and K.R. Chan (1989) In-situ measurements of total reactive nitrogen, total water, and aerosol in polar stratospheric cloud in the Antarctic, J. Geophys. Res., 94, in press.

Hartmann, D. L., M. R. Schoeberl, P. A. Newman, R. L. Martin, B. L. Gary, K. R. Chan, M. Loewenstein, J. R. Podolske, and S. E. Strahan (1989a) Potential vorticity estimates in the south polar vortex from ER-2 flight data, J. Geophys. Res., 94, in press.

Hartmann, D.L., L.E. Heidt, M. Loewenstein, J.R. Podolske, J.F. Vedder, W.L. Starr and S.E. Strahan, Transport into the south polar vortex in early spring (1989b) J. Geophys. Res., 94, in press.

Heidt, L.E., J. F. Vedder, W.H. Pollock, R.A. Lueb and B.E. Henry (1989) Trace gases in the Antarctic Atmosphere. J. Geophys. Res., 94, in press.

Hoffman, D.J., J.M. Rosen and J. W.Harder (1988) Aerosol measurement in the winter/spring antarctic stratosphere. 1. Correlation measurements with ozone, J.Geophys. Res., 93, 665-676.

Holton, J.R. (1986) Meridional distribution of stratospheric trace constituents, J. Atmos. Sci., 43, 1238-1242.

Kelly, K.K., A.F. Tuck, D.M. Murphy, M.H. Proffitt, D.W. Fahey, R.L. Jones, D.S. McKenna, M. Loewenstein, J.R. Podolske, S.E. Strahan, G.V. Ferry, K.R. Chan, J.F.Vedder, G.L. Gregory, W.D. Hypes, M.P. McCormick, E.V. Browell and L.E. Heidt (1989) Dehydration in the lower antarctic stratospher during late wint and early spring, 1987, J. Geophys. Res., 94, in press.

134

Loewenstein, M., J. R. Podolske, S. E. Strahan and K.R. Chan (1989) N2O as a dynamical tracer in the 1987 Airborne Antarctic Ozone Experiment, J. Geophys. Res., 94, in press.

Mahlman, J.D., D.G. Andrews, D.L. Hartmann, T. Matsuno and R.G. Murgatroyd (1984) Transport of trace constituents in the stratosphere, Dynamics of the Middle Atmosphere, edited by J.R. Holton and T. Matsuno, ppl 387-416, Terra Scientific, Tokyo.

Mahlman, J.D., H.D. Levy II, and W.J. Moxim (1986) Three-dimensional simulations of stratospheric N2O: Predictions for other trace constituents, J. Geophys. Res., 91, 2687-2707.

McIntyre, M.E. (1989) On the Antarctic Ozone Hole, J. Atm. Terr. Phys., 51, 29-43..

Molina, L.T., and M.J. Molina (1987) Production of Cl_2O_2 from the self reaction of the ClO radical, J. Phys. Chem., 91, 433-436.

Plumb, R.A., and J. D. Mahlman (1987) The zonally-averaged transport characteristics of the GFDL general circulation/transport model, J. Atmos. Sci., 44, 298-327.

Podolske, J.R., M. Loewenstein, S.E. Strahan, and K.R. Chan (1989) Stratospheric Nitrous Oxide distribution in the Southern Hemisphere, J. Geophys. Res., 94, in press.

Proffitt, M.H, J.A. Powell, A.F. Tuck, D.W. Fahey, K.K. Kelly, A.J. Krueger, M.R. Schoeberl, B.L. Gary, J.J. Margitan, K.R. Chan, M. Loewenstein and J.R. Podolske (1989a) A chemical definition of the boundary of the antarctic ozone hole, J. Geophys. Res., 94, in press

Proffitt, M.H, M.J. Steinkamp, J.A. Powell, RlJ. McLaughlin, O.A. Mills, A.L.Schmeltekopf, T.L. Thompson, A.F. Tuck, T. Tyler, R.H. Winkler and K.R. Chan (1989b) In situ measurements within the 1987 antarctic ozone hole from a high altitude ER-2 aircraft, J. Geophys. Res., 94, in press.

Rosenfield, J., and M.R. Schoeberl (1986) A computation of stratospheric heating rates and the diabatic circulation for the antarctic spring, Geophys. Res. Lett., 13, 1339-1342.

Rosenfield, J., and M.R. Schoeberl (1988) Personal communication, paper in preparation.

Shi, G.-Y, W.-C. Wang, M.K.W. Ko and M. Tanaka (1986) Radiative heating due to stratospheric aerosols over Antarctica, Geophys. Res. Lett., 13, 1335-1338.

Starr, W.:L. and J.F. Vedder (1989) Measurements of ozone in the Antarctic atmosphere during August and September 1987, J. Geophys. Res., 94, in press.

Tuck, A.F., R.T. Watson, E.P. Condon, J.J. Margitan and O.B. Toon (1989) The planning and execution of ER-2 and DC-8 flights over Antarctica, August and September 1987, J. Geophys. Res., 94, in press.

Tung, K.K. (1982) On the two-dimensonal transport of stratospheric trace gases in isentropic coordinates, J. Atmos. Sci., 39, 2330-2355.

Tung, K.K., and H. Yang (1988) Dynamical component of seasonal and year-to-year changes in antarctic and global ozone, J. Geophys. Res., 93, .

White, F.D., Jr., The radiative factor in the mean meridional circulation of the antarctic atmosphere during the polar night, Ph.D. Thesis, University of Wisconsin, Department of Meteorology, 55pp., 1963

LARGE STRATOSPHERIC SUDDEN WARMING IN ANTARCTIC LATE WINTER AND
SHALLOW OZONE HOLE IN 1988: OBSERVATION BY JAPANESE ANTARCTIC
RESEARCH EXPEDITION

HIROSHI KANZAWA and SADAO KAWAGUCHI
National Institute of Polar Research
1-9-10 Kaga Itabashi-ku, Tokyo 173, Japan

ABSTRACT. There occurred a large stratospheric sudden warming in the
southern hemisphere in late winter of 1988 which competes in suddenness
and size with major mid-winter warmings in the northern hemisphere.
Associated with the dynamical phenomenon of the sudden warming, total
ozone increased over the eastern hemispheric part of Antarctica.   The
sudden warming as well as other warmings which followed it made the
1988 ozone hole shallow in depth and small in area. Long-term observation at Syowa Station (69°S, 40°E) well depict characteristics of the
1988 event.

1. INTRODUCTION

There occurred no large stratospheric sudden warmings over Antarctica
in winter and spring of 1987, in other words, there was a delay in the
spring final warming (Randel, 1988), and the deepest ozone hole was
recorded (e.g., Krueger et al., 1988; Gardiner, B.G., 1988; Kaneto et
al., 1989). Much attention had been paid to the 1988 phenomenon.   In
1988, a large stratospheric sudden warming occurred early for the
Antarctic during late winter, and the ozone hole became shallow
(Schoeberl, 1988; Schoeberl et al., 1989; Krueger et al., 1989).   This
paper describes the behavior of stratospheric temperatures and total
ozone mainly over Syowa Station (69°S, 40°E) in 1988 in comparison with
other 22 years, and some features of vertical profile of ozone by
ozonesonde observations at Syowa Station in 1988.   The data show that
dynamics plays an essential role in many aspects of the Antarctic ozone
hole phenomenon.   A shorter version of this paper which omitted some
figures, some discussions, some references, research plan, etc. is seen
in Kanzawa and Kawaguchi (1989).

2. OBSERVATION AT SYOWA STATION, ANTARCTICA

Main data used in the present study have been obtained at Syowa Station
since the International Geophysical Year by the Meteorological team of

135

A. O'Neill (ed.),
Dynamics, Transport and Photochemistry in the Middle Atmosphere of the Southern Hemisphere, 135–148.
© 1990 *Kluwer Academic Publishers.*

136

the Japanese Antarctic Research Expedition. The data are published by the Japan Meteorological Agency (Japan Meteorological Agency, 1963-1989).

Fig. 1 shows temperature on the 30 mb pressure level (at an altitude of about 22 km) at Syowa Station (69°S, 40°E) from August through November for 23 years since 1966 up to 1988. A large sudden warming occurred in 1988 from the end of August to the beginning of September. Such a large sudden warming in such an early period is the record of Syowa Station since 1966. Moreover, the rate of increase of 30 mb temperature competes with or surpasses that of the well known mid-winter major sudden warming in the northern hemisphere in 1963 (e.g., Finger and Teweles, 1964; Julian and Labitzke, 1965) as shown in Table 1 which compares the 1988 event of Syowa Station with the 1963 event of Edmonton (53°N, 113°W), Canada. After the sudden warming, there occurred a sudden cooling. After this warming and cooling event the temperature fluctuates at the period of about two weeks. Such a large amplitude of two-week oscillation is a characteristic of the year 1988. Roughly speaking, the 1988 temperature shows the warm extreme in Fig. 1 while the 1987 temperature shows the cold extreme. Note that the year 1987 is known as the year when the deepest ozone hole was recorded (Krueger et al., 1988), and the level of total ozone at Syowa Station is also lowest in 1987 as shown in Fig. 2 and in Kaneto et al. (1989).

During the same period as the sudden warming from the end of August to the beginning of September, there occurred a sudden increase

Table 1. Sudden warming and associated sudden increase of total ozone in 1988 compared with those in 1963. Temperature on 30 mb (at an altitude of about 22 km) is shown. Unit of total ozone is Dobson Units (DU).

Syowa Station (69°S, 40°E) 1988	Edmonton (53°N, 113°W) 1963
-83°C (22 Aug.)	-81°C (19 Jan.)
59°C/10days	38°C/9days
-24°C (1 Sep.)	-43°C (28 Jan.)
236 DU (22 Aug.)	260 DU (21 Jan.)
227 DU/9days	180 DU/8days
463 DU (31 Aug.)	440 DU (29 Jan.)

*The values of Edmonton, Canada in 1963 are read from Fig. 3 of Finger and Teweles (1964).
**Warming over the north pole in 1963 is from -86°C (12 Jan.) to -35°C (1 Feb.) at the rate of 51°C/20days after Fig. 3 of Julian and Labitzke (1965).

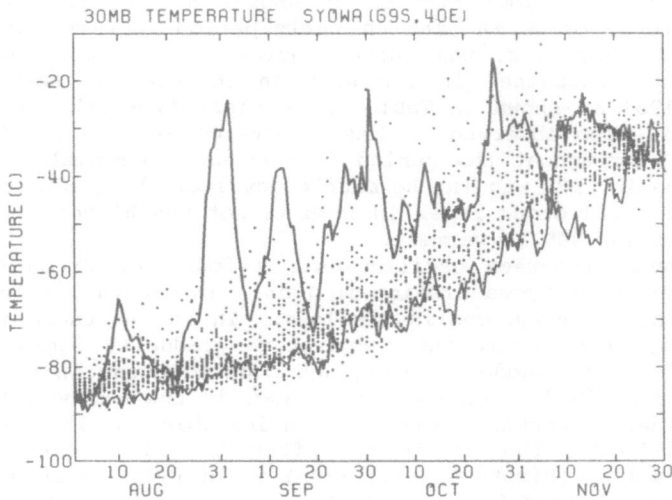

Figure 1. Time change of temperature on 30 mb (about 22 km) over Syowa Station (69°S, 40°E), Antarctica, from August through November. A thick line denotes 1988 and a thin line denotes 1987. Many solid circles denote 21 years for 1966-1986.

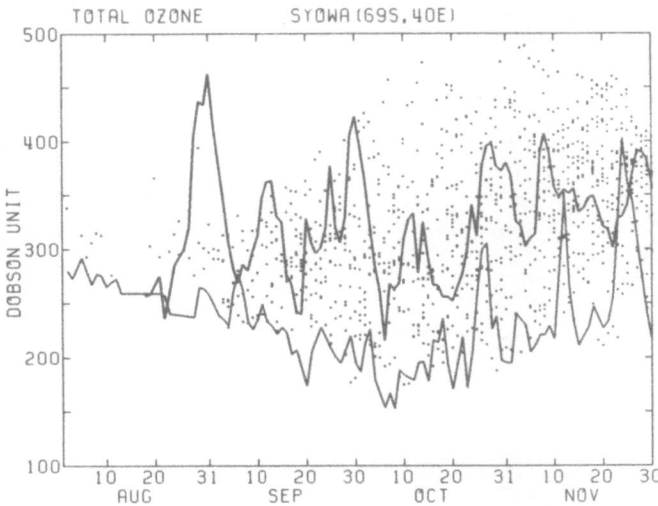

Figure 2. Time change of total ozone over Syowa Station (69°S, 40°E), Antarctica, from August through November. A thick line denotes 1988 and a thin line denotes 1987. Many solid circles denote 21 years for 1962, 1966-1972, 1974-1986.

138

of total ozone over Syowa Station as shown in Fig. 2. Such a large
increase of total ozone in such an early period is also the record of
Syowa Station. Moreover, the sudden increase of total ozone at Syowa
Station in 1988 surpasses in its size an increase of total ozone at
Edmonton in 1963 as shown in Table 1. A clear two-week oscillation of
total ozone is also observed in 1988, corresponding to the temperature
oscillation. Throughout the period from August to November, the level
of total ozone in 1988 was in the middle level of observation values for
23 years; the total ozone level in 1988 is not the highest while the 30
mb temperature in 1988 is warmest.

Strong quasi-two-week oscillation was found in 30 mb temperature
and total ozone over Syowa Station in 1988. In general, after a sudden
warming occurs, a sudden cooling follows. In the northern hemisphere,
there are many cases where the temperature gradually increases to the
summer one after the sudden warming and cooling event in mid-winter as
in 1973 (Kanzawa, 1980). On the other hand, in the southern hemisphere,
in general, the temperature shows a considerable oscillation of about
two weeks to become the summer one after the first event of sudden
warming and cooling (Hirota et al., 1983) as in 1988. A notable characteristic of Fig. 1 (and Fig. 2) is the large amplitude of the oscillation.

Ozonesonde observations at Syowa Station gave vertical profiles of
ozone as well as those of temperature during the first sudden warming
period as shown in Fig. 3. The altitude region where ozone increased
was from about 10 to 25 km in the lower stratosphere. Note that the

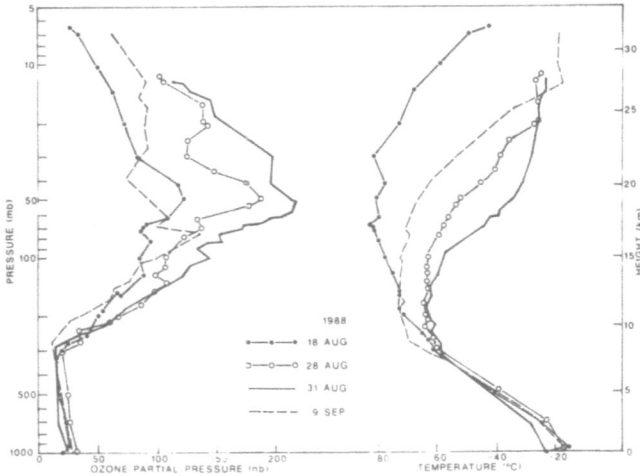

Figure 3. Vertical profiles of ozone (in partial pressure) and temperature at Syowa Station (69°S, 40°E), Antarctica on 18, 28, 31 August, and
9 September in 1988. The profiles were obtained from ozonesonde observations.

ozone loss occurs in just the same altitude region when the ozone depletion is observed, as first shown by Chubachi (1984). Temperature increased in the region from about 10 km at least up to 30 km which is the upper limit of the observation. The ozone profile on 9 September after the event is similar to that on 18 August before the event while temperature on 9 September is considerably warmer than that on 18 August.

3. HORIZONTAL DISTRIBUTION

Time variation of 30 mb temperature maps analyzed by T. Hiraki of the Japan Meteorological Agency is shown in Fig. 4 during the first large sudden warming. The region with temperature warmer than 230 K (-43°C) on this level appeared equatorward of Syowa Station (its location is denoted by a solid circle in the figure) on 26 August. As the season progressed, the warm region became large in area and moved poleward so that nearly the center of it eventually covered Syowa Station on 31 August. After that, the warm region moved eastward and shrank on 5 September. The warm region disappeared on 6 September (not shown). The center of the warm region did not cover the pole as it does in most of mid-winter major warmings in the northern hemisphere. In other words, the breakdown of the polar vortex did not occur. Then this warming is not a major warming if we follow the definition of "a major warming". The size of the 1988 warming, however, was large even if the warm region did not cover the south pole.

Time variation of Nimbus 7/TOMS total ozone maps is shown in Fig. 5 also during the first large sudden warming. As the warming event progressed, the ozone rich region equatorward of Syowa Station (its location is denoted by a solid circle in the figure) with total ozone amount larger than 400 Dobson Units (DU) moved poleward to cover Syowa Station on 29 August. After that, the ozone rich region moved eastward to leave Syowa Station.

The Antarctic ozone hole, the horizontal structure of an ozone poor region over Antarctica surrounded by an ozone rich region with a steep gradient at the boundary of the two regions, did not disappear in September and October even in 1988. However, the ozone hole became shallower in depth and smaller in area (Schoeberl, 1988; Schoeberl et al., 1989). The September temperature at the 20 km altitude over Antarctica in 1988 was about 10-15°C warmer than that in 1987 (Schoeberl, 1988).

4. YEAR TO YEAR VARIATION

Year to year variation of October total ozone at Syowa Station in Fig. 6 shows the October monthly mean value of 1988 (310 DU) extremely in-creased from that of 1987 (204 DU) to become the level of 1970's. As indicated in Hasebe (1980; 1983), Garcia and Solomon (1987), etc., year to year variation of total ozone at high latitudes well correlates with quasi-biennial oscillation (QBO) of the equatorial lower stratosphere. As shown in Fig. 6, during recent 10 years when the ozone depletion is clearly observed, a clear signal of quasi-two-year oscillation of

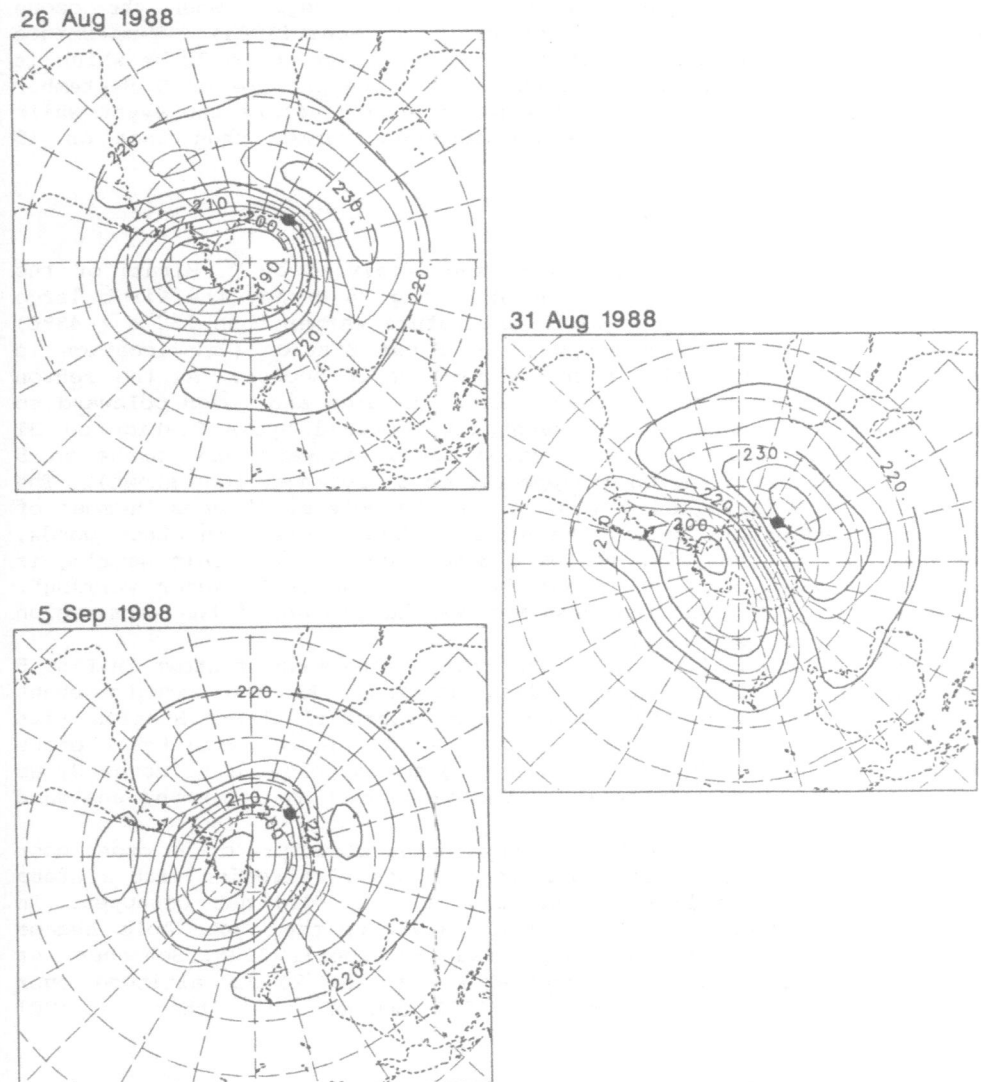

Figure 4. Three-day sequence of temperature maps in Kelvin on 30 mb (about 22 km) in the southern hemisphere on 26, 31 August, and 5 September in 1988. Contours are every 5 K. The solid circle of each panel denotes Syowa Station (69°S, 40°E). Stereographic projection, with both latitude and longitude drawn at intervals of 15 degrees. The Greenwich meridian is toward the top of each panel. The data are analyzed by T. Hiraki and the maps are drawn by M. Shiotani.

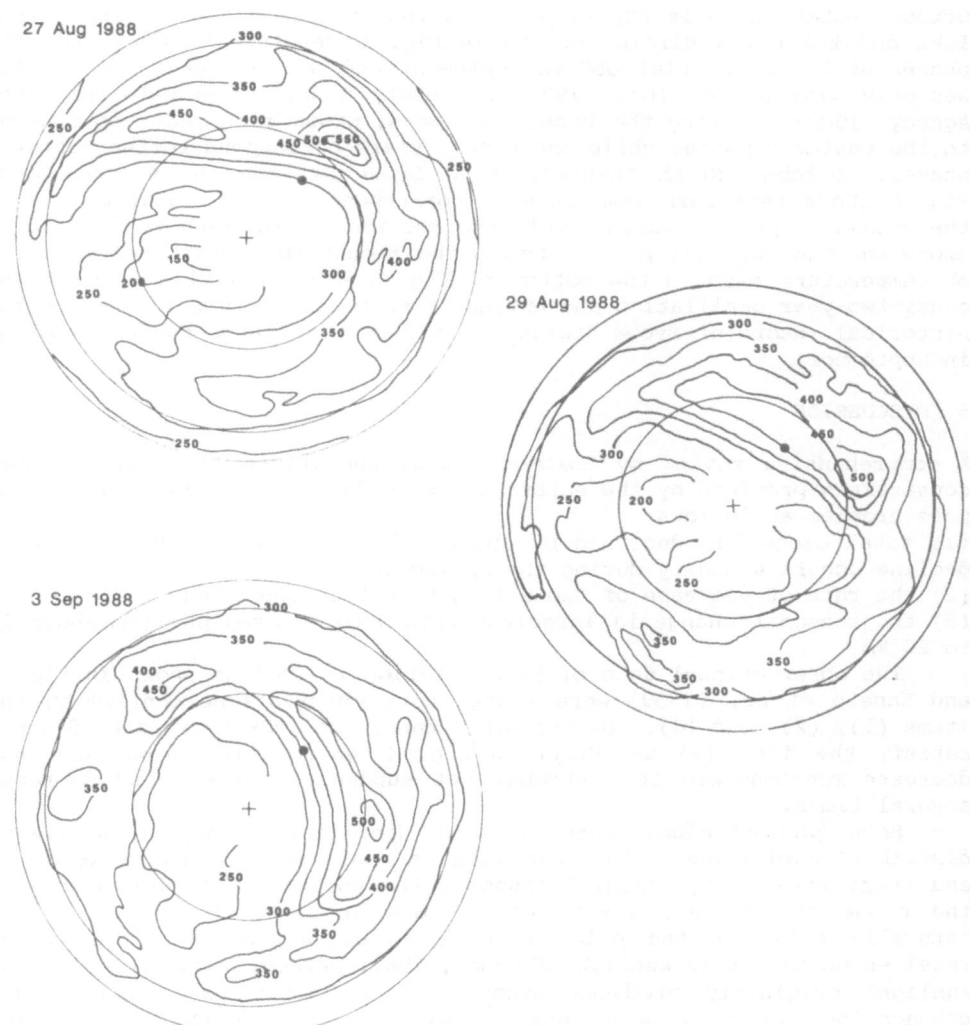

Figure 5. Three-day sequence of total ozone maps in the southern hemisphere on 27, 29 August, and 3 September in 1988. Contours are every 50 Dobson Units (DU). The solid circle of each panel denotes Syowa Station (69°S, 40°E). Orthographic projections, with the pole indicated by a cross and with the equator, 30°S, and 60°S latitudes by three circles. The Greenwich meridian is toward the top of each panel. The data are obtained by the Total Ozone Mapping Spectrometer (TOMS) onboard the Nimbus 7 satellite. The data are analyzed and plotted on real time basis by NASA/GSFC.

October total ozone is superimposed on the depletion trend. The aster-isks and the solid circles denote in Fig. 6 denotes the easterly wind phases of the equatorial QBO in September and October on 30 mb and the westerly wind phases (Coy, 1980; Naujokat, 1986; Japan Meteorological Agency, 1988). During the recent 10 years, ozone rich years correspond to the easterly phases while ozone poor years correspond to the westerly phases. October 30 mb temperature at Syowa Station in the bottom of Fig. 6 shows that warm temperatures correspond to high total ozone and the easterly phases while cold temperatures correspond to low total ozone and the westerly phases during the recent 10 years. September 30 mb temperature also in the bottom of Fig. 6 shows a little sign of the quasi-two-year oscillation and extremely warm temperature in 1988 in the historical record of Syowa Station as a result of large sudden warmings in September.

5. DISCUSSION

A comprehensive review by Solomon (1988) summarizes the observational constraints provided by the Antarctic ozone data for the Antarctic ozone hole problem as follows:
(1) total ozone has declined by about 50% since about 1975; the large decline occurs annually during the spring season;
(2) the rate of decrease of ozone is greatest in September;
(3) the seasonal change is largely confined to the region from about 10 to 25 km.
    The observational data of Syowa Station in 1987 as shown in Fig. 2 and Kaneto et al. (1989) were among those observations confirming the items (1), (2), and (3). On the other hand, the 1988 Syowa data did not satisfy the item (2) as shown in Fig. 2 where total ozone did not decrease monotonously in September but suddenly increase and decrease several times.
    Both photochemical reaction and dynamical transport determine distribution of ozone. To begin with, if only photochemistry operated and there were no dynamical transport, the horizontal structure called the ozone hole, i.e., lower total ozone in the polar area, should naturally exist in the polar area from autumn through spring which receives no or little sunlight (London, 1967; Dütsch, 1971), because the sunlight originally produces ozone. It is noted here that upward propagating planetary waves has a nature of dynamically transporting ozone as well as heat poleward in the stratosphere (e.g., Andrews et al., 1987). In the northern hemisphere where the planetary waves are active, accumulating effects of the transport of ozone bring about the well known spring maximum of total ozone in the Arctic area (e.g., Dütsch, 1971; Bowman and Krueger, 1985). On the other hand, in the southern hemisphere where the planetary waves are generally weak because of weaker contrast of land and sea at the earth's surface and of lower mountains than in the northern hemisphere, it is possible that total ozone has a maximum in middle latitudes and a minimum in the polar area in spring so that the total ozone has a structure of the ozone hole. It is worth noting that climatological temperature in the lower strato-sphere in winter and spring is extremely colder in the Antarctic than in

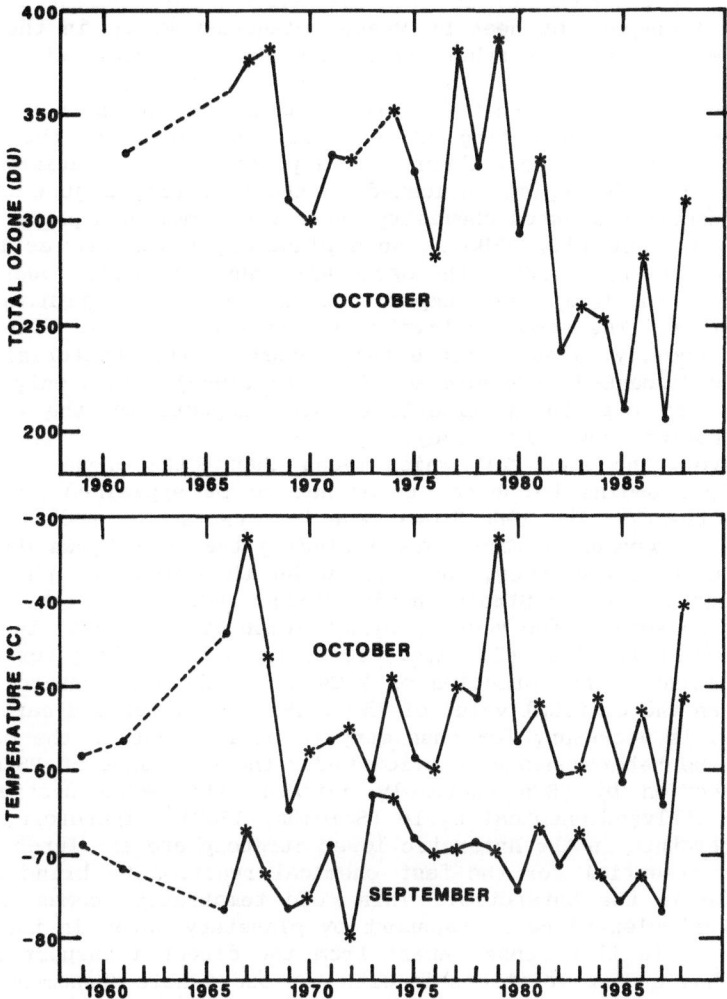

Figure 6. Year to year variations of October monthly means of total ozone, and monthly means of temperature of September and October on 30 mb (about 22 km) at Syowa Station (69°S, 40°E). Broken lines are used if there are missing years. The asterisks and the solid circles denote the easterly wind phase and the westerly wind phase respectively of the equatorial quasi-biennial oscillation (QBO) on 30 mb in September and October. The information on the equatorial wind is taken from Japan Meteorological Agency (1988), Naujokat (1986), and Coy (1980).

the Arctic (e.g., Kanzawa, 1989). The temperature contrast is also due to weaker transport of heat by weaker planetary waves in the southern hemisphere than in the northern hemisphere (e.g., Andrews et al., 1987).

Some observational studies (Nagatani and Miller, 1987; Newman and Randel, 1988) suggested that activity of planetary waves in the southern hemisphere winter and spring is generally decreasing in the recent 10 years. Thus the transport theory has a possibility to satisfy the item (1). The altitude region indicated in the item (3) is just the region where transport dominates chemistry under the normal gas phase chemistry (e.g., Solomon et al., 1985). When planetary waves are active as in 1988 (Hirota et al., 1989), the ozone hole can become shallower in depth and smaller in area, and temperature in the lower stratosphere can become warmer. The strong activity of planetary waves in 1988 might be due to an anomaly in sea surface temperature in the equatorial Southern Pacific as discussed in Kodera and Yamazaki (1989). Thus only dynamical transport can explain qualitatively some aspects of the ozone hole problem (Mahlman and Fels, 1986).

However, the item (2), which means that total ozone decreases as the sunlight begins to shine, seems not to be explained by dynamical transport theory. The 1988 Syowa case is exceptional, and the item (2) is usually observed in many data including the 1987 Syowa data. Thus chroline chemistry emerges. However, under the normal gas phase chemistry without the Polar Stratospheric Clouds (PSCs), chemical time constant may be about a few years (e.g., Solomon et al., 1985) in the altitude region of the item (3), i.e., 10-25 km, and too long for chemistry to be relevant. The formation of PSCs under the temperature condition colder than the critical value of about 195K in winter and early spring, therefore, is necessary for chemistry to be relevant to the ozone hole problem; the heterogeneous reaction under the existence of PSCs and the denitrification by PSCs extremely quickens the ozone destruction by chroline-catalyzed chemical cycle (Solomon, 1988). Therefore extremely cold temperature in the Antarctic lower stratosphere in winter and early spring is essential for the fast chemical reaction to bring about the ozone hole in the Antarctic. This cold temperature comes originally from weaker poleward heat transport by planetary waves in the southern hemisphere. In this sense, apart from the direct transport of ozone, dynamics has a deep relationship with the ozone hole phenomenon through the indirect process. When the lower stratosphere become warm in August and September as a result of large sudden warmings as in the case of over the eastern part of Antarctica in 1988, PSCs cannot exist so that chemical destruction of ozone may be weaker. The relation between total ozone depletion and Antarctic temperature decrease in the lower stratosphere are discussed, for example, in Kawahira and Hirooka (1987, 1989).

The warmings (coolings) are considered to be due to adiabatic compression heating (expansion cooling) by downward (upward) motion induced by vertically propagating planetary waves in the density stratified atmosphere (Matsuno and Nakamura, 1979). The good correlation between temperature and total ozone oscillations with two weeks as shown in Fig. 1 and 2 indicates that the ozone increase and decrease of this time scale are due to transport effects by planetary waves. Whether the ozone variation is due to "advection" or "diffusion" in the sense of terminology of Plumb and Mahlman (1987) is not clear. The event of the temperature oscillation with about two-weeks is intensively discussed by Hirota et al. (1989) in terms of interference between wavenumber 1 stationary planetary waves and wavenumber 2 eastward travelling planetary waves: Wavenumber 1 waves amplified just when the phase of the wavenumber 1 waves coincided with the phase of the wavenumber 2 waves.

Dynamical processes are also considered to play an essential role in interannual variation of Antarctic total ozone. The good correlation between interannual variation of October total ozone with lower stratospheric temperature in the Antarctic and the quasi-biennial oscillation of the equatorial lower stratosphere (see Fig. 6) may mean that under the easterly (westerly) phase of the equatorial QBO planetary wave activity is strong (weak) to bring about high (low) total ozone and warm (cold) lower stratosphere in the Antarctic. The relation between interannual variation of stratospheric temperature in the Arctic and the equatorial QBO is discussed in Holton and Tan (1982), Labitzke (1982), Labitzke and van Loon (1988), etc.

The 1988 phenomenon means that warming and ozone transport by dynamical process can deeply affect the depth and area of the Antarctic ozone hole.

6. JAPANESE RESEARCH PLAN FOR THE ANTARCTIC OZONE HOLE

Meridional distributions of ozone from the equatorial region to the Antarctic region at intervals of about every 5 degrees for November through December of 1987 and 1988 were obtained with ozonesonde observations made by the Japanese Antarctic Research Ship, Shirase, on the way from Japan to Syowa Station, Antarctica (Matsubara et al., 1988; Kawaguchi et al., 1989a, b). Analysis of the data is now in progress to investigate the effect of ozone hole on the summer distribution of ozone in "extrapolar" region by comparing the data of 1987 with those of 1988. The same observation will be conducted also in 1989.

To understand dynamics and chemistry of the ozone hole problem, the Japanese Antarctic Research Expedition has a plan to conduct in situ observation of ozone and aerosol along path of constant pressure balloon on about 50 mb in austral winter and spring of 1991 (Kanzawa et al., 1989) under the Polar Patrol Balloon (PPB) project (e.g., Nishimura et al., 1985; Fujii et al., 1989). About two balloons will be launched from Syowa Station (69°S, 40°E), data acquisition will be made by using the ARGOS system, and life time of observation will be more than about 2 weeks.

Routine observations of ozone and temperature at Syowa Station and observational projects as referred to above will give much information on the Antarctic ozone hole problem.

ACKNOWLEDGEMENTS.

One of the authors (H.K.) wish to thank C.R. Mechoso and A. O'Neill for giving him a chance to participate in this workshop with financial support.

146

The temperature and ozone data were obtained by many members of the
Japan Meteorological Agency (JMA) who wintered by turns in Japanese
Antarctic Station, Syowa. Members of the JMA meteorological team of the
29-th Japanese Antarctic Research Expedition sent the 1988 data from
Syowa Station. T. Hiraki carried out the JMA synoptic analysis of the
southern hemisphere stratosphere especially designed for the 1988
phenomenon. M. Shiotani gave valuable information on synoptic aspects
of the southern hemisphere. Y. Iwasaka contributed to transferring the
Nimbus 7/TOMS total ozone maps from NASA/GSFC to Syowa Station. A.J.
Fleig, A.J. Krueger, members of the TOMS Nimbus Experiment and Ozone
Processing teams made the total ozone map analysis.

K. Kawahira and I. Hirota encouraged us during the course of this
work. T. Yamanouchi gave useful comments on the earlier version of the
manuscript. K. Miyamoto helped to prepare the manuscript.

REFERENCES

Andrews, D.G., Holton, J.R. and Leovy, C.B., 1987: Middle atmosphere
    dynamics. Academic Press, 489pp.
Bowman, K.P. and Krueger, A.J., 1985: A global climatology of total
    ozone from the Nimbus 7 total ozone mapping spectrometer. J.
    Geophys. Res., 90, 7967-7976.
Chubachi, S., 1984: Preliminary result of ozone observations at Syowa
    Station from February 1982 to January 1983. Mem. Natl. Inst. Polar
    Res., Spoc. Issue, 34, 13-19.
Coy, L., 1980: Corrigendum. J. Atmos. Sci., 37, 912-913.
Dütsch, H.U., 1971: Photochemistry of atmospheric ozone. Adv. Geophys.,
    19, 219-322.
Finger, F.G. and Teweles, S., 1964: The mid-winter 1963 stratospheric
    warming and circulation change. J. Appl. Meteorol., 3, 1-15.
Fujii, R., Miyaoka, H., Kadokura, A., Ono, T., Yamagishi, H., Sato, N.,
    Ejiri, M., Hirasawa, T., Nishimura, J., Yazima, N., Yamagami, T.,
    Ohta, S., Akiyama, H., Tsuruda, K., Kodama, M., Fukunishi, H.,
    Yamanaka, M.D., and Kokubun, S., 1989: Polar Patrol Balloon experi-
    ment during 1991-1993. Antarct. Rec., 33, 323-331 (in Japanese with
    English abstract).
Garcia, R.R. and Solomon, S., 1987: A possible relationship between
    interannual variability in Antarctic ozone and the quasi-biennial
    oscillation. Geophys. Res. Lett., 14, 848-851.
Gardiner, B.G., 1988: Comparative morphology of the vertical ozone
    profile in the Antarctic spring. Geophys. Res. Lett., 15, 901-904.
Hasebe, F., 1980: A global analysis of the fluctuation of total ozone
    II. Non-stationary annual oscillation, quasi-biennial oscillation,
    and long-term variations in total ozone. J. Meteorol. Soc. Japan,
    58, 104-117.
Hasebe, F., 1983: Interannual variations of global total ozone revealed
    from Nimbus 4 BUV and ground-based observations. J. Geophys. Res.,
    88, 6819-6834.
Hirota, I., Hirooka, T. and Shiotani, M., 1983: Upper stratospheric
    circulation in the two hemispheres observed by satellites. Q.J.R.
    Meteorol. Soc., 109, 443-454.

147

Hirota, I., Kuroi, K., Shiotani, M., 1989: Mid-winter warmings in the southern hemisphere stratosphere in 1988. Q.J.R. Meteorol. Soc. (submitted)

Holton, J.R. and Tan, H.C., 1982: The quasi-biennial oscillation in the northern hemisphere lower stratosphere. J. Meteorol. Soc. Japan, 60, 140-148.

Japan Meteorological Agency, 1963-1989: Antarctic Meteorological Data Vol. 1-28, Spec. Vol. No. 1-4.

Japan Meteorological Agency, 1988: Monthly Report on Climate System, No. 88-12 (in Japanese).

Julian, P.R. and Labitzke, K., 1965: A study of atmospheric energetics during the January-February 1963 stratospheric warming. J. Atmos. Sci., 22, 597-610.

Kaneto, S., Yamamoto, A., Ogihara, H., Sugawara, H., and Yamanouchi, T., 1989: Ozone observations at syowa station from February 1987 to January 1988. Proc. NIPR Symp. Polar Meteorol. Glaciol., 3 (to appear).

Kanzawa, H., 1980: The behavior of mean zonal wind and planetary-scale disturbances in the troposphere and stratosphere during the 1973 sudden warming. J. Meteorol. Soc. Japan, 58, 329-356.

Kanzawa, H., 1989: Warm stratopause in the Antarctic winter. J. Atmos. Sci., 46, 435-438.

Kanzawa, H. and Kawaguchi, S., 1989: Large stratospheric sudden warming in Antarctic late winter and shallow ozone hole in 1988. Geophys. Res. Lett. (submitted)

Kanzawa, H., Kondo, Y., Iwasaka Y., Makino Y., Yamanaka, M.D., Yamanouchi, T. and Fujii, R. 1989: A research plan for the Antarctic ozone hole under the Polar Patrol Balloon (PPB) project. The 12th Symposium on Polar Meteorology and Glaciology, National Institute of Polar Research, Tokyo, 18-19 July 1989.

Kawaguchi, S., Matsubara, K., and Kanzawa, H., 1989a: Meridional distribution of ozone in the austral summer by ozonesondes. Middle Atmosphere Sciences Symposium, IAMAP89 Fifth Scientific Assembly, Reading, U.K., 31 July - 12 August 1989.

Kawaguchi, S., Shudo, Y., Fukuyama, Y., Kato, Y., Miyamoto, H., and Murayama, S., 1989b: Ozone observation on the Antarctic research ship "Shirase" in 1988. The 12th Symposium of Polar Meteorology and Glaciology, National Institute of Polar Research, Tokyo, 18-19 July 1989.

Kawahira, K. and Hirooka, T., 1987: Interannual variations of zonal mean temperature in the southern hemisphere stratosphere: With reference to the Antarctic ozone depletion. Proc. NIPR Symp. Polar Meteorol. Glaciol., 1, 31-38.

Kawahira, K. and Hirooka, T., 1989: Interannual temperature changes in the Antarctic lower stratosphere-a relation to the ozone hole. Geophys. Res. Lett., 16, 41-44.

Kodera, K. and Yamazaki, K., 1989: A possible influence of sea surface temperature variation on the recent development of ozone hole. J. Meteorol. Soc. Japan, 67, 465-472.

148

Krueger, A.J., Schoeberl, M.R., Stolarski, R.S. and Sechrist, F.S., 1988: The 1987 Antarctic ozone hole: A new record low. Geophys. Res. Lett., 15, 1365-1368.

Krueger, A.J., Stolarski, R.S. and Schoeberl, M.R., 1989: Formation of the 1988 Antarctic ozone hole. Geophys. Res. Lett., 16, 381-384.

Labitzke, K., 1982: On the interannual variability of the middle stratosphere during the northern winters. J. Meteorol. Soc. Japan, 60, 124-139.

Labitzke, K. and van Loon, H., 1988: Associations between the 11-year solar cycle, the QBO, and the atmosphere. Part I: The troposphere and stratosphere in the Northern Hemisphere in winter. J. Atmos. Terr. Phys., 50, 197-206.

London, J., 1967: The average distribution and time variation of ozone in the stratosphere and mesosphere. Space Res., 7, 172-185.

Mahlman, J.D. and Fels, S.B., 1986: Antarctic ozone decreases: A dynamical cause? Geophys. Res. Lett., 13, 1316-1319.

Matsubara, K., Doi, M., Uekubo, T., Okada, K., Kawaguchi, S., and Aoki, S., 1988: Ozone observation on ship from the equatorial region to the Antarctic region (II): Latitudinal variation of vertical distribution of ozone. Fall meeting of the Meteorological Society of Japan, Sendai, 26-28 October 1988.

Matsuno, T. and Nakamura, K., 1979: The Eulerian- and Lagrangian-mean meridional circulations in the stratosphere at the time of a sudden warming. J. Atmos. Sci., 36, 640-654.

Nagatani, R. and Miller, A., 1987: The influence of lower stratospheric forcing on the October Antarctic ozone decrease. Geophys. Res. Lett., 14, 202-205.

Naujokat, B., 1986: An update of the observed quasi-biennial oscillation of the stratospheric winds over the tropics. J. Atmos. Sci., 43, 1873-1877.

Newman, P. and Randel, W., 1988: Coherent ozone-dynamical changes during the southern hemisphere spring, 1979-1986. J. Geophys. Res., 93, 12585-12606.

Nishimura, J., Kodama, M., Tsuruda, K., Fukunishi, H. and Co-Members of PPB Working Group, 1985: Feasibility Studies of "Polar Patrol Balloon". Adv. Space Res., 5, 87-90.

Plumb, R.A. and Mahlman, J.D., 1987: The zonally averaged transport characteristics of the GFDL general circulation/transport model. J. Atmos. Sci., 44, 298-327.

Randel, W.J., 1988: The anomalous circulation in the southern hemisphere stratosphere during spring 1987. Geophys. Res. Lett., 15, 911-914.

Schoeberl, M.R., 1988: Dynamics weaken the polar hole. Nature, 336, 420-421.

Schoeberl, M.R., Stolarski, R.S. and Krueger A.J., 1989: The 1988 Antarctic ozone depletion: Comparison with previous year depletions. Geophys. Res. Lett., 377-380.

Solomon, S., 1988: The mystery of the Antarctic ozone "hole". Rev. Geophys., 26, 131-148.

Solomon, S., Garcia, R.R., and Stordal, F., 1985: Transport processes and ozone perturbations. J. Geophys. Res., 90, 3850-3868.

A POSSIBLE INFLUENCE OF SEA SURFACE TEMPERATURE VARIATION ON THE RECENT DEVELOPMENT OF OZONE HOLE

K. KODERA and K. YAMAZAKI
Meteorological Research Institute
Nagamine, Tsukuba, Ibaraki 305, Japan

ABSTRACT. A possible cause of recent decrease of planetary wave activity and Antarctic ozone depletion in spring is proposed. Tropospheric stationary wave structure in the southern hemisphere shows substantial variation during 1979 to 1988. Before the 1980's, especially in 1979, a well-developed tropospheric ridge was situated in the South Pacific and stationary wave propagated into the stratosphere, while after 1985 the tropospheric ridge became weaker and split into two ridges, and no vertical propagation was clearly seen. In 1988, however, the vertically propagating wave number 1 structure was recovered. This change of planetary wave structure in the South Pacific seems to be related to sea surface temperature anomaly in low latitude South Pacific. A possible role of the sea surface temperature variation in the recent development of ozone hole is also discussed.

1. Introduction

Farman et al. (1985) showed spectacular decreasing of October Antarctic total ozone starting at the end of 1970's. They associated this quasi-monotonous decreasing of ozone with the influence of increasing man-made halocarbons. Since large concentration of cloline monooxide (ClO) at low altitude of the spring stratosphere over Antarctica was detected (Barrett et al.,1988), the primary importance of chemical mechanism on the development of ozone hole has become widely admitted.

In 1988, however, one year after the deepest Antarctic ozone depletion (Krueger et al.,1988), TOMS data showed a great recovery from the deepening ozone hole. The weakening of the ozone hole in 1988 is related to the fact that the polar vortex was broken up very early at the end of October (Schoeberl,1988).

A possibility of the influence of planetary wave activity on the interannual variation of the ozone hole has been pointed out by Mahlman and Fels(1986) and a large decrease of lower stratospheric planetary wave activity since 1979 until 1987 was reported (Nagatani and Miller,1987; Newman and Randel,1988). The cause of the decrease, however, has not been explored.

In the present study, first we examine the interannual variation of planetary wave structure from 1979 to 1988 in the troposphere and stratosphere. Next, sea surface temperature (SST) variation is discussed as a possible cause of the interannual variation of tropospheric planetary waves, and finally a possible relationship between the recent ozone decrease and SST variation is briefly discussed.

149

A. O'Neill (ed.),
Dynamics, Transport and Photochemistry in the Middle Atmosphere of the Southern Hemisphere, 149–158.
© 1990 *Kluwer Academic Publishers.*

150

2. Evolution of stationary wave structure

In the present section, interannual changes of horizontal and vertical structure of planetary waves in the troposphere as well as in the stratosphere are examined using the National Meteorological Center (NMC) and the Japan Meteorological Agency (JMA) monthly mean geopotential data over the period from 1979 to 1988.

To see planetary wave structure, latitude-longitude section of zonally asymmetric component of geopotential height is juxtaposed according to pressure levels. This may be primitive but is straightforward to trace three-dimensional propagation of planetary waves, especially near their source regions.

Fig. 1 shows October mean zonally asymmetric component of geopotential height in the southern hemisphere from 30°S to 90°S, at 50, 100, 200, and 500 mb levels. Fig. 1a is for 1979, 1980 and 1981, from left to right, respectively. Fig. 1b is the same as in Fig. 1a but for 1985, 1986 and 1987. (1982-1984 are not shown).

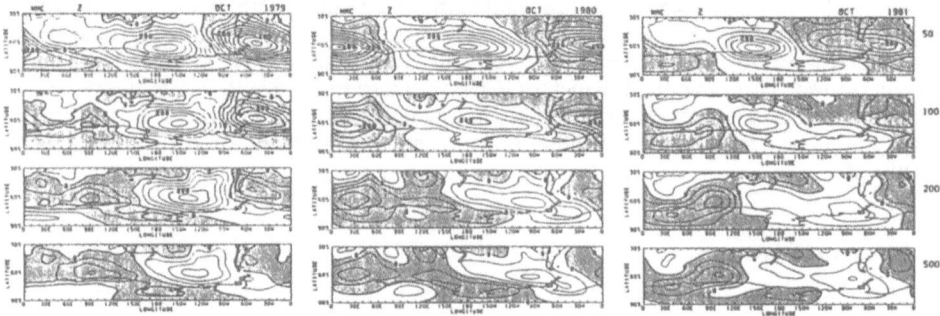

Fig. 1a October mean zonally asymmetric component of NMC geopotential height, at 50,100, 200   and 500 mb (from top to bottom) for 30-90°S, 0-360°E for 1979,1980 and 1981 (from left to right), respectively. Contour interval is 50 g.p.m. and negative values are shaded.

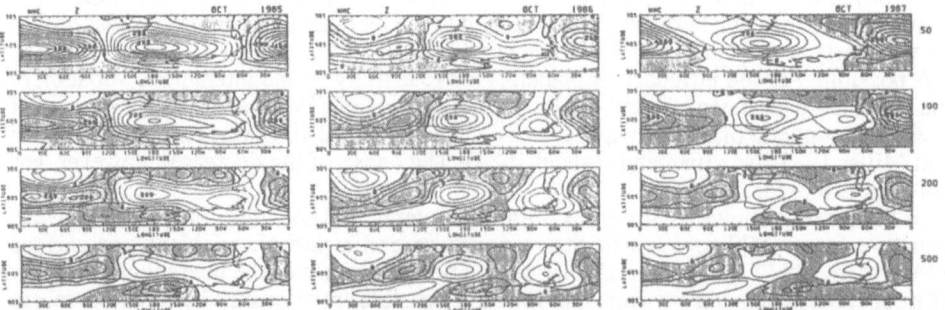

Fig. 1b Same as in Fig. 1a , except for 1985, 1986 and 1987 (from left to right).

In the case of 1979 (Fig. 1a left), well-developed stationary zonal wave number 1 structure is found throughout the troposphere and stratosphere. A tropospheric ridge west of Chile can be traced to a stratospheric ridge south of New Zealand and Australia at 50 mb. This westward tilt of phase with increasing altitude is the characteristics of vertical propagation of forced planetary waves. Although the tropospheric ridge is much weaker, planetary wave propagation from the troposphere into the stratosphere is still seen until 1982. After 1983, especially after 1985 (Fig. 1b), although wave number 1 pattern is seen at the 50 mb level, there is no clear wave number 1 structure in the troposphere, where the ridge is split in two. Note also that the meridional extent of ridges is smaller after 1985. A tropospheric ridge south of New Zealand is traceable up to the stratosphere with no tilt of phase, showing that no apparent propagation of stationary waves from the troposphere occurs.

In the above discussion, we have examined long-term variation from 1979 to 1987 using NMC data. Our analyses are extended to 1988 using JMA data. Fig. 2 shows zonally asymmetric components of the October mean geopotential height maps of JMA data in 1987 (left) and in 1988 (right). Zonally asymmetric component of 1987 October mean geopotential height of JMA analysis (Fig. 2 left) is very similar to that of NMC analysis (Fig. 1b right); the tropospheric ridge is completely split in two. In October 1988 (Fig. 2 right), the tropospheric ridge recovered the early 1980's wave structure; the ridge is situated to the west of Chile and propagates upward and westward.

These examples demonstrate that the vertical propagation is well related to the horizontal structure in the troposphere. When tropospheric wave is clear wave number 1 structure with a developed ridge west of Chile, vertical propagation is clear, while the vertical propagation is not clear when ridge is split in two.

3. Long-term variation of geopotential heights

Fig. 3 shows the results of trend analysis over 1979-1987 for zonally asymmetric component of the October NMC geopotential height at the 50, 100, 200 and 500 mb levels, from top to bottom, respectively.

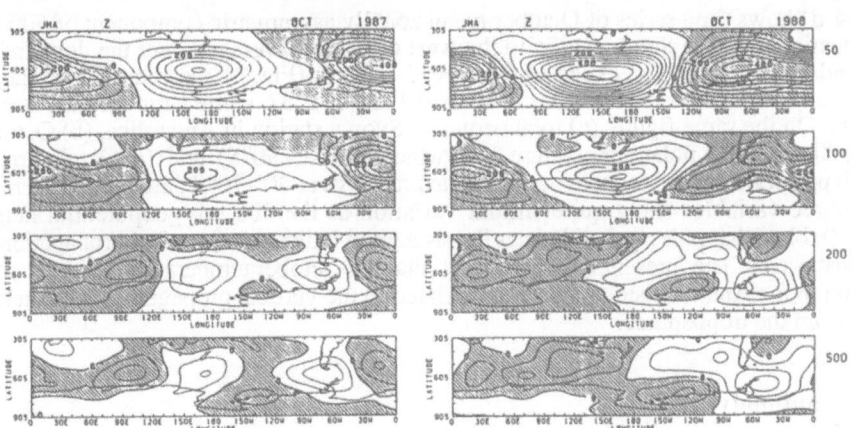

Fig. 2 Same as in Fig. 1, except for JMA geopotential height in October 1987 (left) and 1988 (right).

152

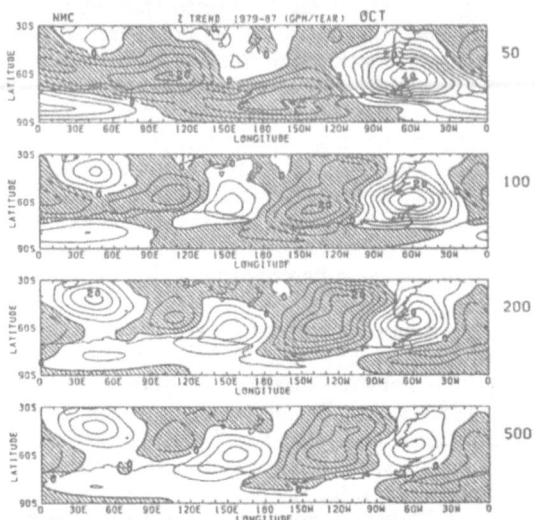

Fig. 3 Linear trend maps (30°-90°S, 0°-360°E) of mean October zonally symmetric
component of NMC geopotential height over 1979-1987 at 50, 100, 200 and 500 mb
level (from top to bottom). Contours are every 5 g.p.m./year and negative values are
shaded.

At 500 mb level (Fig. 3, bottom), as was expected from the difference between 1979
and 1987 (Fig. 1), large negative trend is found in the South Pacific west of Chile, and
positive trend south of South America and south of New Zealand. This negative trend
continues to 50 mb level, shifting westward. The trend in September is similar to that in
October.

Fig. 4-d shows time series of October mean zonally asymmetric component of 500 mb
geopotential height in South pacific to the west of Chile (averaged over the domain 30-
60°S and 110°-130°W). As expected from trend analysis (Fig. 3), the geopotential height
decreases rather monotonically from 1979 until 1986 or 1987, while large increase occurs
in 1988. On the same figure (a) total ozone at Syowa station in Antarctica, (b) October
mean 100 mb south polar temperature (averaged over 60°- 90°S) and (c) October mean
200 mb poleward stationary eddy heat flux averaged over 40°-70°S , are also presented.
These three variables show quite similar variation as the 500 mb geopotential height
(curve d). Note that poleward eddy heat flux is a good indicator of vertical propagation of
planetary waves. The present results suggest that the October temperature and the ozone
variation in the Antarctic stratosphere are related to the vertical propagation of planetary
waves from the troposphere.

4. SST anomaly

In this section we consider a possible cause of the long-term geopotential height variation
in the South Pacific. It is well admitted that lower latitude sea surface temperature (SST)

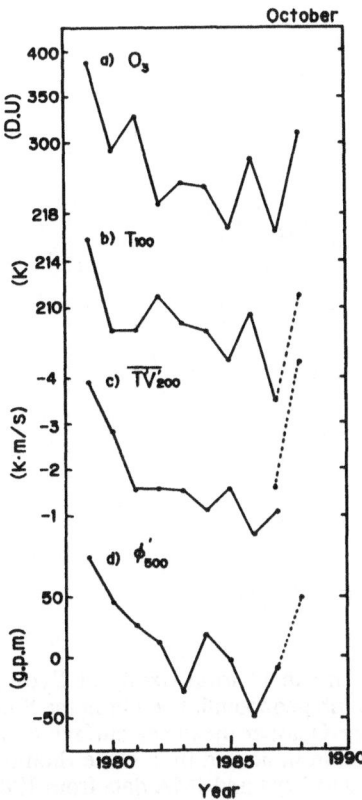

Fig. 4 Time series of October mean of a) total ozone at Syowa station in Antarctica (69°00'S, 39°35'E), b)100 mb south polar temperature averaged over 60°-90°S, c) northward stationary eddy heat flux at 200 mb averaged over 40°-70°S and d) zonally asymmetric component of geopotential height at 500 mb in South Pacific averaged over the domain of 30°-60°S and 110°-130°W. For b), c), and d) solid lines (from 1979-1987) are NMC data and dashed lines (from 1987-1988) are JMA data.

anomaly has large influence on mid-latitude tropospheric circulations (e.g. Blackmon et al.,1983).

Fig. 6 shows linear trend of September-October mean SST over the period from 1974 to 1987. An important increasing trend is seen in the lower latitude South Pacific SST. Fig. 5 shows time series of SST in the South Pacific averaged over the domain indicated by a close rhomboid in Fig. 7 (bottom), and the 500 mb geopotential height anomaly to the west of Chile of Fig. 4-d (middle). A negative correlation is seen between the two curves (correlation coefficient over 1979 to 1988 is -0.73). This coherent increasing of SST in the South Pacific may be a cause of the variation of geopotential height in the South Pacific.

Unfortunately available data is very limited to study the causal relationship between the SST variation and the tropospheric planetary wave structure. In this respect a numerical

154

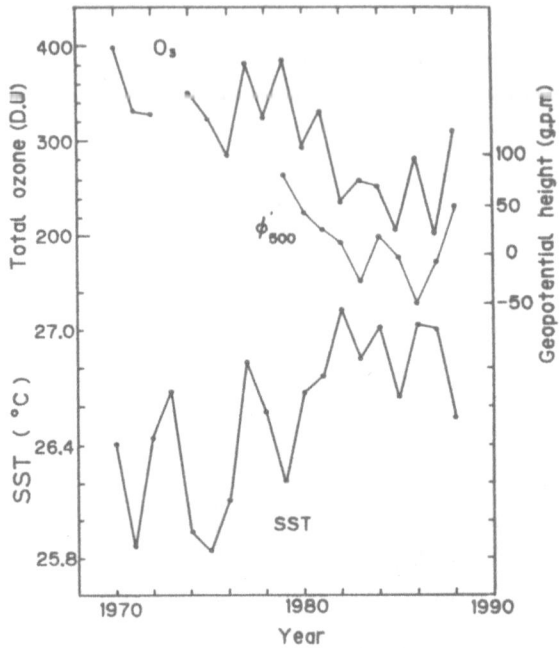

Fig. 5 Time series of October mean a) total ozone at Syowa station, b) zonally asymmetric component of 500 mb geopotential height in the South Pacific (same as in Fig. 4-d), and of c) September-October mean sea surface temperature in the South Pacific averaged over the domain indicated by a close rhomboid in the Fig 7. SST data are NOAA data from 1979 to 1984 and JMA data from 1985 to 1988.

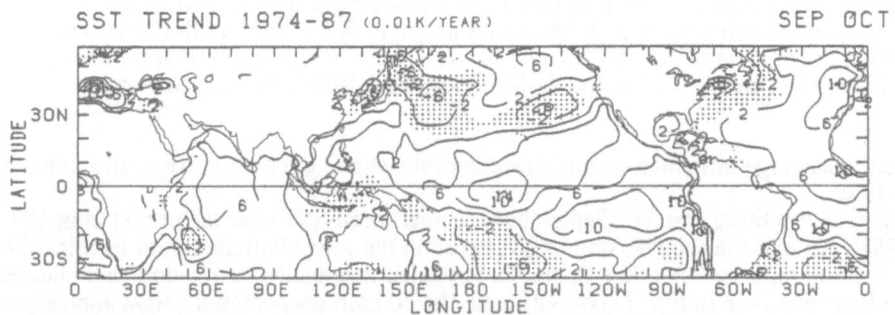

Fig. 6 Linear trend of September-October mean sea surface temperature during the period from1974 to 1987. Unit is 0.01K/year. Contour interval is 0.04K/year. Negative(cooling trend) values are stippled.

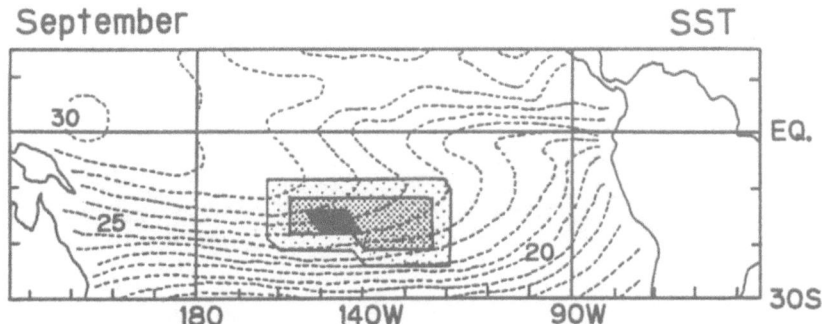

Fig. 7 15 year mean(1970-1984) sea surface temperature (SST) for September. Contour interval is 1°C. Stippled region denotes SST anomaly region used in the GCM experiment. A close rhomboid indicate the region representing the SST in the Fig. 5.

experiment is performed with use of the tropospheric MRI General Circulation Model (GCM), which uses the modified Arakawa-Schubert cumulus parameterization (Tokioka et al.,1988).

From the same atmospheric initial condition on September 1, two integrations are made through the October. In C run, SST anomalies of -1°C (-2°C) are imposed over the lightly(heavily) stippled region in Fig 7. In W run, anomalies of +1°C and +2°C are imposed. The climatological SST used is the 15 year mean (1970-1984) data. Note that the region is near the South Pacific Convergence Zone and the climatological SST over the region is the warmest at the same latitude.

The results show that precipitation over the SST anomaly region is enhanced in W run compared with that in C run. The maximum value of precipitation difference is about 5 mm/day at 130°W, 18°S. The October mean zonally asymmetric components of geopotential height at 500 mb are shown in Fig. 8 for C run, W run and their difference. In C run, wave number 1 structure is dominant at about 60°S and it is similar to that in 1979. In W run, wave number 1 structure split in two. The decrease of geopotential height to the west of Chile and splitting of tropospheric ridge is simulated as expected. The experiment, however, is only one integration for each case hence further study is needed to confirm the results.

5. Summary and discussion

The results obtained in the present studies are:
- A substantial change of planetary wave structure in the troposphere occurs in the South Pacific region after 1979. A clear wave 1 structure in 1979 disappears completely after 1985 (Fig. 1). In 1988, however, the wave 1 structure is recovered (Fig. 2).
- Vertical propagation of planetary waves is related to the tropospheric wave structure. No clear propagation is found when   geopotential height to the west of Chile is lower and tropospheric ridge is split in two .
- 200 mb poleward stationary eddy flux, 100 mb polar temperature and total ozone at Syowa station show similar long-term variation as 500 mb geopotential anomaly in the South Pacific to the west of Chile (Fig. 4).

156

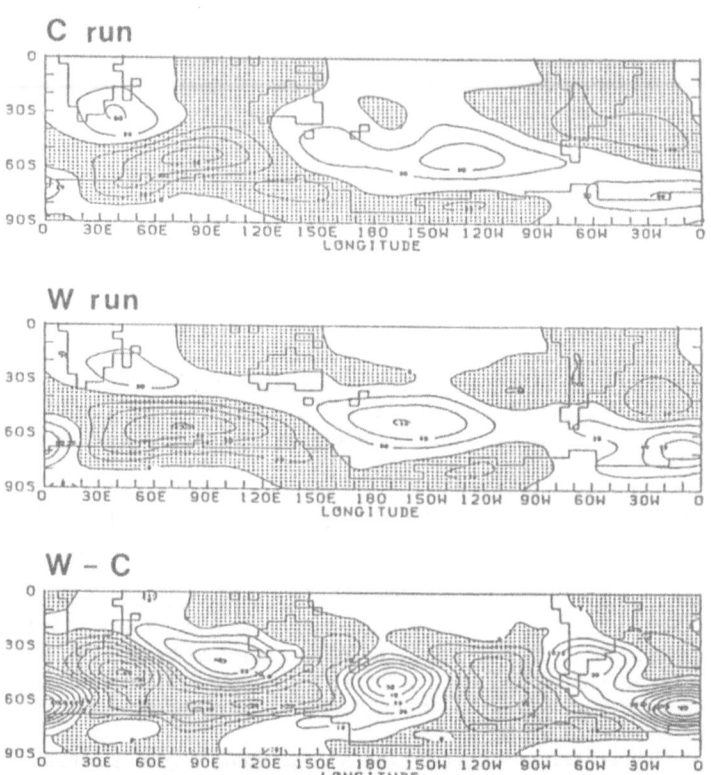

Fig. 8 October mean zonally asymmetric component of geopotential height at 500 mb for C run, W run and W-C (from top to bottom). Contour interval is 30 g.p.m for upper two panels and 10 g.p.m for the difference. Negative values are stippled.

- September-October mean SST in the low latitude South Pacific has a negative correlation with mid-latitude 500 mb geopotential height anomaly in the South Pacific to the west of Chile(Fig. 5).
- The results of the numerical experiment shows that the warming of lower latitude South Pacific SST can produce the depletion of 500 mb geopotential height to the west of Chile.

In the Fig.5, October total ozone at Syowa station(top) is plotted with the lower latitude South Pacific SST(bottom). Except for short-term variation, a good correspondence between the SST and the ozone variation is clearly seen on the figure: the SST starts to increase from the end of 1970's, while ozone decreases from the same epoch. The correlation coefficient between the September-October mean SST and the total ozone at Syowa is -0.53 in October and -0.69 in November for the period from 1970 to 1988(1973 missing).

From these results we may speculate a possible relationship between SST and ozone as follows (Fig. 9);

157

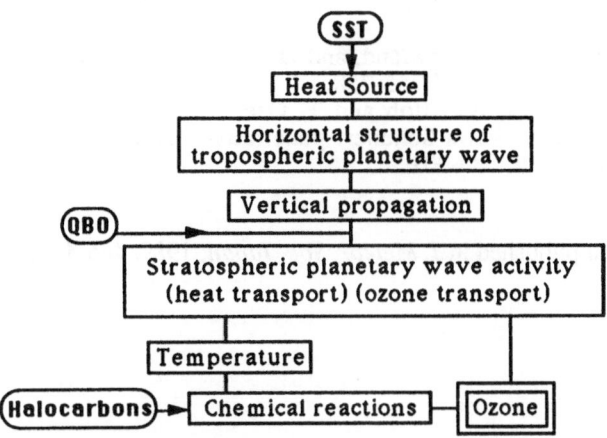

Fig. 9 A possible mechanism for the influence of sea surface temperature variation on the stratospheric ozone content.

Anomaly of SST modulates heat source to change horizontal and vertical structure of planetary waves. The variation of wave activity in the stratosphere influences heat and ozone transport. As a consequence temperature field and ozone content change. The change of stratospheric temperature also affects ozone by chemical process, especially through the formation of polar stratospheric clouds which are crucial for heterogeneous reaction involved in a proposed mechanism of ozone destruction by halocarbons (e.g. Solomon et al.,1986).

References

Barrett, J.W., P.M. Solomon, R.L. de Zafra, M. Jaramillo, L. Emmons, and A. Parrish, 1988: Formation of Antarctic ozone hole by the ClO dimer mechanism, *Nature*, **336**, 455-458.

Blackmon, M.L., J.E. Geisler, and E.J. Pitcher,1983: A general circulation study of January climate anomaly patterns associated with interannual variation of equatorial Pacific sea surface temperatures, *J. Atmos. Sci.*, **40**,1410-1425.

Farman, J.C., B.G. Gardiner, and J.D. Shanklin, 1985: Large losses of total ozone in Antarctica reveal seasonal ClO_x/NO_x interaction, *Nature*, **315**, 207-210.

Krueger, A. J. , M.R. Schoeberl, R.S. Stolarski, and F.S. Sechrist, 1988; The 1987 Antarctic ozone hole: A new record low, *Geophys. Res. Lett.*,**15**,1365-1368.

Mahlman, J.D., and S.B. Fels, 1986: Antarctic ozone decrease: A dynamical cause? , *Geophys. Res. Lett.*, **13**, 1316-1319.

Nagatani, R.M., and A.J. Miller, 1987: The influence of lower stratospheric forcing on the October Antarctic ozone decrease, *Geophys. Res. Lett.*, **14**, 202-205.

Newman, P.A., and W.J. Randel, 1988: Coherent ozone-dynamical changes during the southern hemisphere spring, 1979-1986, *J. Geophys. Res.*, **D10**, 12585-12606.

Schoeberl, M.R, 1988: Ozone depletion: Dynamics weaken the polar hole, *Nature*, **336**, 420-421.

Solomon, S., R.R. Garcia, F.S. Rowland, and D.J. Wuebbles, 1986: On the depletion of Antarctic ozone, *Nature,*,**321**,755-758.

Tokioka, T., K. Yamazaki, A. Kitoh and T. Ose, 1988: The equatorial 30-60 day oscillation and the Arakawa-Schubert penetrative cumulus parameterization, *J. Meteor. Soc. Japan*, **66**, 883-901.

(This paper has been published in *J. Meteor. Soc. Japan*, 1989, 465-472)

GRAVITY WAVES IN THE SOUTHERN HEMISPHERE MIDDLE ATMOSPHERE: A REVIEW OF OBSERVATIONS

R. A. VINCENT
Department of Physics and Mathematical Physics
University of Adelaide
GPO Box 498
Adelaide
Australia 5001

ABSTRACT. Although studies of atmospheric gravity waves in the Southern Hemisphere are limited in number, the waves are found to be ubiquitous in the middle atmosphere. This paper summarizes what information is available about wave activity, with particular emphasis on short and long term variability, anisotropies, fluxes and wave sources. All studies indicate that gravity wave energy and momentum fluxes decrease with height. Estimates of wave drag inferred from both direct and indirect estimates of momentum flux convergences range from tens of ms^{-1} day^{-1} in the mesosphere to less than a few ms^{-1}day^{-1} in the stratosphere. The observations suggest that gravity waves may play an important role in heat and constituent transport throughout the Southern Hemisphere middle atmosphere.

1. Introduction

Satellite measurements have led to significant improvements in our knowledge of the large scale circulation of the middle atmosphere and it is now possible to investigate in broad outline the similarities and differences in the climatologies of the temperature structure and dynamics of the Northern and Southern hemispheres (e.g. Barnett and Corney, 1985; Andrews, 1989). While hemispheric differences can be ascribed partly to differences in the degree of planetary wave activity between the hemispheres, some of the differences must be due to disparities in the degree of atmospheric gravity wave activity, because a growing body of evidence shows that gravity waves play an important role in determining the large scale structure of the middle atmosphere. The momentum flux convergences associated with breaking or saturating waves act to accelerate the mean flow, force meridional circulations and to drive the atmosphere away from radiative equilibrium. Improvements in our understanding of atmospheric processes, and especially hemispheric differences, will therefore result from better knowledge of the temporal and geographical variability of gravity waves in the middle atmosphere.

159

A. O'Neill (ed.),
Dynamics, Transport and Photochemistry in the Middle Atmosphere of the Southern Hemisphere, 159–170.
© 1990 *Kluwer Academic Publishers.*

   The study of atmospheric gravity waves is complicated by their
small spatial scales and relatively high temporal frequencies which
inhibit their observation by satellite techniques. The requirement for
good spatial and/or time resolution means that gravity wave studies have
been generally limited to ground-based radar measurements, and balloon
and rocket soundings. As might be expected, the amount of information is
better in the Northern Hemisphere where, for example, data acquired
through the Meteorological Rocket Network (MRN) has enabled the
development of a climatology of wave activity in the stratosphere (e.g.
Hirota, 1984). These measurements are now being complemented by Rayleigh
scatter lidar observations (e.g. Hauchcorne et al., 1987). Northern
hemisphere mesospheric observations have come from a number of radar
sites which span a range of latitudes (e.g. Balsley and Carter, 1982;
Meek et al., 1985a,b; Maekawa et al., 1987; Yamamoto et al., 1987).
Smaller wave amplitudes make observations more difficult in the
troposphere and lower stratosphere, but nevertheless there has been
impressive progress in wave studies using highly detailed measurements
with rockets (e.g. Dewan et al, 1984), balloons (e.g. Sidi and Barat,
1986), radars (e.g. Fritts et al., 1988) and instrumented aircraft
(Nastrom et al., 1987).
   The lack of information is particularly acute in the Southern
Hemisphere where observations are, in general, limited to only a few
locations. Even where long sequences of mesospheric wind observations
have been made, such as at Christchurch, New Zealand, (44°S, 173°E) and
Mawson, Antarctica, (68°S, 63°E), the emphasis has been on studies of
prevailing winds and tides (e.g. Manson et al., 1985; Fraser, 1989;
Phillips and Vincent, 1989). Only at Adelaide, Australia, (35°S, 138°E)
have comprehensive gravity wave measurements been made in the
mesosphere, although some limited observations were reported for the
lower latitude sites of Townsville (19°S, 147°E) by Vincent and Ball
(1981) and Jicamarca (12°S, 77°W), results for which have been reviewed
by Rastogi (1981). With regard to the mid- to upper-stratosphere, a
gravity wave study has been reported using rocket data taken in
Australia by Eckermann and Vincent (1989). The situation is a little
better in the troposphere and lower stratosphere where there have been
some balloon measurements, including a sequence of constant pressure
balloon observations which covered almost the entire hemisphere
(Massman, 1981). More recently, intensive studies of wave motions in the
troposphere have commenced at Adelaide using a VHF radar of the ST type.
   It is the purpose of this paper to summarize these Southern
Hemisphere gravity wave observations. In an accompanying paper, Fritts
(this volume) compares Northern and Southern Hemisphere measurements and
theory as it relates to the observations. Mesospheric observations made
with both ground-based and optical techniques are discussed in Section
2, while rocket observations of the 30 to 60 km region are summarized in
Section 3. Studies of gravity waves in the lower atmosphere are then
examined in Section 4.

2. Mesospheric Gravity Waves

Some of the earliest observations of mesospheric gravity waves were made with the high-power VHF radar at Jicamarca. In general, only vertical motions have been measured and observations have been of relatively short duration (Rastogi, 1981). The vertical oscillations are often large in amplitude (> 1 ms^{-1}) and show little coherence with height, which suggests a wide bandwidth. Waves with periods in the range 5–15 min often dominate.

Radiowave measurements in the MF/HF frequency band have been used for a number of years to observe gravity wave motions in the 60–100 km altitude range at Adelaide. The investigations have been facilitated by the use of a large antenna array which operates at a frequency of 2 MHz. The large size and versatile nature of the array means that it can be used to measure atmospheric motions with a number of different techniques. Observations over a number of years using the spaced antenna method have enabled a climatology of wave motions to be assembled, and the Doppler beam swinging method has been used to measure vertical momentum fluxes and horizontal scales and phase velocities. Doppler measurements of turbulence dissipation rates, which are thought to be related to gravity wave activity, have also been made at Adelaide. Optical measurements of moving airglow patterns made at a nearby location provide complementary information on horizontal wavelengths and speeds (Freund and Jacka, 1979).

2.1. GRAVITY WAVE AND TURBULENCE CLIMATOLOGIES

Climatologies of gravity wave and turbulence activity are necessary in order to determine how gravity wave dissipation affects the thermal balance and constituent transport in the upper middle atmosphere. Turbulence produced by breaking gravity waves will act to transport constituents vertically through diffusion, as has been proposed to explain the seasonal variation of ozone in the upper mesosphere (Thomas et al., 1984). On the other hand, it has been argued that diffusion is not as effective in transporting heat and constituents as advection by vertical and meridional winds induced by gravity wave body forces (see review by Strobel, 1989; Holton and Schoeberl, 1988).

Climatologies of mesospheric wave activity measured at Adelaide in single years have been presented by Vincent and Fritts (1987) and Vincent (1987). The continuation of these observations over a number of years enables a more complete picture of seasonal variations in activity to be obtained and for interannual variability to be examined. Results from three years of continuous wind measurements from 1984–1987 confirm these earlier studies. The main features are:

(i) An increase in wave amplitudes with height, but with a growth rate much smaller than would be expected for non-dissipating waves. Averaged over the full spectrum, the rms amplitudes are about 25–30 ms^{-1} at 85 km, which agrees with earlier estimates for Adelaide and Townsville (Vincent and Ball, 1981; Vincent, 1984).

(ii) A distinct semi-annual variation of wave activity at heights below about 85 km, with minimum amplitudes attained near the times of

equinoctial transitions in the zonal circulation. Above 85 km the
seasonal variations are small, but the variations seem more annual in
character.

   A weak semiannual variation in turbulence dissipation rates with
equinoctial minima is also evident at heights below about 85 km,
although there is considerable interannual variability (Hocking, 1988).

2.2 SHORT TERM VARIATIONS AND WAVE ANISOTROPIES

   Further investigation shows that there are strong, short-term
variations in mesospheric wave activity at Adelaide which are
superimposed on the regular, seasonal, variations discussed above
(Vincent and Fritts, 1987). Figure 1 shows daily mean values of the
quantity $(\overline{u'^2} + \overline{v'^2})$, a measure of the kinetic energy per unit mass,
measured in the winter of 1988 for waves in the 1-24 hr period range at
85 km, where u' and v' are the zonal and meridional perturbation
velocities. The short-term variations are manifest as bursts in which
the wave energy can more than double over time scales of 5 to 10 days,
especially in winter. These bursts are most probably due to time
variations in the source regions (see Section 4) and to the filtering
effects of the strong zonal winds in the middle atmosphere; changes in
the prevailing wind field due to planetary wave activity will act to
modulate the gravity wave fluxes, removing those which reach critical
levels where their phase speed matches the wind speed.

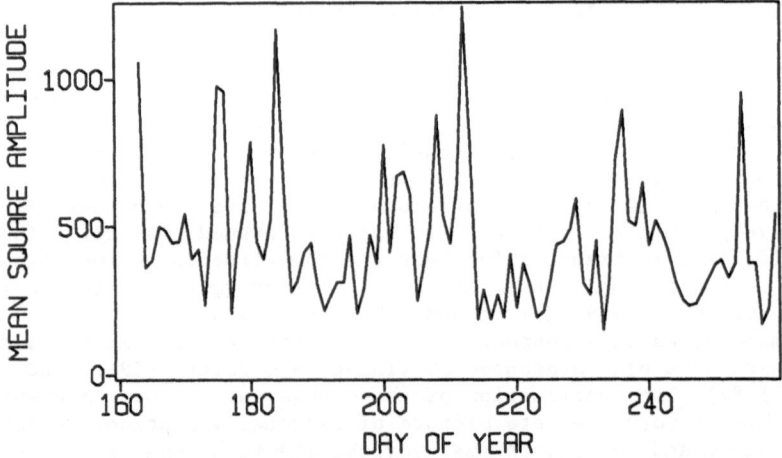

Figure 1.   Daily values of $(\overline{u'^2}+\overline{v'^2})$ in $m^2 s^{-2}$ at 85 km altitude in the
period May to August (winter) at Adelaide (35°S).

   Fluctuations in wave fluxes are also observed on time scales
shorter than one day (Vincent and Reid, 1983; Reid and Vincent, 1987;
Fritts and Vincent, 1987). In the observations of Fritts and Vincent
(1987) strong changes in $\overline{u'w'}$ fluxes were found to be caused by
modulation by the diurnal tide.

Comparison of meridional and zonal amplitudes shows, on both seasonal and short time scales, that v' tends to be larger in magnitude than u', especially in winter. This, together with the non-zero correlation, $\overline{u'v'}$, indicates that wave motions are anisotropic or polarized. Using the so-called Stokes parameter method, Vincent and Fritts (1987) showed that in 1984, the motions were on average polarized in the NS direction in winter and in the NE/SW quadrants in summer.These features are illustrated in Figure 2 which shows the alignments of wave motions in the 1-8 h period band calculated on a daily basis for the winter and summer seasons of 1984. The length of each line indicates the degree to which the wave field is polarized (a value of one means fully polarized and a length of zero means a random field). Significant day-to-day variability of the wave field is evident, but clearly the majority of motions are aligned within a quadrant centered on NS in winter and within the NW/SE quadrants in summer. Similar plots are found for data taken in other years. The horizontal motions of internal gravity waves are polarized such that they are aligned in the direction of horizontal phase progression. Hence, the plots in Figure 2 give an indication of the anisotropic nature of the azimuthal distribution of gravity waves in the mesosphere. Although the direction of wave propagation is unknown it is apparent that the angular distributions in both summer and winter have a strong meridional bias.

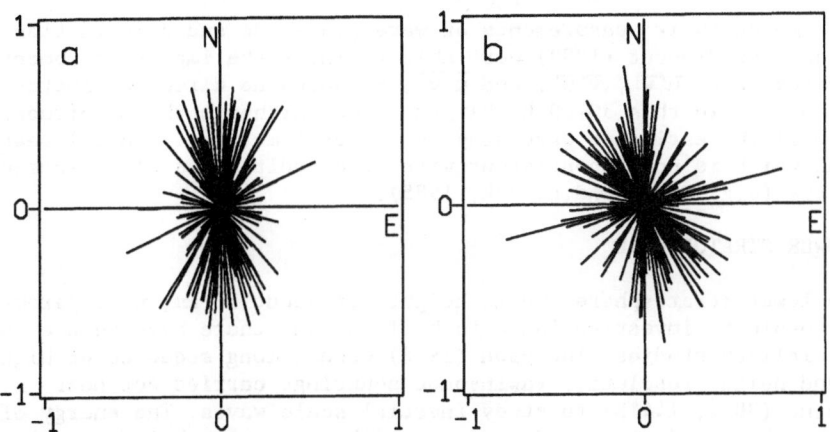

Figure 2. Plots of the polarization of gravity waves in the 1-8 hr period range for each day in (a) winter and (b) summer at an altitude of 85 km at Adelaide in 1984.

2.3 HORIZONTAL PHASE VELOCITIES AND WAVELENGTHS

Direct measurements of the horizontal phase speed (c_h) and wavelength (λ_h) were reported by Reid and Vincent (1987b) who used Doppler observations at Adelaide. In general, λ_h was found to increase from values near 20 km at a ground based period (T) of 10 min up about 60 km for T~100 min. These values are consistent with other observations made in both hemispheres and summarized by Reid (1986). To a good degree of approximation, Reid shows that λ_h is proportional to T, which means that c_h is independent of wave frequency; typically $c_h \sim 30$ ms^{-1} at heights below 90 km, and rises to ~60–100 ms^{-1} near 100 km altitude.

3. Stratospheric Gravity Waves

3.1 UPPER STRATOSPHERE

There are few direct measurements of gravity waves motions in the mid- to upper-stratosphere (30 to 60 km). Eckermann and Vincent (1989), however, used rocket data acquired at Woomera (31°S, 136°E) to infer a wide range of wave characteristics. An interesting feature of these measurements was that they showed that the stratospheric waves had similar characteristics to mesospheric waves observed at Adelaide. Wave amplitudes were found to grow slowly with altitude, implying wave saturation or dissipation, and the mean vertical wavenumber spectrum was broadly consistent with the saturated spectrum postulated by Dewan and Good (1986) and Smith et al. (1987). Eckermann and Vincent (1989) also found that the wave field was polarized, with fluctuations again showing a mean alignment which was predominantly NS in winter, and NW/SE in summer.

Based on their measurements of wave amplitude and polarization Eckermann and Vincent (1989) were able to infer the important momentum flux parameters, $\overline{u'w'}$, $\overline{v'w'}$, and $\overline{u'v'}$, of which no direct estimates have yet been made in this 30–60 km height range. Although the magnitudes were small the estimated wave drag was about 1 ms^{-1}day^{-1} in all seasons, a value which is not inconsistent with more indirect satellite-based estimates (e.g. Smith and Lyjack, 1985).

3.2 LOWER STRATOPSHERE

In the lower stratosphere, up to heights of about 30 km, wave parameters are amenable to investigation with balloons and there have been a number of specialized studies. Thompson (1978) used a long sequence of high time and height resolution rawinsonde soundings carried out near Melbourne (38°S, 144°E) to study inertial scale waves. The energy of the waves in this study appeared to decay with increasing height from near the tropopause and was sharply concentrated at frequencies very close to the inertial frequency and at vertical scales in the range 1.5–2 km at heights between 10 and 20 km. Rather surprisingly, the energy concentration did not seem to be strongly affected by changes in the strength of the background wind speed. This implies that the waves either had very large horizontal wavelengths in order that Doppler shifting not be important or that they were propagating at right angles to the prevailing wind.

By using the so-called rotary decomposition technique, Thompson (1978) found that most of the wave energy resided in components in which the vector wind field rotated counter clockwise with increasing altitude, which indicates that the majority of the waves were

propagating energy upwards, out of the troposphere. Other rotary
spectral studies have also shown that the majority of quasi-inertial
wave energy in the upper stratosphere (Eckermann and Vincent, 1989) and
mesosphere (Vincent, 1984) is upgoing, results which support the view
that most gravity wave sources are in the lower atmosphere.

More recently, Byrne et al. (1989) utilized constant pressure
balloons to study wave motions above the South Pole in the austral
summer. They observed quite strongly polarized wave motions with
amplitudes ~3 ms^{-1}, quite comparable in magnitude to the inferred
intrinsic phase speeds, $|c-u| \sim 5$ ms^{-1}. This strongly suggests that the
waves were close to saturation and so could be important contributors to
the momentum budget of the summertime stratosphere. From the
measurements of Byrne et al. (1989) one can infer $\overline{u'w'}$ fluxes of ~ 0.15–
0.2 m^{2}s^{-2}. Wave fluxes of this magnitude, dissipated over a vertical
region of say one scale height (~ 7 km), would provide an acceleration
of the background flow ~2 ms^{-1}day^{-1}.

From a hemispheric viewpoint, probably the most comprehensive
gravity wave study in the Southern Hemisphere was made by Massman
(1981). He used data taken with constant pressure balloons which floated
at a mean altitude of about 14 km, that is in the upper troposphere at
low-latitudes and in the lower stratosphere at mid- to high-latitudes.
Massman was able to measure a number of important wave properties,
including intrinsic periods (which ranged from 30 min to 3 hr), and
pressure and vertical velocity amplitudes. From the latter, Massman was
able to deduce the vertical energy fluxes and the magnitude of the
vertical flux of horizontal momentum.

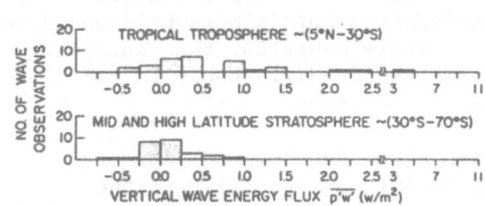
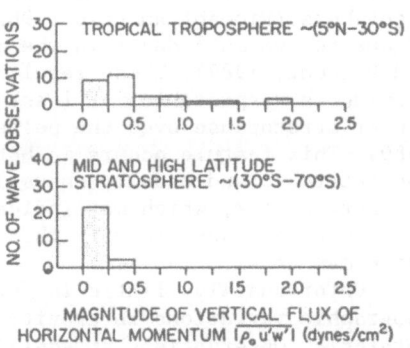

Figure 3. Histograms of vertical energy fluxes (left) and the vertical
fluxes of horizontal momentum (right) in the tropical troposphere and
mid- and high-latitude lower stratosphere (after Massman, 1981).

The most interesting results from Massman's study were that energy
and momentum fluxes were larger in the troposphere than in the
stratosphere (Figure 3) and that waves in the troposphere were mostly
upgoing, whereas waves in the stratosphere were as likely to be
propagating upward as downward. These results strongly suggest that
there is a flux divergence at the tropopause and support the hypothesis

that gravity waves affect the large-scale circulation of the lower
atmosphere in a similar manner to their effect in the mesosphere (e.g.
Lindzen, 1985, Palmer et al., 1986).

4. Discussion

The gravity wave studies discussed above, although limited in number,
show that the waves are ubiquitous in the Southern Hemisphere middle
atmosphere. Those observations which have sufficient time and height
resolution to enable frequency and vertical wavenumber spectra to be
calculated show that the spectra are 'red', so that the most energetic
motions are associated with low-frequency waves and low wavenumbers. The
dominant vertical scales increase with height from values of ~ 1-2 km
near the tropopause to wavelengths ~ 20 km at the mesopause. In this
sense, the observations are comparable to similar studies in the
Northern Hemisphere and with the theoretical predictions of Smith et al.
(1987).

Most importantly, all the results reviewed here suggested that the
amplitudes of the wave motions were often comparable to the intrinsic
phase speed so that the waves satisfied the linear breaking criterion
(i.e. $u' \sim |c-u|$, Lindzen, 1981). In general the density-weighted energy
and momentum fluxes are found to decay with increasing height in the
stratosphere and mesosphere, which implies that gravity wave driven
circulations and diffusion act to redistribute heat and constituents
throughout the middle atmosphere in the Southern Hemisphere. The
inferred mean flow accelerations range from tens of ms^{-1} day^{-1} in the
mesosphere (Vincent an Reid, 1983; Reid and Vincent, 1987) to the order
of one to two ms^{-1} day^{-1} in the middle and lower stratosphere (Eckermann
and Vincent, 1989). These results support the idea that gravity wave
driving is responsible, at least in part, for the elevated and warm
winter stratopause over the poles (Hitchman et al., 1989; Kanzawa,
1989). This feature occurs in both hemispheres but is more pronounced in
the Antarctic and usually shows less variable behaviour than in the
Northern winter, which may indicate hemispheric differences in the
gravity wave transmissivity through the stratospheric wind jets
(Hitchman et al., 1989).

Unfortunately, little is yet known about the temporal and
geographic variations in gravity wave activity, especially in the
Antarctic. Observations at Adelaide show that there is a consistent
semiannual variation in wave activity and turbulence in the mesosphere
up to heights near 85 km, but above this height the seasonal variation
is small. There is also significant short-term variability on time
scales of hours to days (e.g. see Figure 1). These fluctuations are
possibly source related.

Some confirmation for this viewpoint comes from recent studies made
with an ST radar at Adelaide which show substantial temporal variations
in wave activity in the troposphere. Wave energies can increase by one
or two orders of magnitude in a matter of hours prior to the passage of
cold fronts. These bursts of activity appear to be associated with
convective regions propagating ahead of the front, although shear

regions connected with jet streams at the tropopause may also be a
significant source of the waves. Further investigations into the
relative importance of the various sources is continuing. Nevertheless
the burst-like nature of wave activity associated with frontal passages,
and the rather regular weekly occurrence of cold fronts over Southern
Australia in winter, may account for the 5-10 day quasi-periodicity of
mesospheric activity noted in Figure 1.

The anisotropic nature of gravity wave motions in the stratosphere
and mesosphere (Figure 2) is also significant because it implies a
strong contribution from meridionally propagating waves. This bias may
be due to strong source regions to either the north or south of
Australia, to the filtering effects of the strong zonal mean flow in the
lower middle atmosphere, or to refraction effects (S. Eckermann, Private
Communication). Numerical modelling studies made with the high-
resolution "SKYHI" GCM model suggest that equatorial convection could be
a strong source of gravity waves at mid-latitudes (Miyahara et al.,
1986). Whatever the reasons, the effects of a strong meridional
component to the wave drag are poorly understood.

Improved knowledge of the distribution of gravity wave phase speeds
and amplitudes in the lower atmosphere is fundamental to better
parameterization of gravity wave effects in numerical models. The
initial VHF radar observations made at Adelaide, which show that moving
convective bands and/or jet streaks may be the predominant sources at
mid-latitudes, lend weight to the belief that there is a significant
hemispheric difference in phase speed distributions. The large land
masses in the Northern Hemisphere are likely to produce a phase speed
distribution which is dominated by waves with c~0; the absence of
significant topography, except in the Andes and in Antarctica, suggests
that the speed distribution could be dominated by relatively high phase-
speed waves in the Southern Hemisphere. Domination by sources associated
with the cold fronts which sweep regularly across the southern oceans,
together with strong source regions over the equator and Antarctica
could produce a wave field at mid-latitudes in the Southern Hemisphere
middle atmosphere which is highly transient and contains strong
contributions from meridionally propagating waves. In contrast, the wave
field in the Northern Hemisphere may be more homogeneous and isotropic.
It is interesting to note in this context, that Eckermann and Hocking
(1989) inferred from rocket data that the wavefield becomes more
azimuthally isotropic as one moves poleward in the Northern Hemisphere.

Confirmation, or otherwise, for the above speculations can only
come from improved and more widely dispersed observations. The
geographic distribution of gravity wave observations in the Southern
Hemisphere is particularly poor. There is a great need for observations
at other longitudes in the Southern Hemisphere in order to compare with
the measurements made in the Australian sector. Constant pressure
balloon measurements of the type discussed by Massman (1981) seem an
effective way of studying waves in the lower middle atmosphere over the
large oceanic regions of the Southern Hemisphere.

Acknowledgements

The helpful comments of B. H. Briggs, S. D. Eckermann and D. C. Fritts
on early drafts of this paper are gratefully acknowledged.

5. References

Andrews, D.G. (1989) Some comparisons between the middle atmosphere
    dynamics of the Southern and Northern Hemispheres, Pure and App.
    Geophys., 130, 213-232.
Balsley, B.B., and Carter, D.A. (1982) The spectrum of atmospheric
    velocity fluctuations at 8 and 86 km, Geophys. Res. Lett., 9, 465-
    468.
Barnett, J.J., and Corney, M. (1985) Middle atmosphere reference model
    derived from satellite data, Handbook for MAP, 16, 47-85.
Byrne, G.J., Benbrooke, J.R., Bering, E.A., Liao, B., and Theall, J.R.
    (1989) Summertime stratospheric wind measurements above the South
    Pole, J. Atmos. Terr. Phys., 51, 51-60.
Dewan, E.M., Grossbard, N., Quesada, A.F., and Good, R.E. (1984)
    Spectral analysis of 10 m resolution scalar velocity profiles in the
    stratosphere, Geophs. Res. Lett., 11, 80-83, and Correction to
    "Spectral analysis of...", Geophs. Res. Lett., 11, 624.
Dewan, E.M., and Good, R.E. (1986) Saturation and the "Universal"
    spectrum for vertical profiles of horizontal scalar winds in the
    atmosphere, J. Geophys. Res., 91, 2742-2748.
Eckermann, S.D., and Vincent, R.A. (1989) Falling sphere observations of
    anisotropic gravity wave motions in the upper stratosphere over
    Australia, Pure App. Geophys., 130, 509-532.
Eckermann, S.D., and Hocking, W.K. (1989) Effect of superposition on
    measurements of atmospheric gravity waves: A cautionary note and
    some reinterpretations, J. Geophys. Res., 94, 6333-6339.
Fraser, G.J. (1989) Monthly mean winds at 44S and 78S, Pure App.
    Geophys., 130, 291-301.
Freund, J.T, and Jacka, F. (1979) Structure in the λ557.7 nm [OI]
    airglow, J. Atmos. Terr. Phys., 41, 25-51.
Fritts, D.C., and Vincent, R.A. (1987) Mesospheric momentum flux studies
    at Adelaide, Australia: Observations and gravity wave/tidal
    interaction model, J. Atmos. Sci., 44, 605-619.
Fritts, D.C., Tsuda, Sato, T., Fukao, S., and Kato, S. (1988)
    Observational evidence of a saturated gravity wave spectrum in the
    troposphere and lower stratosphere, J. Atmos. Sci., 45, 1741-1759.
Hauchecorne, A., Chanin, M.L., and Wilson, R. (1987) Mesospheric
    temperature inversions and gravity wave breaking, Geophys. Res.
    Lett., 14, 933-936.
Hirota, I. (1984) 'Climatology of gravity waves in the middle
    atmosphere', J. Atmos. Terr. Phys., 46, 767-773.
Hitchman, M.H., Gille, J.C., Rodgers, C.D., and Brasseur, G. (1989) The
    separated polar winter stratopause: A gravity wave driven
    climatological feature, J. Atmos. Sci., 46, 410-422.

Hocking, W.K. (1988) 'Two years of continuous measurements of turbulence parameters in the upper mesosphere and lower thermosphere made with a 2-MHz radar', J. Geophys. Res., 93, 2475-2491.

Holton, J.R., and Schoeberl, M.R. (1988) The role of gravity wave generated advection and diffusion in transport of tracers in the mesosphere, J. Goephys. Res., 93, 11075-11082.

Kanzawa, H. (1989) Warm stratopause in the Antarctic winter, J. Atmos. Sci., 46, 435-438.

Lindzen, R.S. (1985) Multiple gravity wave breaking levels, J. Atmos. Sci., 42, 301-305.

Maekawa, Y., Fukao, S., Hirota, I, Sulzer, M.P., and Kato, S. (1987) Some further studies on long-term mesospheric and lower thermosphere wind observations by the Arecibo radar, J. Atmos. Terr. Phys., 49, 63-71.

Manson, A.H., Meek, C.E., Smith. M.J. and Fraser, G.J. (1985) Direct comparison of prevailing winds and tidal wind fields (24-, 12-h) in the upper middle atmosphere (60-105 km) during 1978-1980 at Saskatoon (52°N, 107°W) and Christchurch (44°S, 173°E), J. Atmos. Terr. Phys., 47, 463-476.

Massman, W.J. (1981) 'An investigation of gravity waves on a global scale using TWERLE data', J. Geophys. Res., 86, 4072-4082.

Meek, C.E., Reid, I.M., and Manson, A.H. (1985 a) Observations of mesospheric wind velocities: I Gravity wave horizontal scales and phase velocities determined from spaced antenna wind observations, Radio Sci., 20, 1363-1382.

Meek, C.E., Reid, I.M., and Manson, A.H. (1985 b) Observations of mesospheric wind velocities. II Cross sections of power spectral density for 48-8h, 8-1 h, 1h-10 min over 60-110 km for 1981, Radio Sci., 20, 1383-1402.

Miyahara, S., Hayashi, Y., and Mahlman, J.D. (1986) Interactions between gravity waves and planetary scale flow simulated by the GFDL "SKYHI" general circulation model, J. Atmos. Sci., 43, 1844-1861.

Nastrom, G.D., Fritts, D.C., and Gage, K.S. (1987) An investigation of terrain effects on the mesoscale spectrum of atmospheric motions, J. Atmos. Sci., 44, 3087-3096.

Palmer, T.N., Shutts, G.J., and Swinbank, R. (1986) Alleviation of a systematic bias in general circulation and numerical weather prediction models through an orographic gravity wave drag parameterization, Quart. J. Roy. Met. Soc., 112, 1001-1040.

Phillips, A., and Vincent, R.A. (1989) Radar observations of prevailing winds and waves in the Southern Hemisphere mesosphere and lower thermosphere, Pure App. Geophys., 130, 303-318.

Rastogi, P.K. (1981) Radar studies of gravity waves and tides in the middle atmosphere: A review, J. Atmos. Terr. Phys., 43, 511-524.

Reid, I.M. (1986) Gravity wave motions in the upper middle atmosphere (60-110 km), J. Atmos. Terr. Phys., 48, 1057-1071.

Reid, I.M., and Vincent, R.A. (1987 a) 'Measurements of mesospheric gravity wave momentum fluxes and mean flow accelerations at Adelaide, Australia', J. Atmos. Terr. Phys., 49, 443-460.

Reid, I.M., and Vincent, R.A. (1987 b) 'Measurements of horizontal scales and phase velocities of short period mesospheric gravity

waves at Adelaide, Australia', J. Atmos. Terr. Phys., 49, 1033-1048.
Sidi, C., and Barat, J. (1986) Observational evidence of an inertial
wind structure in the stratosphere, J. Geophys. Res., 91, 1209-1217.
Smith, A.K., and Lyjak, L.V. (1985) An observational estimate of gravity
wave drag from the momentum balance of the middle atmosphere, J.
Geophys. Res., 90, 2233-2241.
Smith, S.A., Fritts, D.C., and VanZandt, T.E. (1987) Evidence of a
saturated spectrum of atmospheric gravity waves, J. Atmos. Sci., 44,
1404-1410.
Strobel, D.F. (1989) Constraints on gravity wave induced diffusion in
the middle atmosphere, Pure App. Geophys., 130, 533-546.
Thomas, R.J, Barth, C.A., and Solomon, S. (1984) Seasonal variations of
ozone in the upper atmosphere and gravity waves, Geophys. Res.
Lett., 11, 673-676.
Thompson, R.O.R.Y (1978) Observation of inertial waves in the
stratosphere, Quart. J. Roy. Met. Soc., 104, 691-698.
Vincent, R.A., and Ball, S.M. (1981) Mesospheric winds at low and mid-
latitudes in the southern hemisphere, J. Geophys. Res., 86, 9159-
9169.
Vincent, R.A., and Reid, I.M. (1983) HF Doppler measurements of
mesospheric gravity wave momentum fluxes, J. Atmos. Sci., 40, 1321-
1333.
Vincent, R.A. (1984) Gravity wave motions in the mesosphere, J. Atmos.
Terr. Phys., 46, 119-128.
Vincent, R.A., and Fritts, D.C. (1987) A climatology of gravity wave
motions in the mesopause region at Adelaide, Australia, J. Atmos.
Sci., 44, 748-760.
Vincent, R.A. (1987) Radar observations of gravity waves in the
mesosphere, in G. Visconti and R. Garcia (eds.), Transport Processes
in the Middle Atmosphere, D. Reidel, Dordrecht, pp 47-56.
Yamamoto, M., Tsuda, T., Kato, S., Sato, T., and Fukao, S. (1987) A
saturated inertia gravity wave in the mesosphere observed by the
Middle and Upper atmosphere radar, J. Geophys. Res., 92, 11993-
11999.

GRAVITY WAVES IN THE MIDDLE ATMOSPHERE
OF THE SOUTHERN HEMISPHERE

DAVID C. FRITTS

Geophysical Institute and Department of Physics
University of Alaska
Fairbanks, AK 99775-0800

ABSTRACT. Gravity waves in the southern hemisphere exhibit many of the same characteristics as those in the northern hemisphere in terms of their scales, amplitude growth with height, and influences on the lower and middle atmosphere. Yet there are likely to be significant differences in the wave spectra and their effects as well, due to differing source distributions and strengths in the two hemispheres. Presented here will be a brief review of our current understanding of gravity wave saturation processes and effects, and of the evidence for and likely sources of wave variability, in both hemispheres. We will then examine some of the evidence for hemispheric differences in gravity wave sources, energies, drag, and induced diffusion.

1.   Introduction

Our knowledge of middle atmosphere gravity waves and their effects has increased dramatically during the last few years following the recognition of their important role in driving the large-scale circulation and structure of this region. Observational studies have contributed important measurements of gravity wave scales, amplitudes, growth with height, variability, and fluxes of energy and momentum as well as information on the processes responsible for wave excitation, filtering, and dissipation. Theoretical studies also have made progress in addressing these processes and their influences on the large-scale flow. Yet there remain major uncertainties about many of these areas, including the distributions, character, and variability of different wave sources and the roles of wave-wave and wave-mean flow interactions in the evolution of the wave spectrum. Our understanding of gravity waves amd their influences in the Southern Hemisphere is particularly deficient because of the paucity of observations in that area. The purposes of this paper are to review those features of the middle atmosphere gravity wave spectrum common to both hemispheres and to identify the likely areas of major hemispheric differences. A survey of the observations that have been made in the Southern Hemisphere is provided by Vincent (1989).

We will begin by reviewing in section 2 our current understanding of the processes by which gravity waves are saturated, or amplitude limited, and the implications of wave saturation for the wave spectrum and the large-scale flow. This is not intended as a comprehensive review, as this has already been provided (Fritts, 1984a, 1989; Fritts and Rastogi, 1985; Dunkerton, 1989), but rather as a survey of recent work. Evidence of the temporal and geographic variability of

A. O'Neill (ed.),
Dynamics, Transport and Photochemistry in the Middle Atmosphere of the Southern Hemisphere, 171–189.
© 1990 *Kluwer Academic Publishers.*

the gravity wave field and of the processes contributing to this variability will be reviewed in section 3. We consider in section 4 the implications of these and other results for hemispheric differences in the gravity wave spectrum. Our conclusions are presented in section 5.

2. Gravity Wave Saturation Processes and Effects

Numerous observations have shown the velocity and temperature or density variance on time scales less than a day to increase with height in the atmosphere, but at a rate approximately half that expected for conservative wave motions, $E(z) \sim e^{z/H_E}$, with $H_E \sim 12 - 15$ km (Balsley and Carter, 1982; Fritts et al., 1989a). This suggests that the motion spectrum, if it is due largely to gravity waves, is amplitude limited, either by wave reflection or dissipation. An example of this growth with height obtained using space shuttle re-entry data is shown in Figure 1.

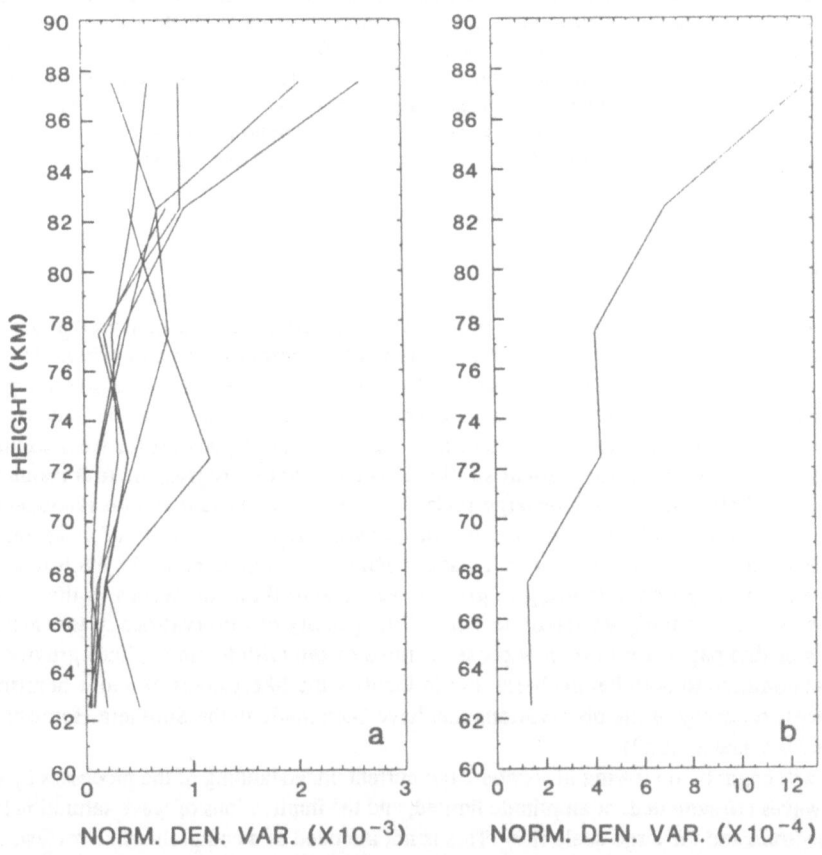

Figure 1. Individual (a) and mean (b) estimates of the variance of (ρ'/ρ) in 5-km height intervals from space shuttle re-entry data for seven flights over the Pacific illustrating the growth of fluctuation amplitudes with height.

173

In addition, there is evidence of processes that act to prevent the attainment of velocity shears or temperature gradients larger than some limiting value. This implies that the observed variance increase occurs at larger vertical scales that have not yet reached these limiting shears (Smith et al., 1987). In the stratosphere, where vertical scales, $\lambda_z \sim 2$ - 3 km, and the vertical flux of wave energy are small, energy dissipation rates are small and wave amplitudes appear to be maintained near nominal saturated values by a slow, systematic extraction of energy from the wave field (Fritts et al., 1988a). This results in bounds on the fluctuations of N^2 with height, several profiles of which are shown in Figure 2. In the mesosphere, where vertical scales and energy fluxes (per unit mass) may be considerably larger, limiting amplitudes are still observed, but other, more vigorous processes may be required to account for the necessary energy dissipation.

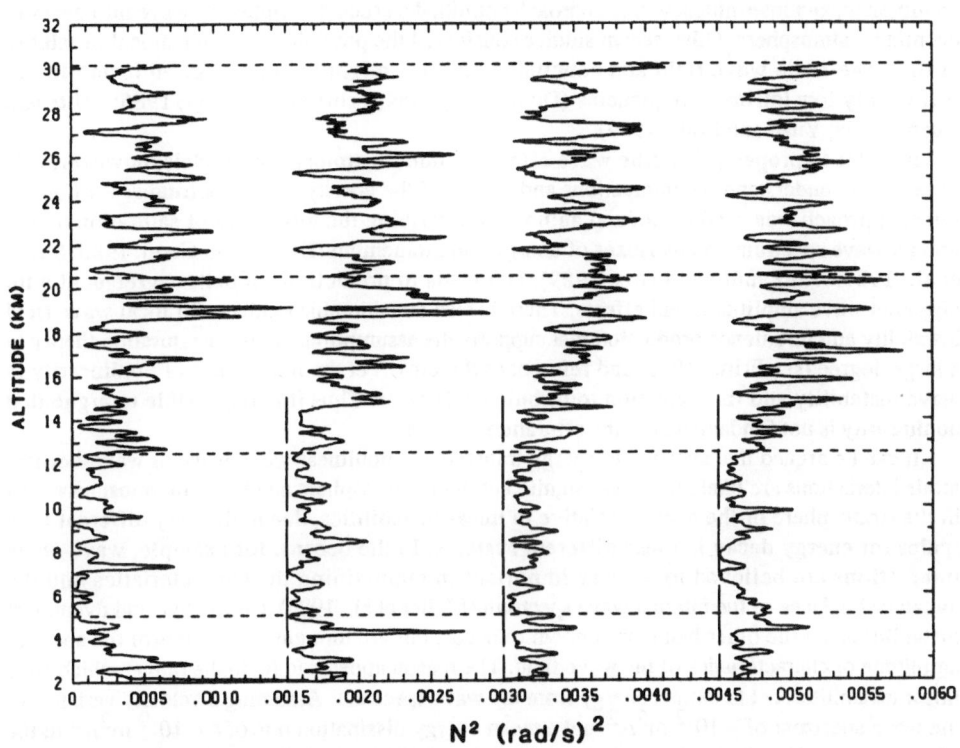

Figure 2. N^2 profiles obtained by high-resolution balloon soundings at the MU radar facility in Shigaraki, Japan (35°N) during March 1986. Although wave amplitudes and vertical scales increase with height, there is evidence of processes limiting temperature gradients to near-adiabatic values.

2.1 GRAVITY WAVE SATURATION MECHANISMS

A number of mechanisms have been proposed to account for the saturation of atmospheric gravity waves. These include convective and dynamical instabilities occurring locally within the wave field (Hodges, 1967; Hines, 1971, 1988; Dunkerton, 1984; Fritts and Rastogi, 1985;

Walterscheid and Schubert, 1989; Fritts and Yuan, 1989a) and nonlinear wave-wave and wave-vortical mode interactions that may act in a more systematic, global manner to extract energy from the more energetic wave motions (Yeh and Liu, 1981, 1985; Weinstock, 1982; Fritts, 1985; Inhester, 1987; Dunkerton, 1987; Dong and Yeh, 1988). There is at present, however, no single mechanism that is believed by most researchers to account for wave saturation in all cases. Indeed, there are reasons to believe that different mechanisms may be important for different components of the wave spectrum (Fritts and Rastogi, 1985; Fritts and Yuan, 1989a).

Early theoretical studies and most recent parameterizations (Hodges, 1967; Lindzen, 1981; Holton, 1982; Garcia and Solomon, 1985; Miyahara et al., 1986) assumed for simplicity that wave amplitudes would be limited by convective instabilities within the wave field. And while this may be somewhat unrealistic, given the neglect of other effects like wave superposition, nonlinearity, and intermittency, this approach permitted a crude description of wave influences in the middle atmosphere. Other recent studies considered the possibility of a dynamical instability of the large-scale wave field and concluded that this might be the preferred instability at sufficiently low intrinsic frequencies (Dunkerton, 1984; Fritts and Rastogi, 1985; Fritts and Yuan, 1989a; Yuan and Fritts, 1989).

In order to properly describe wave effects in middle atmosphere models, however, it is necessary to understand the interactions and effects of the gravity wave spectrum as a whole. A linear approach was used by several authors in describing the influences of saturation on the gravity wave spectrum and its fluxes of energy and momentum (Dewan and Good, 1986; Smith et al., 1987; VanZandt and Fritts, 1989), and leads to predictions in better agreement with observed wave amplitudes and effects. There is also considerable evidence of local wave field instability and turbulence production that supports the assumptions of linear saturation theory to a large degree (see Fritts, 1989, and references therein). Yet the importance of nonlinearity in wave instability and the transition to turbulence is clear. Thus it is impossible to argue that nonlinearity is not fundamental to the saturation process.

It can be argued that systematic energy transfers via nonlinear wave-wave or wave-vortical mode interactions are likely to be less significant in the mesosphere and lower thermosphere than in the stratosphere or the oceans, relative to linear instabilities, due to the very different time scales for energy decay in these different systems. In the oceans, for example, wave-wave interactions are believed to be very important in maintaining the characteristics and the dynamical balance of the internal wave spectrum (Muller et al., 1986). Convective and dynamical instabilities, on the other hand, are known to occur, but are thought not to control the spectral amplitude or characteristics of the wave field. The reason appears to lie in the rate at which such linear instabilities act to extract energy from the wave spectrum. Assuming a velocity variance of the wave spectrum of $\sim 10^{-2}$ m^2/s^2 and a mean energy dissipation rate of $\varepsilon \sim 10^{-9}$ m^2/s^3 in the oceans, we obtain a dissipation time due to such instabilities of ~ 100 days.

The velocity variance in the mesosphere and lower thermosphere, in contrast, is $\sim 10^2$ to 10^3 m^2/s^2, with corresponding energy dissipation rates $\varepsilon \sim 10^{-2}$ to 10^{-1} m^2/s^3, implying a dissipation time scale of a few hours. We obtain an equivalent estimate if we assume characteristic periods and vertical scales of a few hours and ~ 10 km. In the stratosphere, the velocity variance and energy dissipation rates are ~ 10 m^2/s^2 and 10^{-4} m^2/s^3, implying a dissipation time scale of ~ 1 day.

Thus, oceanic internal waves may persist, essentially unaffected by linear dissipation processes, for many wave periods, and may permit significant exchanges of energy via systematic nonlinear transfers. Atmospheric gravity waves at greater heights, however, must shed a large fraction of their energy each wave period. But this rapid rate of energy loss suggests strongly that energy removal must be vigorous and rapid rather than systematic and slow. Gravity waves in the stratosphere are more sensitive to dissipation processes, but may survive for sufficiently long periods to permit wave-wave interactions to play some role in their evolution.

The above arguments suggesting a dominance of gravity wave saturation by linear instability processes in the middle atmosphere are supported by an increasing number of observational and theoretical studies. Radar, balloon, and high-resolution rocket studies of small-scale structure in this region reveal apparent occurrences of dynamical and/or convective instability and a clear correlation of enhanced turbulence activity with the most unstable phase of the large-scale wave field (Tsuda et al., 1985; Sidi and Barat, 1986; Reid et al., 1987; Thrane et al., 1987; Fritts et al., 1988b; Muraoka et al., 1988; Fritts, 1989). Numerical studies by Fritts (1985) and Dunkerton (1987) found nonlinearity to reduce wave amplitudes and momentum fluxes somewhat, but not to limit wave amplitudes effectively. More recently, Walterscheid and Schubert (1989) performed a nonlinear numerical study of gravity wave breaking and concluded that wave amplitudes were limited by convective overturning. As a whole, these results suggest a rapid, local extraction of energy from the gravity wave field rather than a systematic, global transfer via wave-wave interactions.

2.2 EFFECTS OF GRAVITY WAVE SATURATION

Gravity wave saturation is recognized to have two principal effects of relevance in the middle atmosphere. The most important of these is the body force applied to the environment resulting

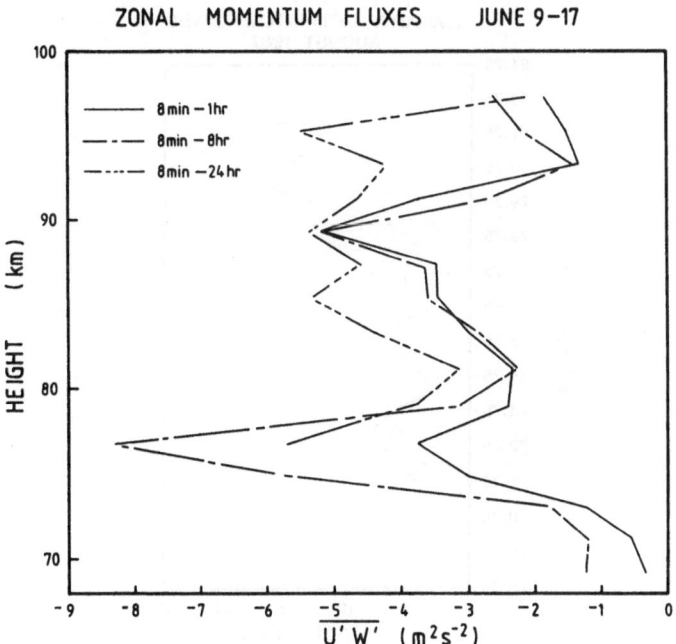

Figure 3. Zonal momentum fluxes measured with the Adelaide MF radar (35°S) during June 1984. Fluxes were estimated for three period bands and reveal that ~ 70 % of the total flux is due to wave motions with observed periods < 1 h.

from the divergence of the vertical flux of horizontal momentum by the wave motion. In the absence of dissipation, of course, gravity waves cause no permanent changes to the mean flow and the wave momentum flux for steady motions is constant with height (Eliassen and Palm, 1960; Andrews and McIntyre, 1978). But if wave amplitudes are reduced by dissipation or other interactions, the resulting momentum flux divergence causes an acceleration towards the phase speed of the wave motion. For localized forcing, the response may be extended in height due to secondary circulation effects arising through departures from balanced flow.

Both studies of the mean atmospheric circulation and thermal structure (Holton, 1982, 1983; Garcia and Solomon, 1985; Palmer et al., 1986; McFarlane, 1987) as well as direct and indirect estimates of wave momentum fluxes (Lilly and Kenedy, 1973; Vincent and Reid, 1983; Smith and Lyjak, 1985; Fritts and Vincent, 1987; Reid and Vincent, 1987; Labitzke et al., 1987; Reid et al., 1988; Fritts and Yuan, 1989b; Fritts et al., 1989b) suggest that the mean wave drag ranges from ~ 2 ms^{-1}/day in the lower stratosphere to ~ 100 ms^{-1}/day near the mesopause. The momentum fluxes giving rise to these body forces vary from $\overline{u'w'}$ ~ 0.1 - 10.0 m^2/s^2 in the troposphere and lower stratosphere to ~ 1 - 60 m^2/s^2 in the mesosphere and lower thermosphere for short intervals. Mean mesospheric momentum fluxes, on the other hand, range from ~ 2 - 20 m^2/s^2, with some evidence for larger values occurring at higher latitudes and greater heights (Reid et al., 1988; Fritts and Yuan, 1989b).

An 8-day mean momentum flux estimate obtained during June 1984 at Adelaide (35°S) is shown in Figure 3. These results indicate, consistent with predictions, that the dominant contribution to the total flux is often provided by the high-frequency component of the gravity

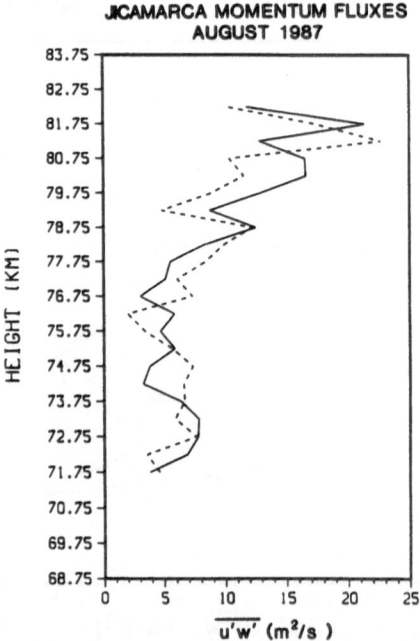

Figure 4. Zonal (——) and meridional (– –) momentum fluxes measured with the Jicamarca MST radar (12°S) during daylight for 10 days in August 1987. The large fluxes suggest a strong gravity wave forcing of the mesopause semiannual oscillation.

wave field. Because of the influence of Doppler shifting on the intrinsic wave frequencies, however, it is possible for much of the momentum flux to occur at lower observed frequencies. A daytime estimate of the momentum flux using the Jicamarca radar near Lima, Peru (12°S) for 10 days during August 1987 is shown in Figure 4 and exhibits the gradual growth with height seen at a number of locations. This likely does not represent the mean momentum flux at this location, however, due to the presence of a significant semidiurnal tidal motion and the effects of such a temporally varying environment on the gravity wave momentum fluxes (Fritts and Vincent, 1987).

The second major consequence of gravity wave saturation in the middle atmosphere is the generation, largely via convective or dynamical instabilities, of significant levels of turbulence. Depending on the height, rate of energy dissipation, and kinematic viscosity, the inertial range of quasi-three dimensional turbulence may extend for two decades or more. This turbulence acts to diffuse heat and constituents vertically and horizontally, depending on the local gradients and its location within the wave field, and thus contributes to a diffusion across mean gradients as well. Initially, this diffusion was thought to be large (Lindzen, 1981; Garcia and Solomon, 1985). However, more recent studies have shown this effect to be lessened considerably when the variation of turbulence intensity within the wave field is taken into account (Chao and Schoeberl, 1984; Fritts and Dunkerton, 1985; Coy and Fritts, 1988; McIntyre, 1989). This results in an effective turbulent Prandtl number that may be substantially larger than 1, Pr ~ 3 - 10, and inefficient turbulent mixing. Other studies have shown that large levels of diffusion are not required to account for the observed thermal and constituent structure of this region (Strobel et al., 1987).

The induced drag and diffusion due to gravity wave saturation processes are also expected to have other local effects as a result of nonuniform source distributions or filtering of the gravity wave spectrum. These include planetary wave excitation or damping (Dunkerton and Butchart, 1984; Schoeberl and Strobel, 1984; Holton, 1984), forced variations in the observed tidal structures (Fritts and Vincent, 1987), and/or systematic changes in large-scale weather patterns (Palmer et al., 1986; McFarlane, 1987; Miyahara et al., 1986; Nastrom et al., 1987). However, the scope of these effects is not known at this time.

2.3 A SATURATED GRAVITY WAVE SPECTRUM

Gravity wave saturation has several implications for the wave spectrum as a whole that are now becoming better understood. The tendency for velocity shears and temperature gradients to be limited imposes constraints on the amplitude and shape of the spectrum at large vertical wavenumbers (Dewan and Good, 1986; Smith et al., 1987). Assuming an amplitude consistent with the onset of convective instability, the saturated spectral amplitudes for horizontal velocity and normalized temperature (T'/ $\overline{T}$) or density fluctuations may be written $(p/10)N^2/m^3$ and $(1/10)N^4/g^2m^3$, respectively. Here p is the slope of the intrinsic frequency spectrum of horizontal velocity fluctuations, N is the buoyancy frequency, g is the gravitational acceleration, and m is the vertical wavenumber in rad/m. These predicted spectral amplitudes and shapes are seen to be in excellent agreement with observations at several locations and heights, suggesting that the convective instability provides a convenient framework in which to view the saturation process (Smith et al., 1987; Chanin and Hauchecorne, 1987; Fritts et al., 1988a; Tsuda et al., 1989). The close agreement between theory and observations made at the MU radar in Shigaraki, Japan during November 1986 is shown in Figure 5. For both velocity and normalized temperature fluctuations, the observed spectral densities in the stratosphere at large m are virtually

178

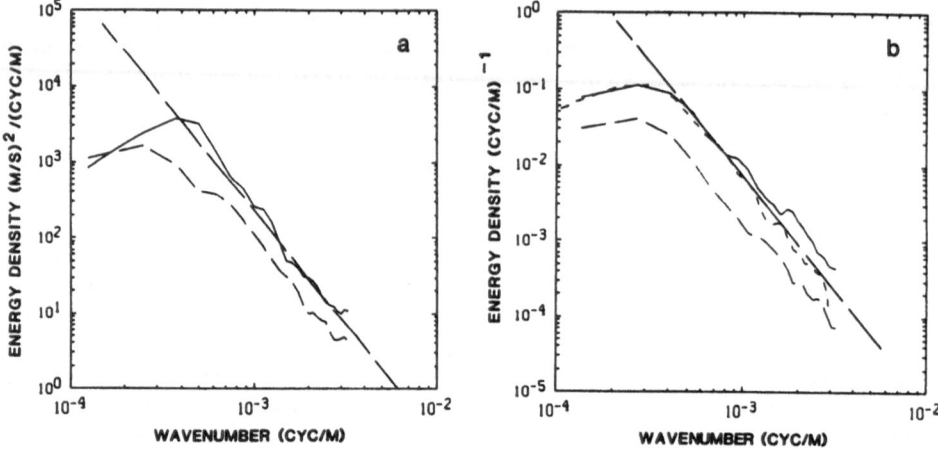

Figure 5. Predicted (long dashed) and observed vertical wavenumber spectra of horizontal velocity and normalized temperature fluctuations from the MU radar (35°N) and high-resolution balloon soundings during November 1986. The short dashed and solid lines are lower and middle stratospheric data, the lower curves are tropospheric data and reflect the smaller N in this region.

indistinguishable from predicted values. Tropospheric spectra fall below stratospheric values because of the dependence of the saturated amplitudes on N.

The saturated spectrum interpretation of the observed motion field offers a more general and physical view of the saturation process than monochromatic saturation theory in a number of respects. Perhaps the most important of these is the gradual evolution of the spectrum with increasing height. Horizontal velocity variances are observed to increase with height with an e-folding depth of $H_E \sim 12 - 15$ km throughout the lower and middle atmosphere (Balsley and Carter, 1982; Vincent, 1984; Balsley and Garello; 1985; Meek et al., 1985; Vincent and Fritts, 1987; Fritts et al., 1989a). But because motions at large m have already achieved saturated amplitudes, this variance increase must occur primarily at smaller wavenumbers that have not yet reached saturated values. This implies a characteristic (most energetic) vertical wavenumber that varies as $m_* \sim e^{-z/2H_E}$ and a spectral slope at small wavenumbers that is positive in order to insure a finite vertical energy flux (VanZandt and Fritts, 1989). With a spectrum of the form $F(m) \sim (E/m_*)\mu^s/(1 + \mu^{s+3})$, where $\mu = m/m_*$ and s is the limiting slope at small m, we infer an amplitude growth for $m < m_*$ of $F(m) \sim (E(z)/m_*)\mu^s \sim e^{(s+3)z/2H_E}$. But this likewise limits s since this growth cannot exceed that expected for conservative wave motions, $(s + 3) < 2H_E/H$, where $H \sim 7$ km is the density scale height. Thus s cannot exceed ~ 1. The gradual spectral evolution also implies a smooth transition from unsaturated to saturated wave amplitudes and a gradual increase in wave drag and diffusion with height. This suggests that the saturated spectrum may provide a useful basis for a more physical parameterization of gravity wave effects than is presently used in large-scale models.

The saturated spectrum concept can also be applied in anticipating the evolution of the gravity wave spectrum with changing atmospheric stratification. It is well known that wave amplitudes must be limited in some sense near the intrinsic phase speed of the wave motion (see Lindzen, 1981; Fritts and Rastogi, 1985). However, an increase in N likewise causes a constriction of wave amplitudes due to the changing character of the wave field. If we assume a wave spectrum as above with s = 1, $F(m) \sim (E/m_*)\mu/(1 + \mu^4)$, composed of conservative motions with $\bar{u}(z)$ constant, then the vertical flux of wave action (or energy) must be constant with height. As N increases the vertical group velocity decreases, requiring that the wave energy density increases accordingly. But this causes the wave motions to achieve amplitudes that are larger than nominal saturated values. The result is enhanced saturation of the wave spectrum throughout some characteristic depth above the transition in N (VanZandt and Fritts, 1989) as well as increased levels of drag and diffusion. This scenario is illustrated in Figure 6 with the initial, WKB, and final spectra in energy content form, mF(m). The difference in wave energy between the WKB and final spectra, and its associated fluxes of energy and momentum, is the additional increment available for local forcing and diffusion due to the change in stratification.

Fractional losses of energy and momentum flux are seen to vary as $(1 - R^{-1/4})$ and $(1 - R^{-1/2})$, respectively, where R is the ratio of N above and below the transition. These losses are expected to be distributed over a depth comparable to the vertical scale of the wave motions undergoing enhanced dissipation. For representative spectra near the tropopause with R = 2, this leads to predictions of an enhanced energy dissipation rate of $\Delta\varepsilon \sim 10^{-3}$ W/kg and an enhanced body force (per unit mass) of $\Delta \bar{u}_t \sim$ -50a ms^{-1}/day $\sim$ -5 ms^{-1}/day for a $\sim$ 0.1, where a is a measure of the excess of westward over eastward propagating wave energy (Fritts et al., 1989b). Near the high-latitude summer mesopause with R = 3, the predicted enhancements of energy dissipation and drag are $\Delta\varepsilon \sim 0.2$ W/kg and $\Delta \bar{u}_t \sim$ -50 ms^{-1}/day, again using a $\sim$ 0.1. Near the tropopause these estimates are comparable to mean values, while they appear to be significantly

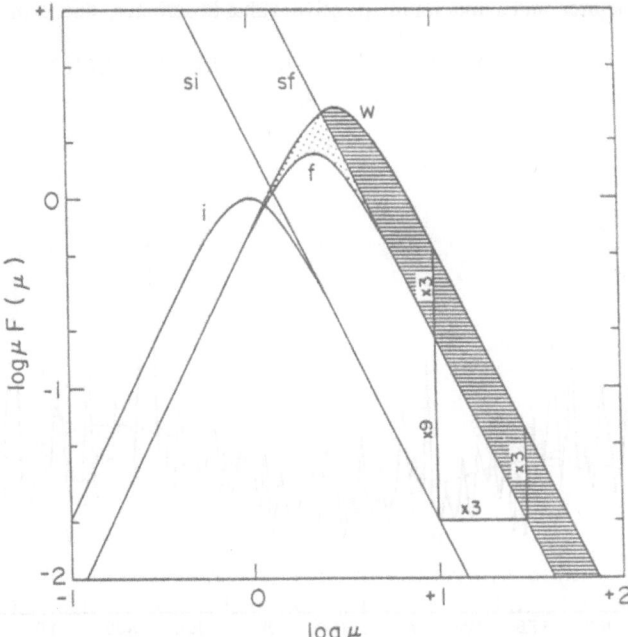

Figure 6. Initial (i), WKB (w), and final, adjusted (f) energy content spectra, $\mu F(\mu)$, illustrating the energy dissipation due to enhanced saturation at an increase in N by a factor R = 3. The increase in total energy from (i) to (w) required to preserve wave energy flux causes the new spectrum to exceed saturated amplitudes in the hatched region.

larger than mean values near the summer mesopause. This suggests that the predicted enhancements may be indistinguishable from the background values at lower levels, but may be major contributors to the totals, and to mesopause dynamics as a whole, at greater heights.

3.   Factors Influencing Gravity Wave Variability

Increasing evidence has appeared during the last few years of the considerable variability of the gravity wave spectrum, its lower and middle atmosphere effects, and the processes contributing to this variability. Our purpose in this section is not to review this evidence, but to focus on those processes believed to be responsible for it.

It is now known that variability of the wave spectrum occurs on a wide range of temporal and spatial scales in response to a variety of factors. Temporal variability occurs on scales ranging from the shortest wave periods to the annual cycle or longer. Spatial variability spans scales from ~ 10 to 10,000 km.

Annual variability of the middle atmosphere wave spectrum has been observed at a number of locations using various techniques (Hirota, 1984; Meek et al., 1985; Vincent and Fritts, 1987; Reid and Vincent, 1987) and is believed to be a result of filtering of upward propagating wave motions by the intervening winds. Gravity wave energies generally undergo a semiannual variation with maxima in summer and winter and minima corresponding to times of smallest mean winds at lower levels. Momentum fluxes, on the other hand, experience a maximum negative value in winter and a maximum positive value in summer, due to the reversal of the

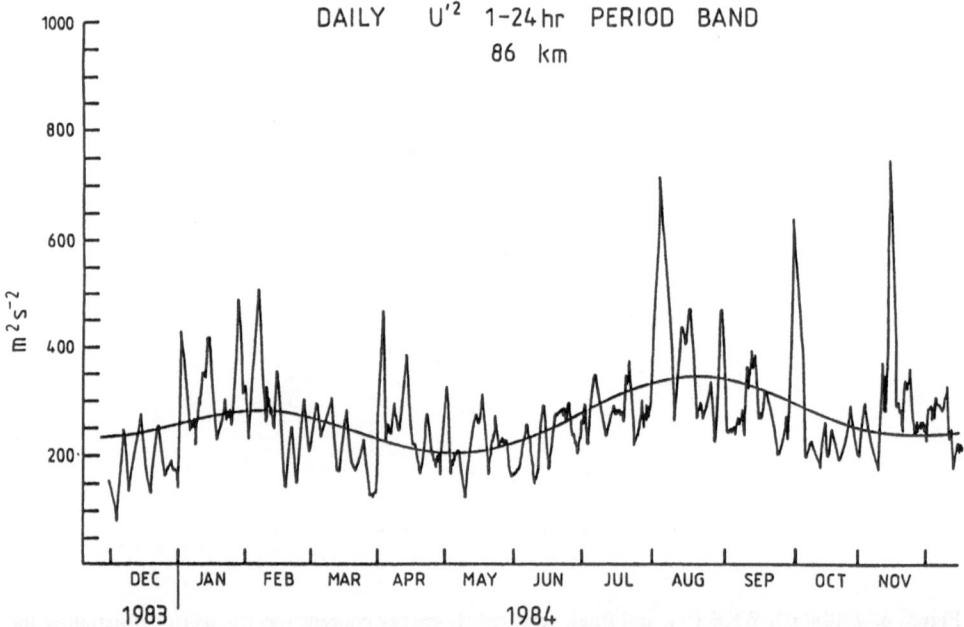

Figure 7. Zonal velocity variances at 86 km obtained with the Adelaide MF radar (35°S) and smoothed over three days. The semiannual oscillation arises in response to seasonal mean wind changes at lower levels. Shorter term variability is likely due to planetary wave modulation.

zonal mean wind from winter to summer. Numerical models using crude gravity wave parameterizations succeed in capturing the gross aspects of this annual mean structure, but are at present unable to treat realistic distributions of gravity wave sources (Holton, 1982; Garcia and Solomon, 1985; Miyahara et al., 1986; Palmer et al., 1986; McFarlane, 1987; Rind et al., 1988).

Evidence of filtering at smaller temporal scales is also provided by recent observations. The studies by Meek et al. (1985) and Vincent and Fritts (1987) found a 5- to 10-day variability in the horizontal wind variance in the mesosphere and lower thermosphere, while Fritts and Vincent (1987) presented evidence of filtering and modulation of gravity wave momentum fluxes by a large-amplitude diurnal tidal motion. The seasonal and short-term variations in horizontal wind variance are illustrated with data from Adelaide in Figure 7. These data reveal both a clear semiannual variation of the mean variance as well as large variations that may correlate with planetary wave modulation of wave sources or filtering conditions. Likewise, geographic modulations of gravity wave momentum fluxes or propagation directions due to filtering by planetary wave wind fields were noted in the winter hemisphere stratosphere and mesosphere in modeling studies by Dunkerton and Butchart (1984) and Miyahara et al. (1986).

The filtering processes leading to short-term momentum flux modulations noted above have also been observed using several other facilities, including the Poker Flat, Alaska MST radar (Fritts and Yuan, 1989b), the SOUSY MST radar in northern Norway (Ruster and Reid, 1989), and the Jicamarca MST radar near Lima, Peru. Data collected at Jicamarca during August 1987 and exhibiting a strong temporal modulation of the gravity wave momentum fluxes due to large fluctuations in the hourly mean wind profiles are shown in Figure 8. These reveal, as proposed by Fritts and Vincent (1987), that short-period waves respond to a favorable environment (with increasing $c - \bar{u}$) with sometimes dramatic increases in the momentum flux due to increasing wave amplitudes and wave field anisotropy. In particular, the large increases in meridional momentum flux at upper heights in Figure 8b correspond to the heights and times of most negative hourly mean meridional motions.

At both large and small scales, this geographic and temporal filtering of the wave variances and momentum fluxes can be understood, qualitatively at least, in terms of the linear response of gravity waves propagating in a shear flow. For waves constrained by the amplitude of incipient convective instability, $u' = (c - \bar{u})$, the covariance of zonal and vertical velocities varies as $\overline{u'w'} \sim (c - \bar{u})^3 \delta_{-}^{1/2}$, where $\delta_{-} = (1 - f^2/\omega^2)$. Thus, in an environment with $\bar{u}$ decreasing with height, waves with c > 0 will experience an increase of $\overline{u'w'} > 0$ while waves with c < 0 will experience a constriction of $\overline{u'w'} < 0$. This growth of $\overline{u'w'}$ ceases where d $\bar{u}$/dz reverses and leads to a large momentum flux divergence at these heights.

Another major source of gravity wave variability is the temporal and geographic variability of various wave sources. There is increasing evidence that orography, convection, and wind shear are significant wave sources, but that other processes, such as geostrophic adjustment and frontal accelerations, may also be important under certain conditions. While extensive data on the geographic variability of the middle atmosphere gravity wave spectrum is not available at present, there are a number of data sets that suggest that significant differences in gravity wave variances may occur over different geographic regions. MF and MST radars, located at present only on or near major land masses, suggest a mean horizontal wind variance near the mesopause of $\sim 1000 \ m^2/s^2$ (Balsley and Carter, 1982; Vincent, 1984, Meek et al., 1985; Fritts and Yuan, 1989b; Ruster and Reid, 1989). In contrast, space shuttle reentry data collected primarily over the central Pacific suggests a mean variance of $\sim 50 - 400 \ m^2/s^2$ between 60 and 90 km (Fritts et al., 1989a). Similar enhancements of velocity and temperature variance over mountainous terrain, relative to that over oceans or plains, were noted in commercial aircraft data obtained near the

182

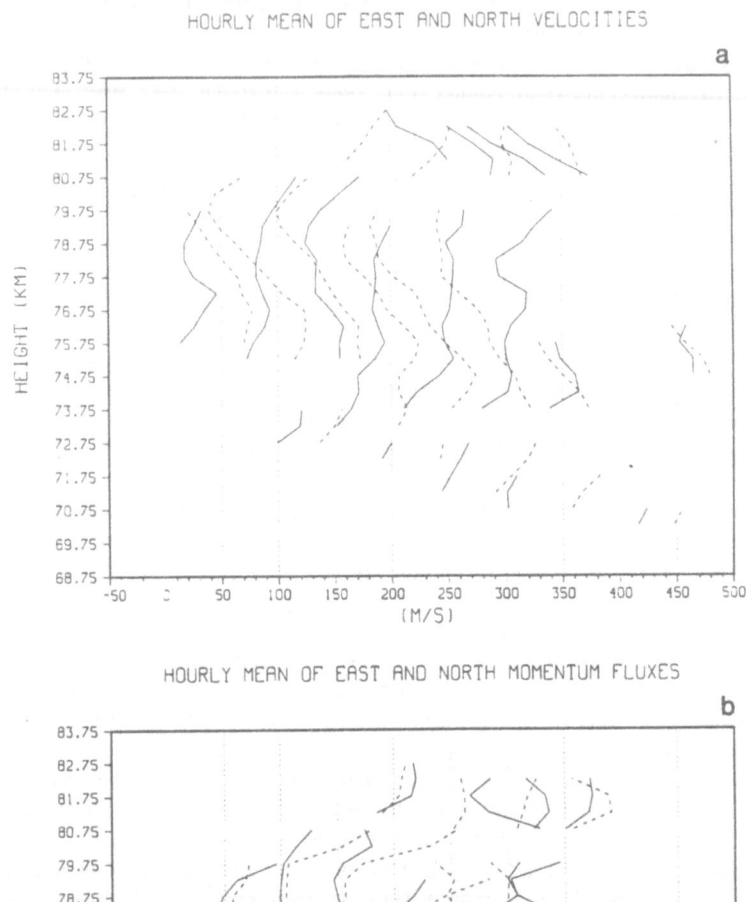

Figure 8. Zonal (————) and meridional (– – –) hourly mean velocities (a) and momentum fluxes (b) measured using the Jicamarca MST radar (12°S) during August 1987. Note the large meridional shears and the correlation of maximum positive meridional fluxes with maximum southward mean velocities.

tropopause (Nastrom et al., 1987). Finally, Ecklund et al. (1982) found the vertical velocity variance in the lee of the Colorado Rockies to be a strong function of mean wind conditions at lower levels, suggesting that mountain wave fluxes of energy and momentum are highly time dependent as well. Thus there is ample evidence that significant orography acts to energize the motion spectrum in the lower and middle atmosphere.

Parameterizations of mountain wave momentum fluxes and their associated divergences have lead to major improvements in mean circulation statistics in large-scale models of the lower atmosphere (Palmer et al., 1986; McFarlane, 1987; Rind et al., 1988). Yet there is also clear modeling evidence that orography is not the only important source. The study by McFarlane (1987), for example, found a mountain wave parameterization to provide a considerable improvement in the northern hemisphere, where there is significant high terrain, but to lead to no such improvement in correcting problems in the southern hemisphere. Assuming that large-scale flow dynamics are well described, this suggests, by analogy with the northern hemisphere, another or other gravity wave sources that may have similar beneficial effects where terrain is not a major factor.

The obvious candidates are convection and wind shear, both of which have received theoretical and observational attention. Convection, primarily thunderstorms, has been demonstrated by a number of authors to initiate gravity waves observed at various heights (Curry and Murty, 1974; Balachandran, 1980; Lu et al., 1984; Einaudi et al., 1987; Taylor and Hapgood, 1988), while recent modeling studies suggest that smaller-scale convection can also lead to appreciable wave activity and GASP data reveal large variance enhancements over frontal zones (T. L. Clark and G. D. Nastrom, private communications, 1989). Likewise wind shear has been shown to lead both to trapped and vertically propagating gravity waves and has been correlated with gravity wave observations at greater heights (Hooke and Hardy, 1975; Lalas and Einaudi, 1976; Lindzen and Tung, 1976; McIntyre and Weissman, 1978; Davis and Peltier, 1976; Fritts, 1984b, Chimonas and Grant, 1984). The challenge with these wave sources, however, is to describe adequately the temporal and geographic distributions and sporadic nature of the forcing.

4. Hemispheric Differences in Gravity Wave Activity and Effects

As noted above, the gravity wave spectrum and its middle atmosphere effects exhibit considerable spatial and temporal variability. This variability arises due to the variable strengths and distributions of sources at lower levels and in response to the different environments through which the waves propagate. The intent of this section is to consider the implications of this variability and the evidence for hemispheric differences in the gravity wave spectrum.

The most obvious potential source of hemispheric differences in the gravity wave spectrum and its effects is the different hemispheric distribution of mountainous terrain (see McFarlane, 1987). In both the lower and middle atmosphere this leads to apparent, significant differences in the energy density of and the likely energy and momentum fluxes by the gravity wave spectrum. GASP data near the tropopause and radar and space shuttle data at greater heights reveal different mean energy densities of $\sim 2 - 3$ over mountains relative to open ocean (Nastrom et al., 1987; Balsley and Carter, 1982; Vincent, 1984; Fritts et al., 1989a), suggesting a major role for orographically generated gravity waves throughout the atmosphere. As noted by Nastrom et al. (1987), the larger wave variances over mountains imply an even greater enhancement in the wave fluxes of energy and momentum because of the larger intrinsic frequencies (and vertical velocities) associated with larger mean variances at fixed horizontal scales. Thus, to the extent that we can extrapolate present data to other geographic locations, the southern hemisphere would appear to have a less energetic mean wave spectrum and smaller mean fluxes of energy and momentum than the northern hemisphere. Because of the different geographic distributions of mountainous terrain and the dependence of strong wave forcing on meteorological conditions (Ecklund et al., 1982; Palmer et al., 1986; McFarlane, 1987), we also anticipate very different responses of the mean circulation to mountain wave activity in the two hemispheres.

A less significant role of orographically generated gravity waves in the southern hemisphere may be offset to some extent by other sources of wave activity. This different mix of sources is likely to impose, as well, a different distribution of phase speeds and possibly varying lower and middle atmosphere responses. For example, convection would appear to be a likely source for the strong gravity wave activity and large momentum fluxes observed at Jicamarca and illustrated in Figure 8. There is too little data at present, however, to assess either the role of other wave sources or their influences at greater heights, and additional observations will be necessary to answer these questions.

Different gravity wave energy and momentum fluxes anticipated for the two hemispheres also have implications for the mean circulation and constituent structures and for the dissipation of other motions. A larger mean momentum flux and divergence will lead to stronger flow decelerations and weaker mean circulations in regions of strong wave dissipation, while a larger energy flux implies greater turbulent diffusion and more rapid damping of other wave motions. As with a gravity wave source evaluation, there is at present too little data to permit an assessment of the differences in momentum fluxes between the two hemispheres. There is, however, some evidence to support the view that there are different levels of diffusion. This is provided primarily by conjugate tidal observations at Adelaide (35°S) and Kyoto (35°N) which reveal a large difference in the tidal amplitudes at the two stations. Observations of the diurnal tide, which generally has a smaller vertical structure and is more sensitive to dissipation, suggest that there is a significantly higher mean diffusion and a corresponding reduction in tidal amplitudes in the northern hemisphere (Vincent, private communication, 1988).

Finally, we should anticipate that the different planetary and tidal wave environments in the two hemispheres arising from different planetary wave excitation and tidal damping may contribute to quantitatively varying filtering environments for gravity waves originating at lower levels. The smaller degree of planetary wave activity in the southern hemisphere winter may contribute to smaller variations of gravity wave momentum fluxes than occur in the northern hemisphere winter (Miyahara et al., 1986), while larger tidal amplitudes are likely to contribute to greater modulation of gravity wave momentum fluxes in the southern hemisphere (Fritts and Vincent, 1987). But as with the distribution of wave sources discussed above, much of this discussion is speculative at this point due to a lack of more complete data on the gravity wave spectrum in the southern hemisphere.

5.   Summary

We have reviewed here the primary processes and effects of gravity wave saturation in the middle atmosphere, the likely causes of gravity wave variability, and the causes of and evidence for hemispheric differences in the wave spectrum and its effects. Because of the large vertical flux of wave energy in the middle atmosphere, the most likely mechanisms for wave instability are the processes of convective and dynamical instability occurring locally within the wave field. Convective instability appears to be favored for waves with intrinsic frequencies substantially larger than f, while dynamical instability seems to be possible for wave frequencies very near f. Unlike the oceans or the lower atmosphere, where wave dissipation times are longer, the middle atmosphere wave spectrum appears not to be amplitude limited by the slower, more systematic nonlinear wave-wave or wave-vortical mode interactions. It is clear, however, both from observations and modeling studies, that one result of local convective and/or dynamical instabilities is the rapid nonlinear transfer of energy from the wave motions into small-scale turbulence.

The consequences of gravity wave saturation for the mean circulation and constituent structures, namely wave drag and turbulent diffusion, are becoming better understood, with observational, modeling, and theoretical work now converging to a common view. Gravity wave drag is believed to be a fundamental contributor to the overall structure of the middle atmosphere and to play a smaller but important role near the tropopause as well. Turbulent diffusion arising from wave saturation processes likewise plays a role in middle atmosphere transports, but now appears not to be as important as was originally believed. In terms of the wave spectrum itself, saturation appears to establish a high vertical wavenumber region with an amplitude of the velocity spectrum that scales with N^2 and a slope of -3 which gradually extends to smaller and smaller wavenumbers as the spectral components propagate vertically. This gradual evolution of the wave spectrum leads to smooth variations in wave drag and diffusion with height and permits a description of the wave field that allows for a quantitative prediction of its response to changes in the environment.

Several factors are believed to contribute to the variability of the gravity wave spectrum. These include the temporal and spatial variability of various wave sources, primarily orography, convection, and wind shear, and filtering by intervening wind fields due to mean, planetary, and/or tidal and low-frequency gravity wave motions. Orography and convection appear to be major sources of gravity wave activity, with mountain waves strongly modulated by the mean flow at lower levels and frontal zones and especially thunderstorms acting as significant sources of wave variance throughout the lower and middle atmosphere. Filtering also modulates the wave field and its orientation and momentum flux strongly, resulting in a preferred sense of propagation against the local mean flow and large fluxes at levels where the wave motions achieve the maximum intrinsic phase speeds.

Finally, we have presented evidence of the geographic variability of the wave spectrum which suggests that there may be significant differences in the gravity wave spectrum and its effects between the northern and southern hemispheres. These differences include wave energy densities and the dominant wave sources, and may include as well vertical fluxes of energy and momentum. Some evidence of different energy fluxes is provided by very different tidal amplitudes at conjugate locations, suggestive of different turbulent diffusion rates in the two hemispheres.

acknowledgments This work was supported by the Air Force Office of Scientific Research (AFSC) under contract F49620-87-C-0024 and by the National Aeronautics and Space Administration under grant NAGW-1179.

6. References

Andrews, D.G., and M.E. McIntyre, 1978: An exact theory of nonlinear waves on a Lagrangian mean flow, J. Fluid Mech., 89, 609-646.

186

Balachandran, N.K., 1980: Gravity waves from thunderstorms, Mon. Wea. Rev., 108, 804-816.

Balsley, B.B., and D.A. Carter, 1982: The spectrum of atmospheric velocity fluctuations at 8 and 86 km, Geophys. Res. Lett., 9, 465-468.

Balsley, B.B., and R. Garello, 1985: The kinetic energy density in the troposphere, stratosphere and mesosphere: A preliminary study using the Poker Flat radar in Alaska, Radio Sci., 20, 1355-1362.

Chanin, M.-L., and A. Hauchecorne, 1987: Lidar sounding of the structure and dynamics of the middle atmosphere. A review of recent results relevant to transport processes, Transport Processes in the Middle Atmsophere, NATO Advanced Studies Workshop, G. Visconti and R. Garcia, eds., 459-477.

Chao, W.C., and M.R. Schoeberl, 1984: A note on the linear approximation of gravity wave saturation in the mesosphere, J. Atmos. Sci., 41, 1893-1898.

Chimonas, G., and J.R. Grant, 1984: Shear excitation of gravity waves. Part II: Upscale scattering from Kelvin-Helmholtz waves, J. Atmos. Sci., 41, 2278-2288.

Coy, L., and D.C. Fritts, 1988: Gravity wave heat fluxes: A Lagrangian approach, J. Atmos. Sci., 45, 1770-1780.

Curry, J.M., and R.C. Murty, 1974: Thunderstorm-generated gravity waves, J. Atmos. Sci., 31, 1402-1408.

Davis, P.A., and W.R. Peltier, 1976: Resonant parallel shear instability in the stably stratified planetary bondary layer, J. Atmos. Sci., 33, 1287-1300.

Dewan, E.M., and R.E. Good, 1986: Saturation and the "universal" spectrum for vertical profiles of horizontal scalar winds in the atmosphere, J. Geophys. Res., 91, 2742-2748.

Dong, B, and K.C. Yeh, 1988: Resonant and nonresonant wave-wave interactions in an isothermal atmosphere, J. Geophys. Res., 93, 3729-3744.

Dunkerton, T.J., 1984: Inertio-gravity waves in the stratosphere, J. Atmos. Sci., 41, 3396-3404.

Dunkerton, T.J., 1987: Effect of nonlinear instability on gravity wave momentum transport, J. Atmos. Sci., 44, 3188-3209.

Dunkerton, T.J., 1989: Theory of internal gravity wave saturation, Pure Apl. Geophys., in press.

Dunkerton, T.J., and N. Butchart, 1984: Propagation and selective transmission of gravity waves in a sudden warming, J. Atmos. Sci., 41, 1443-1460.

Ecklund, W.L., K.S. Gage, B.B. Balsley, R.G. Strauch, and J.L. Green, 1982: Vertical wind variability observed by VHF radar in the lee ofthe Colorado Rockies, Mon. Wea. Rev., 110, 1451-1457.

Einaudi, F., W.L. Clark, D. Fua, J.L. Green, and T.E. VanZandt, 1987: Gravity waves and convection in Colorado during July 1983, J. Atmos. Sci., 44, 1534-1553.

Eliassen, A., and E. Palm, 1960: On the transfer of energy in stationary mountain waves, Geophys. Publ., 22, 1-20.

Fritts, D.C., 1984a: Gravity wave saturation in the middle atmosphere: A review of theory and observations, Rev. Geophys. Space Phys., 22, 275-308.

Fritts, D.C., 1984b: Shear excitation of atmospheric gravity waves. Part II: Nonlinear radiation from a free shear layer, J. Atmos. Sci., 41, 524-537.

Fritts, D.C., 1985: A numerical study of gravity wave saturation: Nonlinear and multiple wave effects, J. Atmos. Sci., 42, 2043-2058.

Fritts, D.C., 1989: A review of gravity wave saturation processes, effects, and variability in the middle atmosphere, Pure Appl. Geophys., 130, 343-371.

Fritts, D.C., R.C. Blanchard, and L. Coy, 1989a: Gravity wave structure between 60 and 90 km inferred from space shuttle re-entry data, J. Atmos. Sci., 46, 423-434.

Fritts, D.C., and T.J. Dunkerton, 1985: Fluxes of heat and constituents due to convectively unstable gravity waves, J. Atmos. Sci., 42, 549-556.

Fritts, D.C., and P.K. Rastogi, 1985: Convective and dynamical instabilities due to gravity wave motions in the lower and middle atmosphere: Theory and observations, Radio Sci., 20, 1247-1277.

Fritts, D.C., S.A. Smith, B.B. Balsley, and C.R. Philbrick, 1988b: Evidence of gravity wave saturation and local turbulence production in the summer mesosphere and lower thermosphere during the STATE experiment, J. Geophys. Res., 93, 7015-7025.

Fritts, D.C., T. Tsuda, T. Sato, S. Fukao, and S. Kato, 1988a: Observational evidence of a saturated gravity wave spectrum in the troposphere and lower stratosphere, J. Atmos. Sci., 45, 1741-1759.

Fritts, D.C., T. Tsuda, T.E. VanZandt, S.A. Smith, T. Sato, S. Fukao, and S. Kato, 1989b: Studies of velocity fluctuations in the lower atmosphere using the MU radar: II. Momentum fluxes and energy densities, J. Atmos. Sci., in press.

Fritts, D.C., and R.A. Vincent, 1987: Mesospheric momentum flux studies at Adelaide, Australia: Observations and a gravity wave/tidal interaction model, J. Atmos. Sci., 44, 605-619.

Fritts, D.C., and L. Yuan, 1989a: A stability analysis of inertio-gravity wave structure in the middle atmosphere, J. Atmos. Sci., in press.

Fritts, D.C., and L. Yuan, 1989b: Measurement of momentum fluxes near the summer mesopause at Poker Flat, Alaska, submitted to J. Atmos. Sci.

Garcia, R.R., and S. Solomon, 1985: The effect of breaking gravity waves on the dynamical and chemical composition of the mesosphere and lower thermosphere, J. Geophys. Res., 90, 3850-3868.

Hines, C.O., 1971: Generalizations of the Richardson number criterion for the onset of atmospheric turbulence, Quart. J. Roy. Met. Soc., 97, 429-439.

Hines, C.O., 1988: The generation of turbulence by atmospheric gravity waves, J. Atmos. Sci., 45, 1269-1278.

Hirota, I., 1984: Climatology of gravity waves in the middle atmosphere, Dynamics of the Middle Atmosphere,, J.R. Holton and T. Matsuno, Eds., 65-76.

Hodges, R.R., Jr., 1967: Generation of turbulence in the upper atmosphere by internal gravity waves, J. Geophys. Res., 72, 3455-3458.

Holton, J.R., 1982: The role of gravity wave-induced drag and diffusion in the momentum budget of the mesosphere, J. Atmos. Sci., 39, 791-799.

Holton, J.R., 1983: The influence of gravity wave breaking on the general circulation of the middle atmosphere, J. Atmos. Sci., 40, 2497-2507.

Holton, J.R., 1984: The generation of mesospheric planetary waves by zonally asymmetric gravity wave breaking, J. Atmos. Sci., 41, 3427-3430.

Hooke, W.H., and K.R. Hardy, 1975: Further study of the atmospheric gravity waves over the eastern seaboard on 18 March 1969, J. Appl. Meteor., 14, 31-38.

Inhester, B., 1987: The effect of inhomogeneities on the resonant parametric interaction of gravity waves in the atmosphere, Ann. Geophys., 87, 209-217.

Labitzke, K., A.H. Manson, J.J. Barnett, and M. Corney, 1987: Comparison of geostrophic and observed winds in the upper mesosphere over Saskatoon, Canada, J. Atmos. Terres. Phys., 49, 987-997.

Lalas, D.P., and F. Einaudi, 1976: On characteristics of gravity waves generated by atmospheric shear layers, J. Atmos. Sci., 33, 1248-1259.

Lilly, D.K., and P.J. Kennedy, 1973: Observations of a stationary mountain wave and its associated momentum flux and energy dissipation, J. Atmos. Sci., 30, 1135-1152.

Lindzen, R.S., 1981: Turbulence and stress due to gravity wave and tidal breakdown, J. Geophys. Res., 86, 9707-9714.

Lindzen, R.S., and K.K. Tung, 1976: Banded convective activity and ducted gravity waves, Mon. Wea. Rev., 104, 1602-1617.

Lu, D., T.E. VanZandt, and W.L. Clark, 1984: VHF Doppler radar observations of buoyancy waves induced by thunderstorms, J. Atmos. Sci., 41, 272-282.

McFarlane, N.A., 1987: The effect of orographically excited gravity wave drag on the general circulation of the lower stratosphere and troposphere, J. Atmos. Sci., 44, 1775-1800.

McIntyre, M.E., 1989: On dynamics and transport near the polar mesopause in summer, J. Geophys. Res., in press.

McIntyre, M.E., and M.A. Weissman, 1978: On radiating instabilities and resonant over-reflection, J. Atmos. Sci., 35, 1190-1196.

Meek, C.E., I.M. Reid and A.H. Manson, 1985: Observations of mesospheric wind velocities. II. Cross sections of power spectral density for 48-8h, 8-1h, 1h-10 min over 60-110 km for 1981, Radio Sci., 20, 1383-1402.

Miyahara, S., Y. Hayashi, and J.D. Mahlman, 1986: Interactions between gravity waves and the planetary scale flow simulated by the GFDL "SKYHI" general circulation model, J. Atmos. Sci., 43, 1844-1861.

Muller, P., G. Holloway, F. Henyey, and N. Pomphrey, 1986: Nonlinear interactions among internal gravity waves, Rev. Geophys., 24, 493-536.

Muraoka, Y., T. Sugiyam, K. Kawahira, T. Sato, T. Tsuda, S. Fukao, and S. Kato, 1988: Formation of mesospheric VHF echoing layers due to a gravity wave motion, J. Atmos. Terres. Phys., 50, 819-829.

Nastrom, G.D., D.C. Fritts, and K.S. Gage, 1987: An investigation of terrain effects on the mesoscale spectrum of atmospheric motions, J. Atmos. Sci., 44, 3087-3096.

Palmer, T.N., G.J. Shutts, and R. Swinbank, 1986: Alleviation of a systematic westerly bias in general circulation and numerical weather prediction models through an orographic gravity wave drag parameterization, Quart. J. Roy. Met. Soc., 112, 1001-1040.

Reid, I.M., R. Ruster, P. Czechowsky, and G. Schmidt, 1988: VHF radar measurements of momentum flux in the summer polar mesosphere over Andenes (69°N, 16°E), Norway, Geophys. Res. Lett., 11, 1263-1266.

Reid, I.M., R. Ruster, and G. Schmidt, 1987: VHF radar observations of cat's-eye-like structures at mesospheric heights, Nature, 327, 43-45.

Reid, I.M., and R.A. Vincent, 1987: Measurements of mesospheric gravity wave momentum fluxes and mean flow accelerations at Adelaide, Australia, J. Atmos. Terres. Phys., 49, 443-460.

Rind D., R. Suozzo, N. K. Balachandran, A. Lacis, and G. Russell, 1988: The GISS global climate - middle atmosphere model, Part II: Model variability due to interactions between planetary waves, the mean circulation and gravity wave drag, J. Atmos. Sci., 45, 371-386.

Ruster, R., and I.M. Reid, 1989: VHF radar observations of the dynamics of the summer polar mesopause region, J. Geophys. Res., in press.

Schoeberl, M.R., and D.F. Strobel, 1984: Nonzonal gravity wave breaking in the winter mesosphere, Dynamics of the Middle Atmosphere, J.R. Holton and T. Matsuno, Eds., 45-64.

Sidi, C., and J. Barat, 1986: Observational evidence of an inertial wind structure in the stratosphere, J. Geophys. Res., 91, 1209-1217.

Smith, A.K., and L.V. Lyjak, 1985: An observational estimate of gravity wave drag from the momentum balance in the middle atmosphere, J. Geophys. Res., 90, 2233-2241.

Smith, S.A., D.C. Fritts, and T.E. VanZandt, 1987: Evidence for a saturated spectrum of atmospheric gravity waves, J. Atmos. Sci., 44, 1404-1410.

Strobel, D.F., M.E. Summers, R.M. Bevilacqua, M.T. DeLand, and M. Allen, 1987: Vertical constituent transport in the mesosphere, J. Geophys. Res., 92, 6691-6698.

Taylor, M.J., and M.A. Hapgood, 1988: Identification of a thunderstorm as a source of short period gravity waves in the upper atmospheric nightglow emissions, Planet. Space Sci., 36, 975-985.

Thrane, E.V., T.A. Blix, C. Hall, T.L. Hansen, U. von Zahn, W. Meyer, P. Czechowsky, G. Schmidt, H.-U. Widdel, and A. Neumann, 1987: Small scale structure and turbulence in the mesosphere and lower thermosphere at high latitudes in winter, J. Atmos. Terres. Phys., 49, 751-762.

Tsuda, T., K. Hirose, S. Kato, and M.P. Sulzer, 1985: Some findings on correlation between the stratospheric echo power and the wind shear observed by the Arecibo UHF radar, Radio Sci., 20, 1503-1508.

Tsuda, T., T. Inoue, D.C. Fritts, T.E. VanZandt, S. Kato, T. Sato, and S. Fukao, 1989: MST radar observations of a saturated gravity wave spectrum, J. Atmos. Sci., in press.

VanZandt, T.E., and D.C. Fritts, 1989: A theory of enhanced saturation of the gravity wave spectrum due to increases in atmospheric stability, Pure Appl. Geophys., 130, 399-420.

Vincent, R.A., 1984: Gravity wave motions in the mesosphere, J. Atmos. Terres. Phys., 46, 119-128.

Vincent, R.A., 1989: Gravity wave observations in the Southern Hemisphere, Dynamics, Transport and Photochemistry in the Middle Atmosphere of the Southern Hemisphere, A. O'Neill and C.R. Mechoso, Eds., in press.

Vincent, R.A., and D.C. Fritts, 1987: A climatology of gravity waves in the mesosphere and lower thermosphere over Adelaide, Australia, J. Atmos. Sci., 44, 748-760.

Vincent, R.A., and I.M. Reid, 1983: HF Doppler measurements of mesospheric momentum fluxes, J. Atmos. Sci., 40, 1321-1333.

Walterscheid, R.L., and G. Schubert, 1989: Nonlinear evolution of an upward propagating gravity wave: Overturning, convection, and turbulence, J. Atmos. Sci., in press.

Weinstock, J., 1982: Nonlinear theory of gravity waves: Momentum deposition, generalized Rayleigh friction, and diffusion, J. Atmos. Sci., 39, 1698-1710.

Yeh, K.C., and C.H. Liu, 1981: The instability of atmospheric gravity waves through wave-wave interactions, J. Geophys. Res., 86, 9722-9728.

Yeh, K.C., and C.H. Liu, 1985: Evolution of atmospheric spectrum by processes of wave-wave interaction, Radio Sci., 20, 1279-1294.

Yuan, L., and D.C. Fritts, 1989: Influence of a mean shear on the dynamical instability of an inertio-gravity wave, J. Atmos. Sci., in press.

NITROGEN CHEMISTRY IN ANTARCTICA: A BRIEF REVIEW

S. SOLOMON
Aeronomy Laboratory
National Oceanic and Atmospheric Administration
Boulder, Colorado 80303 USA

ABSTRACT. Measurements of reactive nitrogen species in Antarctica are briefly reviewed and their links to ozone depletion are summarized. Observations of NO, NO_2, HNO_3, particulate nitrate, and total NO_y demonstrate that the composition of the Antarctic stratosphere is greatly perturbed by the presence of clouds. Further, measurements have shown that the clouds themselves are composed in part of HNO_3, and that sedimentation of cloud particles apparently can remove reactive nitrogen from the gas phase altogether. These processes reduce the abundance of stratospheric NO_2, a primary requirement for elevated ClO densities and attendant ozone loss.

1. Introduction

It has long been known that nitrogen species play a critical role in the photochemistry of the Earth's stratosphere through catalytic destruction of ozone (Crutzen, 1970; Johnston, 1971). In recent years, however, the discovery and interpretation of the Antarctic ozone hole has led to new perspectives on the mechanisms whereby nitrogen exerts control over the stratospheric ozone layer. In particular, it has become clear that in addition to the catalytic destruction of ozone in the upper stratosphere, nitrogen chemistry plays a major role in the formation of polar stratospheric clouds and their strong influence upon stratospheric ozone chemistry in polar regions. It is the purpose of this paper to briefly review those findings.

Total active nitrogen is commonly referred to as NO_y, and includes the following major stratospheric species: NO, NO_2, NO_3, N_2O_5, $ClONO_2$, and HNO_3. The oxides of nitrogen are often referred to as NO_x, and include NO, NO_2, NO_3 and N_2O_5 for the purposes of this review. Interconversion of NO and NO_2 can take place via reaction of NO with ozone followed by reaction of NO_2 with atomic oxygen; these two reactions represent the natural catalytic cycle that dominates the chemical destruction of ozone over much of the middle and upper stratosphere.

When the Antarctic ozone hole was first discovered by the British Antarctic Survey (Farman et al., 1985), many questions regarding its veracity and origin were posed. Modest ozone depletions of 5-8% in winter and early spring were also observed in the Arctic (Ozone Trends Report, 1989), but these are much smaller than the extensive ozone loss of about 50% found in the Antarctic spring. The reality of the ozone depletion phenomenon was quickly established by satellite, balloon-borne and ground-based sensors that confirmed the evidence presented by Farman et al., and interpretive studies quickly followed (see the review by Solomon, 1988). Dynamical mechanisms involving upward transport of ozone-poor air from the troposphere were proposed by Tung et al. (1986) and by Mahlman and Fels (1986). Two chemical mechanisms were also proposed. Callis and Natarajan (1986) proposed that unusually large quantities of reactive nitrogen might have been produced by the solar maximum of the early 1980's; subsequent transport of these species to the Antarctic lower stratosphere might then have led to rapid ozone depletion via the well-

191

A. O'Neill (ed.),
Dynamics, Transport and Photochemistry in the Middle Atmosphere of the Southern Hemisphere, 191–201.
© 1990 *Kluwer Academic Publishers*.

established NO_x-catalyzed destruction of atmospheric ozone as mentioned above. On the other hand, chlorine chemistry was also suggested as a possible explanation for the ozone decline. Solomon et al. (1986) and McElroy et al. (1986a) suggested that heterogeneous reactions occurring on the surfaces of the polar stratospheric clouds (PSC's) known to abound in the Antarctic stratosphere might drastically alter the composition of this region as compared to an environment where only gas-phase chemical processes occur, setting the stage for substantial ozone destruction in the lower stratosphere through halogen chemistry.

Heterogeneous reactions such as

$$HCl + ClONO_2 \longrightarrow HNO_3 + Cl_2 \tag{1}$$

were suggested (Solomon et al., 1986) and later shown in laboratory studies to be remarkably rapid (Molina et al., 1987; Tolbert et al., 1987). This surface reaction has two important effects: it converts chlorine from the relatively unreactive species ($ClONO_2$ and HCl) into Cl_2, a highly reactive substance that readily photodissociates to produce chlorine atoms. These subsequently react with ozone to form chlorine monoxide. Chlorine monoxide in turn can dimerize readily at the cold temperatures typical of the Antarctic stratosphere, ultimately leading to rapid ozone loss (see Molina and Molina, 1987; Anderson et al., 1989). Reactions of chlorine monoxide with bromine monoxide (McElroy et al., 1986) also contribute to Antarctic ozone destruction. The Antarctic ozone loss due to reaction of chlorine monoxide with the perhydroxyl radical (Solomon et al., 1986; Crutzen and Arnold, 1986) has not been quantified based on direct measurements, but is likely to be relatively small. Both cold temperatures to provide cloud surfaces and sunlight to drive photochemical processes are critical to the chlorine theory (sunlight is required, for example, in the photolysis of the Cl_2 molecule formed in reaction 1 above). Additional heterogeneous reactions currently believed to be of importance are:

$$ClONO_2 + H_2O \longrightarrow HOCl + HNO_3 \tag{2}$$

$$HCl + N_2O_5 \longrightarrow ClONO + HNO_3 \tag{3}$$

$$H_2O + N_2O_5 \longrightarrow 2HNO_3 \tag{4}$$

The reaction of HCl with $ClONO_2$ on particles converts reactive nitrogen from a relatively active species to HNO_3, thereby lowering the abundance of NO_x and slowing the reformation of $ClONO_2$. Thus heterogeneous processes enhance the abundances of reactive chlorine while also suppressing the abundances of reactive nitrogen species. Both effects are critical to establishing the amount of chlorine that can be made available to destroy ozone and, perhaps more importantly, the time scale over which it is able to do so. In the limit of complete removal of NO_x, elevated chlorine monoxide abundances liberated by polar stratospheric clouds can persist for a time scale of months, thereby destroying ozone quite effectively, while for nominal levels of NO_x, chlorine monoxide persists only a few hours before it is reconverted to $ClONO_2$ in the lower stratosphere, limiting the amount of ozone destroyed.

The solar cycle and chlorine theories of Antarctic ozone depletion thus began with opposite notions regarding the abundance of nitrogen oxides in the Antarctic stratosphere: the solar theory required greatly enhanced levels of NO_x, while the chlorine theories suggesting substantial suppression of NO_x through reactions on PSC surfaces. Measurements of nitrogen oxides and chlorine species in Antarctica showed that the chlorine theory was basically correct insofar as its gross predictions regarding the Antarctic stratosphere were concerned: the abundances of NO_x species were found to be remarkably low in the Antarctic spring stratosphere while those of reactive chlorine were greatly enhanced and were broadly consistent with levels required to explain the observed ozone destruction given presently accepted photochemistry (Barrett et al., 1988; Anderson et al., 1989; Jones et al., 1989).

Antarctic data and related theoretical studies also revealed that the chemical mechanisms responsible for NO_x suppression are considerably more complicated than first suggested, and established that the chemical composition of the Antarctic stratosphere plays an important role in cloud microphysics as well as in ozone depletion. Seminal studies by Toon et al. (1986), Crutzen and Arnold (1986), and McElroy et al. (1986b) suggested that PSCs might begin to form at temperatures above the frost point as (probably frozen) binary mixtures of HNO_3 and H_2O. Such a mechanism would further suppress reactive gaseous nitrogen species beyond the heterogeneous conversion processes noted above. Laboratory studies of the HNO_3 - $H2O$ system quickly confirmed these suggestions, demonstrating the thermodynamic stability of the HNO_3 trihydrate ice (Hansen and Mauersberger, 1988a,b). Further, as noted by Toon et al. (1986), such particles might remove reactive nitrogen irreversibly from ("denitrify") the polar stratosphere through sedimentation of large PSC particles containing HNO_3.

Denitrification is potentially of particular importance, since it represents a mechanism whereby the chemical perturbations induced by PSC's could persist long after the clouds themselves are no longer present. Stratospheric warmings lead to generally warmer temperatures in the Arctic than in the Antarctic and hence fewer PSC's, especially in the latter part of the winter and spring (i.e., February through April as compared to August through October). This portion of the annual cycle is of special interest because ozone depletion requires both sunlight and photochemical perturbations due to PSC's. The spring overlap of the two is clearly more limited in the Arctic than in the Antarctic. However, when denitrification is considered, these two factors may no longer need to occur simultaneously for substantial ozone depletion to take place, a possibility that could assume special importance in the warm Arctic spring stratosphere.

The next section describes measurements of the gas-phase abundances of reactive nitrogen species in the Antarctic stratosphere. The following section discusses measurements that revealed the important and perhaps surprising role played by HNO_3 in the microphysics of cloud formation. Finally, the implications of the current understanding of polar odd nitrogen chemistry for polar ozone chemistry are summarized.

2. Measurements of Reactive Nitrogen Species in Antarctica

In retrospect, observations of the column abundance and diurnal cycle of nitrogen dioxide obtained in the late 1970's and early 1980's provided strong indications of perturbed chemistry in the polar winter and spring long before the discovery of the Antarctic ozone hole directed worldwide attention to south polar regions. The NO_2 column abundance in the Arctic winter and spring was found to be significantly lower than theoretical expectations (Noxon, 1979). This phenomenon came to be known as the Noxon "cliff" and was highlighted as a major challenge to our understanding of stratospheric chemistry. Although many of the observations were obtained in the northern hemisphere winter, a similar Antarctic "cliff" was found in the only available latitude survey from the southern hemisphere (Noxon, 1978). Observations from the New Zealand Antarctic research station (Scott Base at 78S) published prior to the discovery of the Antarctic ozone hole also displayed remarkably low levels of NO_2 during Antarctic spring (McKenzie and Johnston, 1984), well below those generally found in the Arctic spring. In the high latitude air, both the diurnal variation and absolute abundance of NO_2 were greatly reduced compared to theoretical expectations.

Observations and interpretation of HNO_3 abundances in the Arctic polar night (Wofsy, 1978; Austin et al., 1986b) provided important evidence for an unexplained source of HNO_3 in polar winter, and the possibility of heterogeneous production on atmospheric aerosol was suggested. In particular, the observations showed larger abundances of HNO_3 inside the heart of the polar vortex than found at lower latitudes. These are difficult to reconcile with purely gas-phase production of HNO_3 in the dark polar winter. A few observations of the column abundance of HNO_3 were also available from Antarctica (Williams et al., 1982), confirming that its behavior there was inconsistent with "standard" gas-phase photochemical schemes, and supporting the very low NO_2 column measurements of McKenzie and Johnston (1984). Solomon et al. (1986) compared these observations of the latitude gradient of total HNO_3 column abundance observed in Antarctica to two-dimensional model calculations both including and neglecting heterogeneous production. They suggested that the observed low NO_2 and high HNO_3 column amounts in Antarctica indicated net production of HNO_3, probably through surface reactions.

Thus early observations of reactive nitrogen species suggested the possibility that heterogeneous reactions affected the partitioning of nitrogen species in the Antarctic winter and spring. While these observations were very important and strongly suggestive of major gaps in our understanding, it was not until further measurements of reactive nitrogen species were carried out following the discovery of the ozone hole that the mechanisms responsible for them and their significance to ozone chemistry became clearer.

Observations of the NO_2 column abundance in Antarctic spring were obtained with both visible (Keys and Johnston, 1986; Mount et al., 1987; Wahner et al., 1989) and infrared absorption methods (Farmer et al., 1987; Coffey and Mankin, 1989; Toon et al., 1989), from ground-based and aircraft-borne instruments. A detailed comparison between techniques and platforms is beyond the scope of this review, but the measurements are in broad agreement, particularly insofar as the very low column abundances of NO_2 within the Antarctic vortex in the spring are concerned. The ground-based measurements allowed study of the diurnal and seasonal variations at a fixed point, while the aircraft measurements provide important latitude gradient information. The airborne measurements demonstrated that a sharp decrease in NO_2 occurred as the Antarctic polar vortex was approached.

The observations of column NO_2 by Keys and Johnston (1986; 1988) revealed that the spring abundances of NO_2 are far smaller than the autumn levels, and the diurnal variation largely absent in spite of the fact that the diurnal variations apparent in the autumn season should be expected to take place in a similar manner during spring. The lower absolute NO_2 abundances and the absence of diurnal variation observed in spring as compared to autumn points towards removal of reactive nitrogen or its sequestration in a reservoir whose lifetime exceeds a few days (e.g., HNO_3). Similar NO_2 observations from the Soviet stations of Molodezhnaya and Mirny (Elokhov and Gruzdev, 1989) showed very low NO_2 column abundances at Molodznaja until late November, 1987, suggesting limited resupply of nitrogen to the polar vortex until about the time of the stratospheric warming in early December in that particular year.

Satellite measurements of NO_2 are also available from the SME, SAGE and SAGE II experiments. These data reveal sharp reductions in NO_2 in the Antarctic winter, with latitudinal gradients that are qualitatively similar to those obtained from the column measurements discussed above. Thomas et al. (1988) described SME observations of the abundance of NO_2. The SME observations are restricted to the altitude range above about 28 km. Thomas et al. (1988) compared these data to observations of the total NO_2 column abundance over McMurdo by Mount et al. (1987), and showed that much of the NO_2 normally expected to be located below 28 km was missing.

HNO_3 column abundances were also measured by infrared absorption methods (Farmer et al., 1987; Coffey and Mankin, 1989; G. Toon et al., 1989). Toon et al. (1989) reported an abrupt drop in the HNO_3 column to very low values within the Antarctic vortex. HNO_3 can be affected by heterogeneous chemistry in several ways. If PSC clouds are present, then a substantial amount of the gas-phase HNO_3 can be incorporated into the clouds and abundances of gas-phase HNO_3 will be low. However, the presence of the clouds is believed to lead to conversion of other forms of reactive nitrogen (N_2O_5, $ClONO_2$ into particulate HNO_3. If the cloud particles then evaporate, HNO_3 abundances should be expected to be unusually high. If, on the other hand, the cloud particles sediment out of the stratosphere, then the HNO_3 and NO_y abundances will remain low until transport processes replace or effectively mix with the 'denitrified' air.

In-situ measurements of NO_y via a chemiluminescence technique demonstrate that substantial, though not complete, denitrification of the Antarctic lower stratosphere (near 18-20 km) occurred in September, 1987 (Fahey et al., 1989a). The very low column abundances of nitric acid observed in 1987 and the vertical extent of very cold temperatures observed in that particular year indicated that this denitrification extended throughout much of the stratosphere. In contrast, the high values of HNO_3 column observed in November, 1978 (Williams et al., 1982) suggest that either a) denitrification was not very extensive in that year (e.g., due to warmer temperatures) or b) that nitric acid was brought in through transport processes, perhaps those associated with the stratospheric warming.

Karcher et al. (1988) report measurements of HNO_3, NO_2, HCl and O_3 column abundances between 64N and 57S in June, 1983 and 1984. The latitude gradients obtained and the similarity with measurements from other studies in Antarctic spring led Karcher et al. to conclude that their observations at 57S revealed significant perturbations due to heterogeneous chemistry. These findings indicate that anomalous chemistry takes place well before springtime and can be evident at latitudes rather far equatorward of the polar vortex. Such perturbations could have important effects on the seasonal and latitudinal variations in ozone depletion.

In-situ measurements of NO and NO_y in the Antarctic vortex were obtained by Fahey et al. (1989a,b). The interpretation of measurements of NO_y can be complicated, since the instrument is more sensitive to the NO_y present in particles than the gas phase. This leads to NO_y "enhancements" when large amounts of NO_y-containing particulate material is present. As discussed in the next section, measured NO_y enhancements during PSC events provide strong evidence that Type 1 PSC's contain a substantial fraction of nitrate and begin to form at approximately the temperatures expected for the nitric acid trihydrate (Fahey et al., 1989b).

Fahey et al. (1989a) also studied the behavior of NO_y on occasions when clouds were not present locally (as indicated by the cloud particle counter measurements). Since N_2O is the source of stratospheric NO_y through the reaction $O(^1D) + N_2O \rightarrow 2NO$, the destruction of N_2O is accompanied by a production of NO_y. Thus, one expects increasing concentrations of NO_y as N_2O decreases, and this indeed occurs for N_2O mixing ratios greater than about 140 ppbv at 450K (about 20 km). However, for N_2O mixing ratios below about 140 ppbv (in the interior of the Antarctic vortex) this behavior breaks down, and substantial decreases in NO_y abundances are found. This strongly suggests denitrification processes. NO_y mixing ratios between 1 and 4 ppbv were observed over a wide range of potential temperatures where values close to 10 ppbv would be anticipated, suggesting that substantial denitrification had occurred over much of the low stratosphere. As noted earlier, these results are consistent with the very low HNO_3 columns observed at the same time by Toon et al. (1989) and Coffey and Mankin (1989).

On two Antarctic flights, Fahey et al. (1989a) measured NO rather than NO_y. They found that the NO mixing ratios decreased as the polar vortex was approached, in contrast to model calculations that predict increasing abundances in the absence of removal processes (e.g., on PSC's). NO mixing ratios inside the chemically perturbed region were on the order of a few tens of parts per trillion by volume or less. Measurements of the diurnal variations of OClO and BrO reported by Solomon et al. (1989) suggest similar values of 10-100 pptv of NO_2. Thus measurements show that very little NO_x was available within the polar vortex, which is a prime requirement for elevated levels of ClO and attendant ozone loss as emphasized earlier.

3. Reactive Nitrogen and Cloud Microphysics

3.1. OBSERVATIONS OF REACTIVE NITROGEN IN PSC'S.

Recent field investigations have shed a great deal of light on the composition of polar stratospheric clouds. A number of studies of PSCs have revealed that the clouds begin to form at temperatures far above the frost point (by as much as 8K), strongly suggesting condensation of trace species other than water. This was first pointed out by Austin et al. (1986a) based on LIMS observations of anomalous infrared radiance, and studied in detail by Poole et al. (1988a,b) using SAM II observations of PSC extinction. Hofmann (1989a,b) and Rosen et al. (1988) also noted the formation of PSCs at temperatures well above the frost point. Arnold and Knop (1989) reported measurements of HNO_3 in the Arctic that set an upper limit for HNO_3 cloud formation temperatures at 195K at 23 km, well below the frost point.

The observations of total reactive nitrogen (NO_y) by Fahey et al. (1989a,b) provide important insights into the composition of PSCs. The measurements of NO_y are somewhat difficult to interpret in the presence of PSC particles, because the instrument has a greater sampling efficiency for NO_y in large particulates than for NO_y in the gas phase. This can lead to substantial, particle size-dependent, NO_y enhancements when large amounts of NO_y-containing particles are present. These enhancements complicate interpretation of the gas-phase component of the signal but also provide important insights to the particle composition, especially when coupled with concurrent observations of the particle size distribution so that their effects can be quantitatively evaluated (see Fahey et al., 1989b).

Airborne measurements of NO_y, total water vapor, temperature, pressure and cloud particles as a function of time on August 17, 1987 by Fahey et al. (1989) demonstrate that cloud particle formation occurred well above the frost point of water vapor. These observations permit identification of cloud "edges", although it should be emphasized that the past temperature history of air parcels also plays a role in determining the onset of cloud formation and should be examined in a detailed analysis. Nonetheless, it is clear that the observed condensation point was far warmer than the frost point, and in general agreement with expectations based on the thermodynamics of the nitric acid trihydrate as proposed by Toon et al. (1986) and quantified by the laboratory study of Hanson and Maursberger (1988a,b). Further, the dramatic NO_y enhancement directly confirms that the particles contain a substantial amount of nitrate. This set of observations therefore provides strong evidence that condensation of HNO_3 plays a critical role in PSC cloud formation.

Observations by Gandrud et al. (1989) support and extend this picture. Gandrud et al. conducted filter sampler measurements of cloud composition from an aircraft platform. Using a dual filter system, Gandrud et al. (1989) deduced the amount of nitrate in both the particulate and gas phases. The instrument is sensitive to HNO_3, $ClONO_2$, N_2O_5, and particulate nitrate. The data obtained on several flights suggest that a substantial fraction of the available nitrate had been taken up on particles.

Pueschel et al. (1989) collected aerosol particles during the same aircraft experiments using a wire impactor technique and subsequently analyzed them for nitrate and chloride. Their results revealed a sharp dependence of condensed nitrate on temperature: nitrate-containing aerosols began to form at temperature below about 193K, in general agreement with expectations based on the thermodynamics of Type 1 nitric acid trihydrate particles.

As expected from theoretical and laboratory studies, in-situ measurements of total NO_y and gas and particulate phase nitrate thus confirm that PSC particles often contain a large amount of nitrate and form at roughly the temperatures expected for the $HNO_3/3H_2O$ system. The formation of nitrate-containing PSC Type 1 clouds is likely to affect the chemistry of the polar lower stratosphere in several important ways: by providing a surface for heterogeneous reactions attainable at temperatures significantly warmer than the frost point and by removing reactive nitrogen from the gas phase (and hence affecting the photochemical balance, at least temporarily). The observations are consistent with suggestions regarding irreversible removal of reactive nitrogen via incorporation onto particles and subsequent sedimentation.

3.2. DENTRIFICATION

The previous section demonstrated that cloud particles containing substantial amounts of reactive nitrogen do form in the Antarctic stratosphere. However, the possibility of denitrification requires not only formation of such particles but also that they grow large enough to sediment fairly rapidly. Thus cloud microphysics plays an important role in the composition of the Antarctic stratosphere. Many recent studies have been concerned with the formation and growth of PSC particles, and the mechanism for dentrification of the stratosphere. These studies have focussed on the formation of two types of polar stratospheric clouds, those composed of the nitric acid trihydrate (formed at temperatures 5-8K above the frost point) and those composed largely of water ice (formed at temperatures below the frost point). Poole (1987) and Poole and McCormick (1988) presented calculations from a two-stage PSC microphysics model which assumed that "Type 1" PSCs form above the frost point as $HNO_3/3\ H_2O$ deposited on frozen background aerosol nuclei, and that "Type 2" PSCs form subsequently (below the frost point) as H_2O ice deposited on Type 1 nuclei. They used vapor pressure relationships extrapolated from laboratory measurements of liquid HNO_3-H_2O mixtures and assumed that no barrier to cloud particle nucleation existed other than the Kelvin vapor pressure elevation factor. Results showed that for slow cooling conditions (0.5K/day), only a fraction (5%) of the background aerosol population would be activated as PSC particles, resulting in a bimodal particle size distribution. For typical Antarctic vapor mixing ratios, the authors reported Type 1 modal radii on the order of 1 μm and Type 2 radii near 4 μm. Poole and McCormick (1988) suggested that sedimentation of these larger Type 2 particles might lead to irreversible removal of HNO_3 from the Antarctic stratosphere. Calculated optical properties for Type 1 PSCs generally agreed well with lidar and SAM II observations. Theoretical calculations of the growth of Type 1 PSCs from the background aerosol were also reported by Hamill et al. (1988), who noted that Type 1 particle radii vary inversely (and markedly) with the fraction of aerosol activated into PSC particles.

The formation and growth of Type 2 PSCs has been discussed in recent papers by Ramaswamy (1988) and O. B. Toon et al. (1989). Ramaswamy specifically addressed the role played by Type 2 PSCs in the dehydration of the Antarctic stratosphere during winter, assuming that the particles form by deposition of H_2O ice directly onto the background aerosol and that the Kelvin effect was the only barrier to particle activation. Although not focusing on particle size, Ramaswamy found that Type 2 particles of radii 2-3 μm formed over the course of several days with temperature decreases below the frost point, and that such a process could lead to extensive irreversible removal of H_2O vapor in the Antarctic. O. B. Toon et al. (1989) presented calculations on the size and lifetime of Type 2 particles forming on Type 1 PSC nuclei assumed to have a modal radius of 0.5 μm. The authors found that an energy barrier to ice nucleation (particle activation) akin to that observed in tropospheric cirrus was necessary in order to explain the dependence of observed Type 2 PSC properties on cooling rate. For cooling rates of several degrees K/day, inclusion of the nucleation barrier led to Type 2 particle radii on the order of 20 μm which could rapidly dehydrate and denitrify the Antarctic stratosphere through sedimentation.

These cloud microphysical modeling studies suggest that $HNO_3/3$ H_2O particles are likely to attain sizes too small for significant sedimentation, while water ice clouds formed below the frost point can probably achieve sizes large enough for rapid sedimentation. This occurs in large part because HNO_3 is present at roughly the 10 ppbv level in the stratosphere, while water vapor abundances are on the order of 5 ppmv. Thus the available HNO_3 vapor leads to limited sizes if a large number of particles grow. Only if a very few particles take up the bulk of HNO_3 mass can their sizes reach levels that will lead to significant sedimentation (Salawitch et al., 1989), a scenario not anticipated based on current cloud microphysical modeling. If only those particles composed largely of water ice are capable of growing large enough to fall out of the stratosphere, then denitrification must be accompanied by dehydration and presumably requires very cold temperatures below the frost point. This would likely allow for extensive dentrification in the Antarctic but limit the dentrification in the relatively warm, Arctic stratosphere and could hence play a significant role in the inter-hemispheric differences in ozone depletion observed. On the other hand, if nitric acid trihydrate particles are capable of attaining sizes sufficient to allow sedimentation, then denitrification is more likely to pervade both polar regions, and would occur without accompanying dehydration. Kelly et al. (1989) report evidence for substantial dehydration of the Antarctic stratosphere along with the denitrification described by Fahey et al. (1989a). These measurements were conducted in September, 1987, and the sedimentation process likely occurred much earlier during the Antarctic winter season. Thus these data do not prove that denitrification and dehydration occurred concurrently and do not definitively establish the microphysical mechanism. Arctic measurements of reactive nitrogen and water vapor under a wide range of temperature regimes are needed to examine this issue.

Recent in-situ measurements of PSC's by Hofmann (1989a) in the Arctic during January 1989 showed in one case a 3-km thick PSC layer near 21 km at temperatures from 186-187 K in which roughly half of the available condensation nuclei grew to radii >0.2 μm, about 1 in 10 grew to radii >1.0 μm, and none grew to radii >5.0 μm. The absence of large ice crystals implied that the observed PSC particles were of the Type 1 class. On a second occasion, thin (300 m thick) PSC layers were observed near 25 km at temperatures near 191 K which consisted predominantly of particles in the 1-2 μm radius range. Further interpretation of these and similar Antarctic data has been presented by Hofmann (1989b), who noted that the small particle mode (r≤0.5 nm) appeared to be associated with fast cooling events such as those due to mountain lee waves or tropospheric anticyclones. In contrast, the thin layers of larger particles contained only a few percent of the available condensation nuclei and were apparently associated with even more rapid cooling events. Salawitch et al. (1989) note that sedimentation of the large particle mode could significantly denitrify the polar stratosphere without much accompanying dehydration if these particles are composed of nitric acid trihydrate. However, Hofmann (1989b) emphasized that such thin layers could not survive long (hours to days) and pointed out that current understanding of cloud microphysics cannot explain their formation. Thus a major current puzzle in polar chemistry is the detailed mechanism responsible for formation of the large particles observed by Hofmann (1989a,b) and a full understanding of their composition and implications for denitrification and dehydration.

4. Summary

Measurements of reactive nitrogen species in Antarctica have been carried out with a variety of techniques and a wide range of platforms. The observational methods include infrared and visible spectroscopy, chemiluminesence and filter sampling. Airborne and ground-based observations concur that substantial perturbations in reactive nitrogen chemistry characterize the Antarctic spring. Observations of NO and NO_2 show greatly decreased abundances. The aircraft data confirms that the in-situ abundances of NO near 15-20 km are much smaller than expected based on gas-phase photochemical models and are broadly consistent with expectations based on PSC chemistry. The suppression of NO_2 inhibits the formation of $ClONO_2$ and HCl, thus allowing any ClO formed by heterogeneous reactions to destroy ozone over prolonged periods.

Theoretical studies suggested that PSC clouds might be composed of binary mixtures of nitric acid and water. The thermodynamic stability of such mixtures was rapidly demonstrated. Measurements of the NO_y and acidic nitrate in the particulate phase confirm that some Antarctic PSC are composed of such compounds, and that they form at temperatures considerably warmer than the frost point. These clouds remove nitric acid from the gas phase and hence further extend the time period over which ClO abundances can remain elevated once formed by PSC's.

Further, observations of the total reactive nitrogen abundance (NO_y) demonstrate that the Antarctic stratosphere has been substantially denitrified. Column measurements of HNO_3 support these findings. The denitrification is accompanied by dehydration in the Antarctic spring and may take place though sedimentation of large PSC particles. The detailed mechanism for denitrification is currently a subject of debate, and it is not known whether it takes place exclusively through nitric acid trihydrate particles, or through water ice particles, or both. Understanding of the mechanism of dentrification and its further implications for ozone depletion in both the south and north polar regions is a focus of current research.

REFERENCES

Anderson, J. G., W. H. Brune, S. A. Lloyd, W. L. Starr, M. Loewenstein and J. R. Podolske, Kinetics of ozone destruction by ClO and BrO within the Antarctic vortex: an analysis based on in-situ ER-2 data, J. Geophys. Res., AAOE special issue, 1989.

Arnold, F., and G. Knop, Stratospheric nitric acid vapour measurements in the cold arctic vortex - implications for nitric acid condensation, Nature, 338, 746, 1989.

Austin, J., E. E. Remsberg, R. L. Jones, and A. F. Tuck, Polar stratospheric clouds inferred from satellite data, Geophys. Res. Lett., 13, 1256, 1986a.

Austin, J., R. R. Garcia, J. M. Russell, S. Solomon and A. F. Tuck, On the atmospheric photochemistry of nitric acid, J. Geophys. Res., 91, 5477-5485, 1986b.

Barrett, J. W., P. M. Solomon, R. L. DeZafra, M. Jaramillo, L. Emmons, and A. Parrish, Formation of the antarctic ozone hole by the ClO dimer mechanism, Nature, 336, 455, 1988.

Callis, L. B., and M. Natarajan, The antarctic ozone minimum: relationship to odd nitrogen, odd chlorine, the final warming, and the 11-year solar cycle, J. Geophys. Res., 91, 10771, 1986.

Coffey, M. T., and W. G. Mankin, Airborne measurements of stratospheric constituents over Antarctica in the austral spring 1987, 2, Halogen and nitrogen trace gases, J. Geophys. Res., AAOE special issue, 1989.

Crutzen, P. J., The influence of nitrogen oxide on the atmospheric ozone content, Q. J. R. Meteorol. Soc., 96, 320, 1970.

Crutzen, P. J., and F. Arnold, Nitric acid cloud formation in the cold antarctic stratosphere: a major cause for the springtime "ozone hole", Nature, 324, 651-655, 1986.

Elokhov, A. S., and A. N. Gruzdev, Total ozone and NO2 observations at Molodeznaja and Mirny stations, Antarctica, in spring 1987 and in summer and fall, 1988, Middle Atmosphere Symposium, Dushanbe, USSR, 1989.

Fahey, D. W., D. M. Murphy, C. S. Eubank, K. K. Kelly, M. H. Proffit, G. V. Ferry, M. K. W. Ko, M. Loewenstein, and K. R. Chan, Measurements of nitric oxide and total reactive nitrogen in the Antarctic stratosphere: observations and chemical implications, J. Geophys. Res., AAOE special issue, 1989a.

Fahey, D.W., Kelly K.K., Ferry G.V., Poole L.R., Wilson J.C., Murphy D.M. and Chan K.R, In situ measurements of total reactive nitrogen, total water and aerosols in Polar Stratospheric Clouds in the Antarctic Stratosphere, J. Geophys. Res, AAOE special issue, 1989b.

Farman, J. C., B. G. Gardiner, and J. D. Shanklin, Large losses of total ozone in Antarctica reveal seasonal ClOx/NOx interaction, Nature, 315, 207, 1985.

Farmer, C. B., G. C. Toon, P. W. Shaper, J. F. Blavier, and L. L. Lowes, Ground-based measurements of the composition of the antarctic atmosphere during the 1986 spring season, I. Stratospheric trace gases, Nature, 329, 126, 1987.

Gandrud, B. W., P. D. Sperry, L. Sanford, K. K. Kelly, G. V. Ferry and K. R. Chan, Filter measurement results from the Airborne Antarctic Ozone Experiment, J. Geophys. Res., AAOE special issue, 1989.

Hamill, P., R. P. Turco, and O. B. Toon, On the growth of nitric and sulfuric acid aerosol particles under stratospheric conditions. J. Atmos. Chem., 7, 287-315, 1988.

Hanson, D. R. and K. Mauersberger, Laboratory studies of the nitric acid trihydrate: Implications for the South Polar stratosphere, Geophys. Res. Lett., 15, 855-858, 1988a.

Hanson, D. R. and K. Mauersberger, Vapor Pressures of HNO3/H2O solutions at low temperatures, J. Phys. Chem., 92, 6167-6170, 1988b.

Hofmann, D. J., Direct ozone depletion in springtime Antarctic lower stratospheric clouds., Nature, 337, 447, 1989a.

Hofmann, D. J., Comparison of stratospheric clouds in the Antarctic and in the Arctic, submitted to Geophys. Res. Lett., 1989b.

Johnston, H. S., Reduction of stratospheric ozone by nitrogen oxide catalysts from supersonic transport exhaust, Science, 173, 517, 1971.

Jones, R. L., J. Austin, D. S. McKenna, J. G. Anderson, D. W. Fahey, C. B. Farmer, L. E. Heidt, K. K. Kelly, D. M. Murphy, M. H. Proffit and A. F. Tuck, Lagrangian photochemical modelling studies of the 1987 Antarctic spring vortex: comparison with observations, J. Geophys. Res., AAOE special issue, 1989.

Karcher, F., M. Amodei, G. Armand, C. Besson, B. DuFour, G. Froment, and J. P. Meyer, Simultaneous measurements of HNO3, NO2, HCl, O3, N2O, CH4, H2O and CO, and their latitudinal variations as deduced from airborne infrared spectrometry, Ann. Geophysicae, 4, 425, 1988.

Kelly, K. K., A. F. Tuck, D. M. Murphy, M. H. Proffitt, D. W. Fahey, R. L. Jones, D. S. McKenna, M. Loewenstein, J. R. Podolske, S. E. Strahan, G. V. Ferry, K. R. Chan, J. F. Vedder, G. L. Gregory, W. D. Hypes, M. P. McCormick, E. V. Browell, and L. E. Heidt, Dehydration in the lower Antarctic stratosphere during late winter and early spring, 1987, J. Geophys. Res., AAOE special issue, 1989.

Keys, J. G., and P. V. Johnston, Stratospheric NO2 and O3 in Antarctica: dynamic and chemically controlled variations, Geophys. Res. Lett., 13, 1260, 1986.

Keys, J. G., and P. V. Johnston, Stratospheric NO2 column measurements from three Antarctic sites, Geophys. Res. Lett., 15, 898, 1988.

Mahlman, J. D., and S. B. Fels, Antarctic ozone decreases: a dynamical cause, Geophys. Res. Lett., 13, 1316, 1986.

McElroy, M. B., R. J. Salawitch, S. C. Wofsy and J. A. Logan, Antarctic ozone: reductions due to synergistic interactions of chlorine and bromine, Nature, 321, 759, 1986a.

McElroy, M. B., R. J. Salawitch, and S. C. Wofsy, Antarctic O3: chemical mechanisms for the spring decrease, Geophys. Res. Lett., 13, 1296, 1986b.

McKenzie, R. L., and P. V. Johnston, Springtime stratospheric NO2 in Antarctica, Geophys. Res. Lett., 11, 73, 1984.

Molina, L. T. and M. J. Molina, Production of Cl2O2 from the self-reaction of the ClO radical, J. Phys. Chem. 91, 433-436, 1987.

Molina, M. J., T.L. Tso, L. T. Molina, and F. C.Y. Wang, Antarctic stratospheric chemistry of chlorine nitrate, hydrogen chloride, and ice: Release of active chlorine, Science, 238, 1253-1257, 1987.

Mount, G. H., R. W. Sanders, A. L. Schmeltekopf, and S. Solomon, Visible spectroscopy at McMurdo Station, Antarctica, 1. Overview and daily variations of NO2 and O3 during austral spring, 1986, J. Geophys. Res., 92, 8320, 1987.

Noxon, J. F., Stratospheric NO2 in the Antarctic winter, Geophys. Res., Lett., 5, 1021, 1978.

Noxon, J. F., Stratospheric NO2: 2, Global behavior, J. Geophys. Res., 84, 5067, 1979.

Ozone Trends Panel (R.T. Watson, M.J. Prather and M.J. Kurylo et al.), Present state of knowledge of the upper atmosphere 1988: an assessment report. NASA reference publ. 1208, available from the National Technical Information Service, Springfield, VA 22161, USA, 1989.

Poole, L. R., Airborne lidar studies of Arctic polar stratospheric clouds, Ph.D. dissertation, University of Arizona, 1987.

Poole, L. R., and M. P. McCormick, Polar stratospheric clouds and the Antarctic ozone hole, J. Geophys. Res., 93, 8423-8430, 1988.

Poole, L. R., M. T. Osborn, and W. H. Hunt, Lidar observations of Arctic polar stratospheric clouds: Signature of small, solid particles above the frost point, Geophys. Res. Lett., 15, 867-870, 1988a.

Poole, L. R., M. P. McCormick, E. V. Browell, C. T. Trepte, D. W. Fahey, K. K. Kelly, G. V. Ferry, R. Pueschel, and R. L. Jones, Extinction and backscatter measurements of Antarctic PSCs, 1987: Implications for particle and vapor removal, NASA CP-10014, 77-79, 1988b.

Pueschel, R. F., K. G. Snetsinger, J. K. Goodman, O. B. Toon, G. V. Ferry, V. R. Oberbeck, J. M. Livingston, S. Verma, W. Fong, W. L. Starr and K. R. Chan, Condensed nitrate, sulfate and chloride in Antarctic stratospheric aerosols, J. Geophys. Res., In press, AAOE special issue, 1989.

Ramaswamy, V., Dehydration mechanism in the Antarctic stratosphere during winter. Geophys. Res. Lett., 15, 863, 1988.

Rosen, J. M., D. J. Hofmann, and J. W. Harder, Aerosol measurements in the winter/spring Antarctic stratosphere, 2, Impact on polar stratospheric cloud theories, J. Geophys. Res., 93, 677-686, 1988.

Salawitch, R. J., G. P. Gobbi, S. C. Wofsy and M. B. McElroy, Dentrification in the Antarctic stratosphere, Nature, 339, 525, 1989.

Solomon, S., R. R. Garcia, F. S. Rowland, and D. J. Wuebbles, On the depletion of antarctic ozone, Nature, 321, 755, 1986.

Solomon, S., The mystery of the Antarctic ozone hole, Rev. Geophys., 26, 131, 1988.

Solomon, S., R. W. Sanders, M. A. Carroll, and A. L. Schmeltekopf, Visible and near-ultraviolet spectroscopy at McMurdo Station, Antarctica, 5. Observations of the diurnal variations of OClO and BrO, J. Geophys. Res., AAOE special issue, 1989.

Thomas, R. J., K. H. Rosenlof, R. T. Clancy, and J. M. Zawodny, Stratospheric NO2 over Antarctica as measured by the solar mesosphere explorer during austral spring, 1986, J. Geophys. Res., 93, 12561, 1988.

Tolbert, M. A., M. J. Rossi, R. Malhotra, and D. M. Golden, Reaction of chlorine nitrate with hydrogen chloride and water at Antarctic stratospheric temperatures, Science, 238, 1258, 1987.

Toon, G.C., C.B. Farmer, L.L. Lowes, P.W. Schaper, J.-F. Blavier and R.H. Norton, Infrared measurements of stratospheric composition over Antarctica during September 1987, J. Geophys. Res., AAOE special issue, 1989.

Toon, O. B., P. Hamill, R. P. Turco, and J. Pinto, Condensation of HNO3 and HCl in the winter polar stratosphere, Geophys. Res. Lett., 13, 1284, 1986.

Toon, O. B., R. P. Turco, J. Jordan, J. Goodman, and G. Ferry, Physical processes in polar stratospheric ice clouds, J. Geophys. Res., AAOE special issue, 1989.

Tung, K. K., M. K. W. Ko, J. M. Rodriguez and N. D. Sze, Are antarctic ozone variations a manifestation of dynamics or chemistry?, Nature, 333, 811, 1986.

Wahner, A., R. O. Jakoubek, G. H. Mount, A. R. Ravishankara and A. L. Schmeltekopf, Remote sensing of daytime column NO_2 during the airborne Antarctic ozone experiment, J. Geophys. Res., AAOE special issue, 1989.

Williams, W. J., J. J. Kosters, and D. G. Murcray, Nitric acid column densities over Antarctica, J. Geophys. Res., 87, 8976, 1982.

Wofsy, S. C., Temporal and latitudinal variations of stratospheric trace gases: a critical comparison between theory and experiment, J. Geophys. Res., 83, 364, 1978.

THE POTENTIAL ROLE OF HO_x AND ClO_x INTERACTIONS IN THE OZONE HOLE PHOTOCHEMISTRY

P.J. CRUTZEN and C. BRÜHL

Max Planck Institute for Chemistry
Air Chemistry Department
POB 3060
D-6500 Mainz
F.R. Germany

ABSTRACT. It is demonstrated by model calculations, that odd hydrogen can play a significant role in ozone destruction in the polar lower stratosphere whenever nitric acid vapor volume mixing ratios are below about 1 *ppbv*. The production of odd hydrogen after sunrise is significantly enhanced by a newly discovered reaction involving ClO and the methyl peroxy radical. Decrease of overhead ozone in the southern hemisphere leads to an increase in OH and hydrogen peroxy radical concentrations and a reduction in tropospheric ozone because of increasing penetration of UV radiation.

1. Introduction

Chlorine is converted from the reservoirs HCl and $ClONO_2$ to its active forms during polar night by heterogeneous reactions on polar stratospheric cloud particles. In the paper it is, however, demonstrated by model calculations, that the speed of ozone depletion after sunrise is not only controlled by the amount of active chlorine produced during polar night; it can also be influenced to a large extent by the enhanced buildup of OH radicals which convert HCl in the gas phase to chemically active $ClOX$ species (Crutzen and Arnold, 1986), setting into effect the ozone destruction cycle involving the ClO dimer

$$
\begin{aligned}
ClO + ClO + M &\rightarrow Cl_2O_2 + M \\
Cl_2O_2 + h\nu &\rightarrow 2\,Cl + O_2 \quad \text{(finally)} \\
Cl + O_3 &\rightarrow ClO + O_2 \quad (2\times)
\end{aligned}
\qquad (1)
$$

$$\text{net}: \ 2\,O_3 \rightarrow 3\,O_2$$

The ozone destruction cycles involving the interaction between HO_x and ClO_x

A. O'Neill (ed.),
Dynamics, Transport and Photochemistry in the Middle Atmosphere of the Southern Hemisphere, 203–212.
© 1990 *Kluwer Academic Publishers*.

species may play also some role in ozone depletion:

$$\begin{aligned} HO_2 + ClO &\rightarrow HOCl + O_2 \\ HOCl + h\nu &\rightarrow OH + Cl \\ OH + O_3 &\rightarrow HO_2 + O_2 \\ Cl + O_3 &\rightarrow ClO + O_2 \end{aligned} \qquad (2)$$

and to a smaller extent

$$\begin{aligned} OH + ClO &\rightarrow HO_2 + Cl \\ Cl + O_3 &\rightarrow ClO + O_2 \\ HO_2 + O_3 &\rightarrow OH + 2O_2 \end{aligned} \qquad (3)$$

During sunlit conditions in polar spring, a newly discovered, fast reaction between ClO and CH_3O_2 ($k = 3\times10^{-12} cm^3$molecule$^{-1} s^{-1}$) detected at our institute (Simon et al, 1989) may lead to high concentrations of OH and HO_2 via reactions in the CH_4 oxidation cycle:

$$\begin{aligned} OH \text{ (or } Cl) + CH_4 + O_2 &\rightarrow H_2O \text{ (or } HCl) + CH_3O_2 \\ ClO + CH_3O_2 &\rightarrow Cl + CH_3O + O_2 \\ CH_3O + O_2 &\rightarrow CH_2O + HO_2 \\ CH_2O + h\nu + 2O_2 &\rightarrow 2HO_2 + CO \quad \text{(30\% probability)} \end{aligned}$$

Starting this reaction chain with OH, a net gain of 0.6 HO_x is provided, starting it with Cl means a net gain of 1.6 HO_x but one active chlorine ($ClOX$) is lost. The reaction cycle can certainly be started by reaction with Cl, as high $OClO$ and ClO concentrations have been discovered during ozone hole conditions and Cl atoms are formed from the photolysis of Cl_2O_2. Furthermore, it is rather likely that the chain involving OH and Cl can be started via the photolysis of $ClOH$.

The buildup of OH and HO_2 is, however, found to be highly sensitive to the level of denoxification, due to the freezing of gaseous HNO_3 into NAT (nitric acid trihydrate, $HNO_3 \cdot 3H_2O$) and ice particles, which is determined by the minimum temperature the air experienced during polar night, and ensuing particle settling. In the lower stratosphere of non-polar latitudes most of the active chlorine from photolysis of the CFCs is converted by nitrogen species and methane to the reservoir species HCl and $ClONO_2$ and is therefore not available for ozone destruction:

$$\begin{aligned} ClO + NO &\rightarrow Cl + NO_2 \\ Cl + CH_4 &\rightarrow HCl + CH_3 \end{aligned}$$

and

$$ClO + NO_2 \xrightarrow{M} ClONO_2$$

Under those conditions the reaction

$$HCl + OH \rightarrow Cl + H_2O$$

is inefficient because odd hydrogen is kept low by nitric acid and nitrogen oxides which destroy OH by catalytic cycles

$$HNO_3 + OH \rightarrow NO_3 + H_2O$$
$$NO_3 + h\nu \rightarrow NO_2 + O$$
$$OH + NO_2 \overset{M}{\rightarrow} HNO_3$$
$$\text{net}: \ 2OH \rightarrow H_2O + O$$

and

$$HO_2 + NO_2 \overset{M}{\rightarrow} HO_2NO_2$$
$$OH + HO_2NO_2 \rightarrow O_2 + H_2O + NO_2$$
$$\text{net}: \ OH + HO_2 \rightarrow H_2O + O_2$$

In the Antarctic, and sometimes also Arctic, during winter and spring, however, the temperatures are low enough, that most of the gaseous nitric acid can freeze out to form polar stratospheric cloud particles. On these particles active chlorine is formed by heterogeneous reactions of HCl with $ClONO_2$ and N_2O_5, also leading to the removal of NO_x. Dependent on temperatures and cloud microphysical processes, individual layers in the lower stratosphere may, therefore, become more or less denoxified , denitrified and dehydrated by settling of the particles. This will also allow for a rapid buildup of OH radicals, and enhanced production of chemically active $ClOX$.

2. Results for the lower stratosphere

In the sensitivity studies we have carried out with our one-dimensional time-dependent chemical model for the latitude of 75°S we do not simulate heterogeneous reactions in detail because of too large uncertainties. Instead, we assume, that $ClONO_2$ and/or $ClOH$, which are transported downward from the middle stratosphere, have reacted with HCl to form Cl_2 during the polar night (at day 210 in the figures), since we are interested in its effects after sunrise. After day 220, transport is switched off in the model, while the amount of gaseous HNO_3 is controlled by the minimum temperature and water vapor concentrations which the air parcel has gone through, using the expression for the HNO_3 vapor pressure over HNO_3-trihydrate and ice particles as given in Hanson and Mauersberger (1988). We, therefore, assumed, that all particles containing HNO_3 have fallen out from a particular layer during polar night. Our study is mainly aimed to show that HO_x chemistry can play a substantial role in the lower post-winter stratosphere, and how sensitive this is to assumed minimum temperatures and remaining HNO_3 concentrations.

The temporal developments of the mixing ratios of the active chlorine species and the HO_x radical concentrations at the $64\,hPa$ level (about $18\,km$) for the mid-1980th are shown in the upper parts of Fig. 1, volume mixing ratios of ozone,

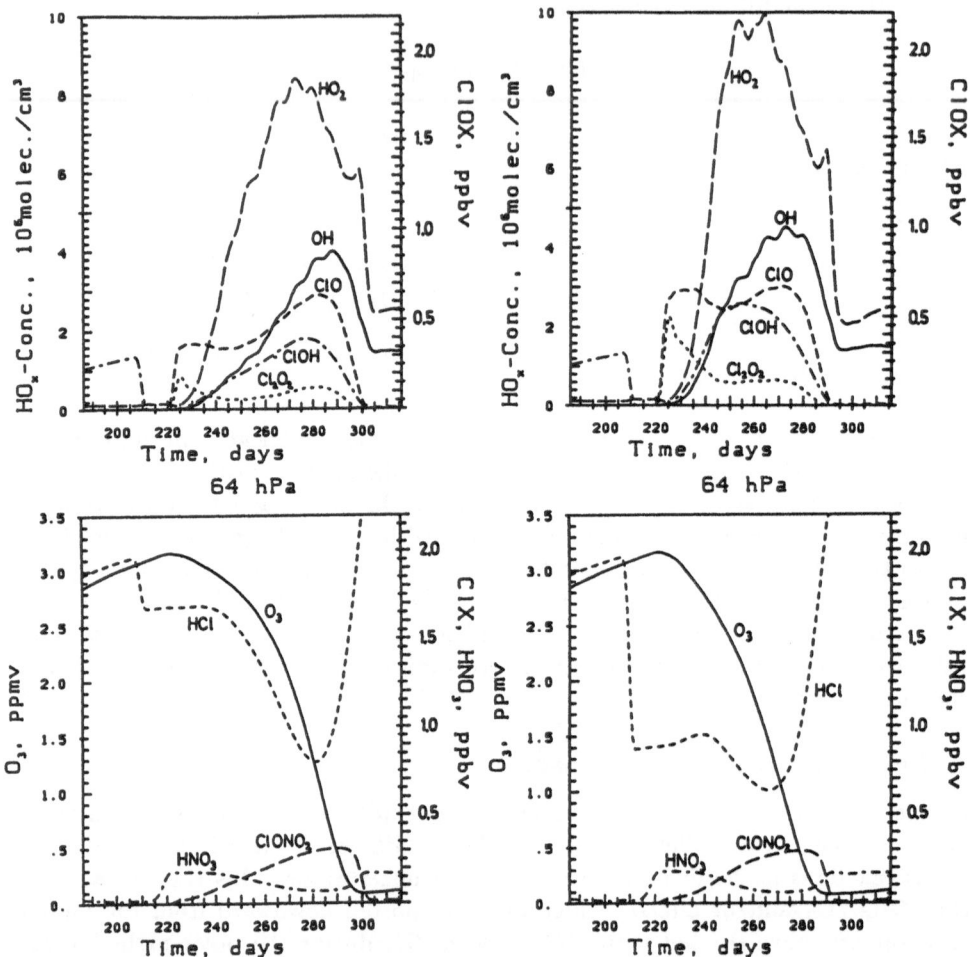

Fig. 1. Chemical development of ozone, HNO_3 and chlorine and odd hydrogen species in polar spring at a latitude of 75°S with the reaction $ClO + CH_3O_2$ and different initial $ClOX$. About 0.3ppb of HCl are assumed to be converted by heterogeneous reactions (day 210, left part), or about 1.1ppb (right part); HNO_3 corresponding to 188.6K and 2.6ppmv H_2O.

gaseous HNO_3 and the chlorine reservoir species in the lower parts. After polar sunrise a rapid buildup of HO_x is calculated. With the concentrations shown for October (upper left part of Fig.1), the two odd hydrogen ozone destruction cycles (Eqs. 2 and 3) together are at least as efficient as the ClO-dimer cycle (Eq. 1) in the particular layer. Very low HNO_3 concentrations of about 0.2ppb (corresponding to 188.6K and 2.6ppm H_2O) are a prerequisite for this fast ozone depletion.

In the scenario in the right part of Fig. 1 is assumed, that 1.1ppbv of HCl

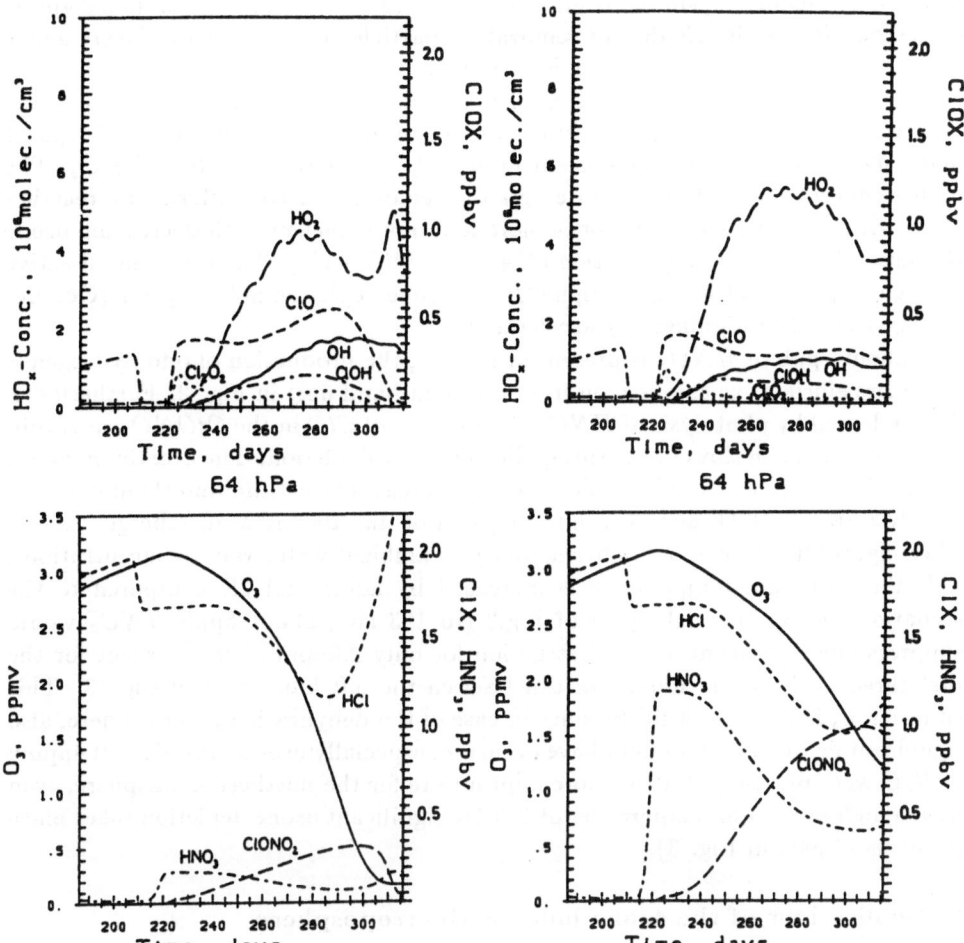

Fig. 2. The effects of gaseous HNO_3 and of the reaction $ClO + CH_3O_2$. Left hand side: About 0.3ppb of HCl artificially converted to $ClOX$ and HNO_3 like in Fig. 1, without the new reaction; right hand side: all reactions, initial $ClOX$ like in Fig. 1, HNO_3 corresponding to 191.1K and 2.6ppmv H_2O.

is converted to active chlorine before sunrise. The amount of 0.8ppbv exceeding the sum of $ClONO_2$ and $ClOH$ is added to simulate the potential effect of the heterogeneous reaction $N_2O_5 + HCl_s \rightarrow ClNO_2 + HNO_{3s}$ and to get high concentrations of ClO just after sunrise. Comparison of the left and right hand parts of Fig. 1 demonstrates, that some weeks after sunrise the concentrations of active chlorine are nearly independent of the initial conditions set by heterogeneous reactions. However, starting with more chlorine, initial ozone depletion is faster.

In the other layers between about 13 and 21km, which might be interesting for 'ozone hole chemistry', similar results are found. In all these layers less than 0.4ppbv of gaseous HNO_3 is left due to removal by particle settling. In the layers above 22km normal gas phase chemistry is assumed.

With the mechanism proposed in Crutzen and Arnold (1986) also some ozone depletion is calculated (Fig. 2), but comparison of the left hand parts of Figs. 1 and 2 shows the significant enhancement of HO_x production by $ClO+CH_3O_2$. The higher concentration of OH counteracts the removal of active chlorine by reaction of Cl with methane which becomes more and more efficient with decreasing ozone (because of the competing reaction $Cl + O_3 \rightarrow ClO + O_2$). The recycling of active chlorine implies, that in case of high OH the ozone depletion in late spring (October in Figs 1 and 2) is significantly accelerated.

If about 1ppb of HNO_3 is present at sunrise, the production of odd hydrogen is much less efficient, as can be seen from the right hand part in Fig. 2. Furthermore, NO_2 released by photolysis of HNO_3 ties up a lot of ClO in the $ClONO_2$ reservoir. This causes a much slower but still significant ozone depletion. The drastic change of chemical development can be achieved by an increase of the (minimum) temperature of only 2.5K to 191.1K as can be seen from Fig. 1 of Hanson and Mauersberger (1988). Their figure shows also the high sensitivity to ambient water vapor concentrations.

If the minimum temperature is increased by additional 2K compared to the scenario given in the right part of Fig.2 (to 193.1K), about 5ppb HNO_3 would suppress any significant ozone destruction for only 2.6ppm of H_2O except for the first three weeks after sunrise, as can be seen the left hand part of Fig. 3. This situation is, however, not likely, since in case of the dehydration assumed here, also significant denitrification might have occurred, especially over Antarctica. If 4ppmv of H_2O were present, which is more appropriate for the northern hemisphere, even with a high minimum temperature of 193.1K significant ozone depletion takes place (right hand part in Fig. 3).

3. Implications of the ozone hole for the troposphere

Total ozone has decreased nearly everywhere in the southern hemisphere when the average values for the years 86/87 and 79/80 are compared (Watson et al, 1988; Sze et al, 1989). The most likely explanation is the dilution effect of the ozone hole. Less ozone in the lower stratosphere enables more UV-B radiation to penetrate into the troposphere which enhances the production of OH there and destruction of O_3 by $O(^1D) + H_2O$, and $HO_2 + O_3 \rightarrow OH + 2O_2$. In our two-dimensional photochemical model (Gidel et al, 1983) we imposed the observed ozone depletions in the lower stratosphere in the optical path calculations needed for the computation of photolysis rates. Comparison with a standard model run shows a significant increase of OH and $O(^1D)$ in high southern latitudes. For example, as shown in Fig. 4, tropospheric OH rises by up to about 15% in polar summer, when the observed ozone depletions are imposed. This is accompanied by an ozone depletion exceeding 10% in the lower troposphere as can be seen from Figs. 4 and 5. In

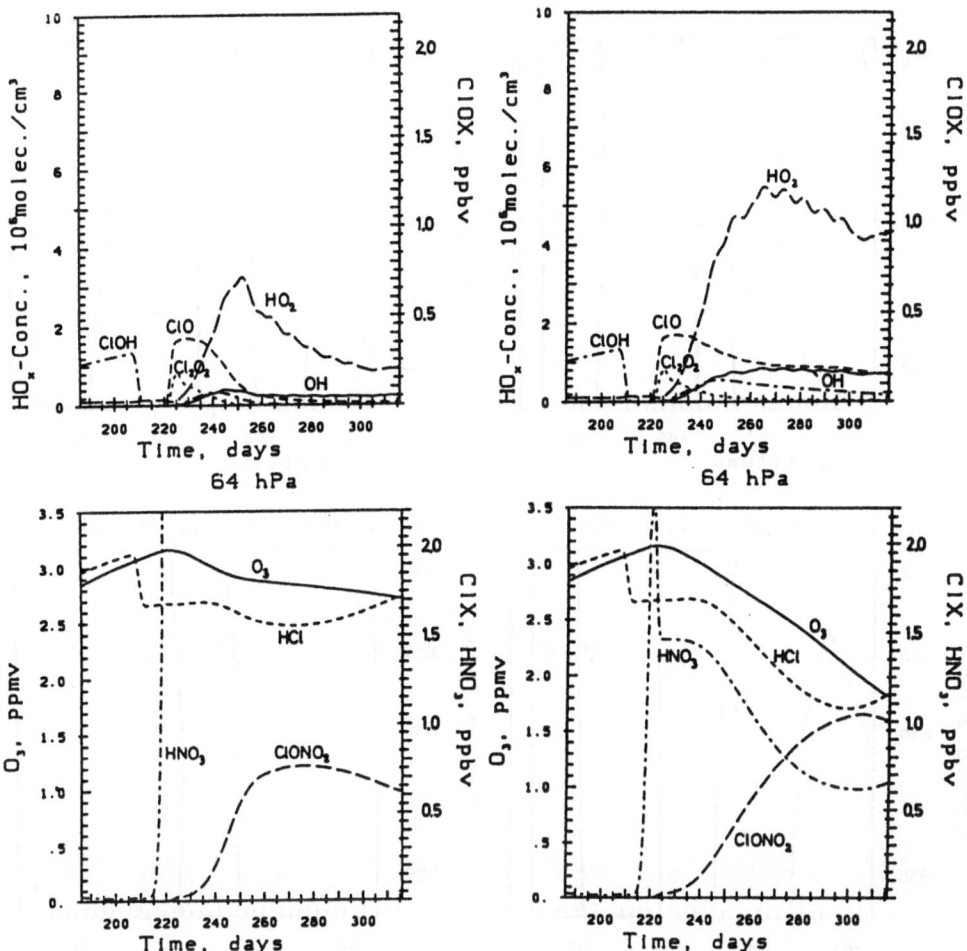

Fig. 3. The effect of ambient H_2O on HNO_3 vapor pressure and on ozone depletion for 193.1K, initial $ClOX$ like in Fig. 1. Left hand side: 2.6ppmv H_2O, right hand side: 4ppmv H_2O.

Fig. 5 the seasonal changes of lower tropospheric ozone in the atmosphere of 1987 with and without imposed stratospheric ozone depletions are intercompared. The results for ozone changes in the southern hemisphere are similar to a time-dependent calculation from 1979 to 1987 where from year to year a linear increase in imposed ozone depletion is assumed. Methane increases during that time as assumed in Brühl and Crutzen (1988).

4. Conclusions

It is shown by model calculations, that strongly enhanced concentrations of odd hy-

210

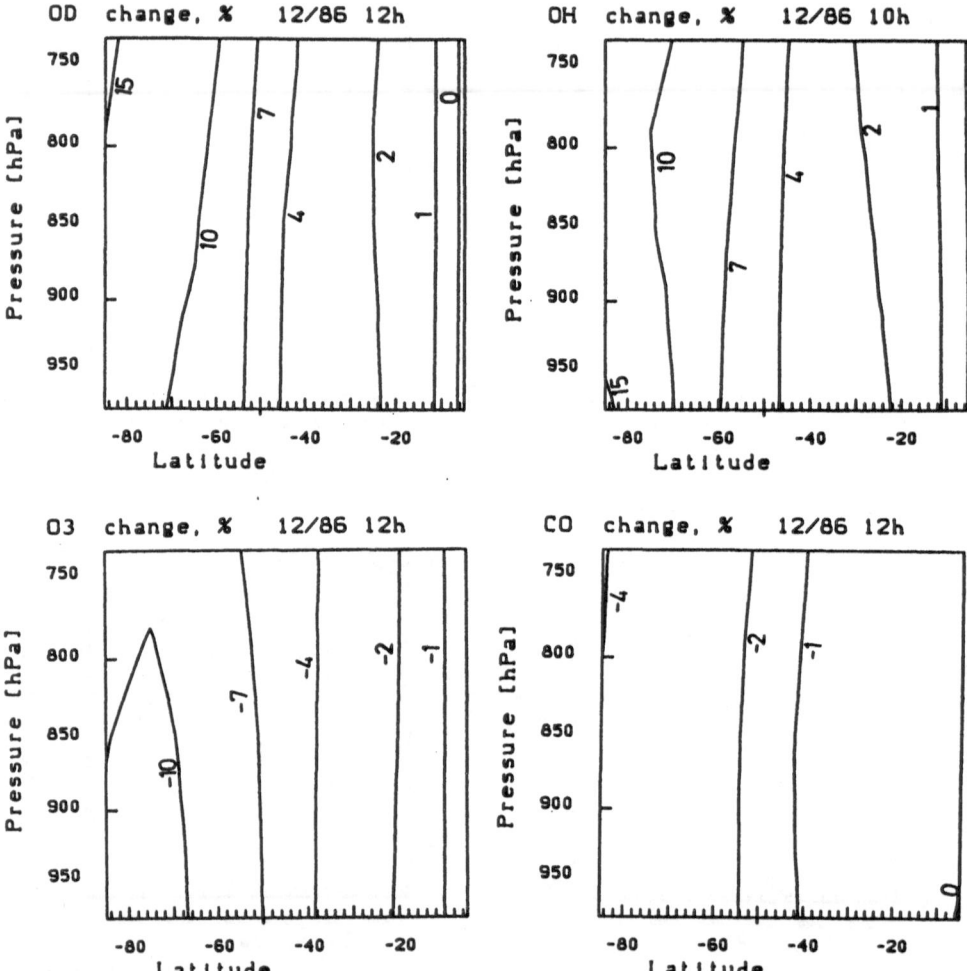

Fig. 4. Changes in tropospheric OH, $O(^1D)$, O_3 and CO in austral summer induced by the ozone hole, as calculated with a two dimensional model.

drogen due to a new reaction involving ClO and CH_3O_2 may significantly contribute to ozone depletion in polar spring by recycling of active chlorine. The buildup of odd hydrogen was found to be very sensitive to the abundance of gaseous HNO_3, i.e. to the amount of denitrification and dehydration of the air during polar night. Increases of odd hydrogen because of more penetrating UV due to observed ozone depletions in the stratosphere may lead to tropospheric ozone depletions by up to 10% in summer in high southern latitudes.

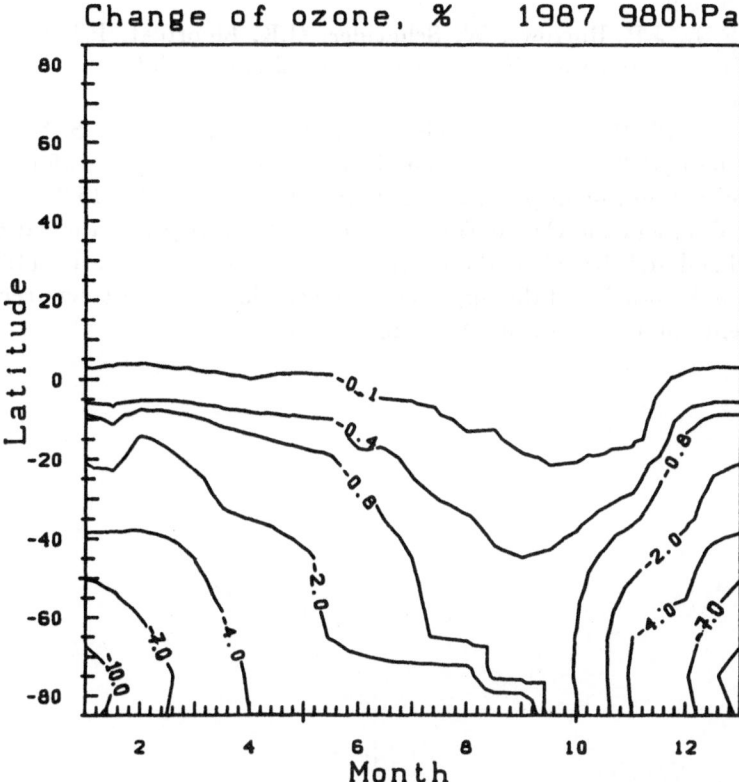

Fig. 5. Seasonal changes in lower tropospheric ozone due to depletions in total ozone as observed between 1979 and 1987 (compared to a model run without imposed depletions).

5. References

Brühl, C. and P.J. Crutzen (1988) 'Scenarios of possible changes in atmospheric temperatures and ozone concentrations due to man's activities, estimated with a one-dimensional coupled photochemical climate model'. *Climate Dynamics 2*, 173-203.

Crutzen, P.J. and F. Arnold (1986) 'Nitric acid cloud formation in the cold Antarctic stratosphere: a major source for the springtime 'ozone hole''. *Nature 324*, 651-655.

Gidel, L.T., P.J. Crutzen, J. Fishman (1983) 'A two-dimensional photochemical model of the atmosphere. I: Chlorocarbon emissions and their effect on stratospheric ozone'. *J. Geophys. Res.* **88**, 6622-6640.

Hanson, D. and K. Mauersberger (1988) 'Laboratory studies of the nitric acid trihydrate: implications for the south polar stratosphere'. *Geophys. Res. Lett.*

15, 855-858.

Simon, F.G., J.P. Burrows, W. Schneider, G.K. Moortgat, P.J. Crutzen (1989) 'Study of the reaction $ClO + CH_3O_2 \rightarrow$ products at 300K'. *J. Phys. Chemistry*, in press.

Sze, N.D., M.K.W. Ko, D.K. Weisenstein, J.M. Rodriguez, R.S. Stolarski, M.R. Schoeberl (1989) 'Antarctic ozone hole: possible implications for ozone trends in the southern hemisphere'. *J. Geophys. Res.* **94**, 11521-11528.

Watson, R.T. and the Ozone Trends Panel, M.J. Prather and the Ad Hoc Theory Panel and M.J. Kurylo and the NASA Panel for Data Evaluation (1988) Present state of knowledge of the upper atmosphere 1988: an assessment report. NASA reference publication 1208, Washington, DC.

OZONE TRANSPORT IN THE SOUTHERN HEMISPHERE

K. K. TUNG and H. YANG
Department of Applied Mathematics
University of Washington, Seattle, WA 98195, USA

Extended Abstract

We discuss the mechanism of tracer transport in the stratosphere, with emphasis on the effect of transport on ozone concentration in the Southern Hemisphere.

In the lower to middle stratosphere, it is likely that large-scale planetary waves are responsible for the global scale transport of chemical tracers, angular momentum and heat, and these waves mix predominantly along isentropic surfaces (Mahlman et al, 1984), with planetary wave "breaking" (McIntyre and Palmer, 1983) being one of the processes of irreversible mixing. One consequence of the wave mixing process is that ozone is transported to high latitudes. Another effect is that because of the transport of heat the high (low) latitude region becomes warmer (colder) than it would have been in the absence of wave transport. High latitude region consequently cools radiatively. Thus air subsides diabatically over high latitudes and rises over low latitudes. This diabatic circulation further transports tracers *across* the isentropes. Isentropic wave mixing and diabatic mean circulation are simply two manifestations of the same transport process (See also Haynes and McIntyre, 1987). In the zonally averaged framework, the former is parameterized as diffusion with an isentropic mixing coefficient K_{yy}, and the latter as a diabatic advection proportional to $\bar{Q}$, the net radiative heating rate. The fact that K_{yy} and $\bar{Q}$ are related to each other has already been pointed out previously. (Mahlman, 1985; Newman et al, 1986; and Plumb and Mahlman, 1987).

The validity of the K-theory can be shown for the case of small amplitude waves (Plumb, 1979; Matsuno, 1980; Tung, 1982), and also appears to hold approximately for moderate amplitude waves in a General Circulation Model (Plumb and Mahlman, 1987). However, for the large amplitude atmospheric waves that "break" and mix in the process, it is only a conjecture that the end result can be mimiced by a K-diffusion process in some gross sense. Some tests are needed.

Assuming downgradient diffusion of Ertel's potential vorticity on an isentropic surface, Yang et al (1990a) deduced the K_{yy}-field corresponding to an observed NMC

A. O'Neill (ed.),
Dynamics, Transport and Photochemistry in the Middle Atmosphere of the Southern Hemisphere, 213–215.
© 1990 *Kluwer Academic Publishers.*

temperature (and hence to the real atmospheric angular momentum) field. The relationships between the mean angular momentum budget, the wave driving (E-P flux divergence), and the isentropic flux of Ertel's potential vorticity have been derived by Tung (1986) and Andrews et al (1987) using isentropic coordinate formulation. The same NMC temperature field is also used, together with model determined ozone, to calculate the net radiative heating, $\bar{Q}$.

A coupled 2-D model is constructed that utilizes the consistent set of K_{yy} and $\bar{Q}$ as transport parameters to simulate simultaneously the distributions of a number of tracer gases including ozone. The simulated results are then compared with satellite observations. These are presented in Yang et al (1990b). The overall favorable comparison of the simulated and the observed appears to indicate the reasonableness of the adopted formulation and the approximate validity of the K-theory in simulating the effect of large-scale mixing process.

Although, as mentioned above, the transport process of K_{yy} and $\bar{Q}$ can not be separated, it is useful for diagnostic purposes to assess the contributions of each separately. It appears that major seasonal and latitudinal variations of column ozone are generated by the mean diabatic circulation. The subpolar ozone maximum in the Southern Hemisphere late winter and spring is related to poleward transport from tropics, which is strongest through the winter accumulating into a maximum in early spring. In the Northern Hemisphere, the poleward transport is even stronger as a consequence of stronger wave driving (and larger K_{yy}, see Yang et al, 1990a). These stronger wave events result in mid-winter warmings and ozone transport into the polar region. As a consequence, the ozone maximum occurs over the pole in the Northern Hemisphere and appears earlier in season than in the Southern Hemisphere.

The K_{yy} field that was deduced from NMC temperature field (Yang et al, 1990a) has a maximum in the midlatitude regions and a minimum over the equatorial region as well as a minimum over both poles. If the K_{yy} field is taken to be a constant instead, at a value of the global average of the variable K_{yy}, one finds that the ozone values will be higher in the tropics, at about 300 Dobson units instead of the 260 Dobson units observed and also modeled using a variable K_{yy} field (Yang et al, 1990b). The polar ozone values also become higher. It is thus seen that the high gradient region across the polar vortex in the Southern Hemisphere is maintained only in the absence of strong wave mixing, which is usually the case in the Southern Hemisphere. There are however years when a moderate increase in wave mixing can transport enough heat and ozone into the polar vortex to disrupt the "containment vessel" and the heteorogenous chemical reactions that require cold temperatures. Also, since the polar area is small in comparison with midlatitude area, year-to-year variations of polar ozone can be quite sensitive to moderate fluctuations of wave mixing process outside the polar vortex. Such variations are possibly further amplified in recent years by enhanced chemical destruction of ozone during cold years.

References

Andrews, D. G., J. R. Holton and C. B. Leovy (1987) Middle Atmosphere Dynamics, Academic press, 489 pp.

Haynes, P. H., and M. E. McIntyre (1987) 'On the evolution of vorticity and potential vorticity in the presence of diabatic heating and frictional or other forces', *J. Atmos. Sci.*, **44**, 828-841.

Mahlman, J. D. (1985) 'Mechanistic interpretation of stratospheric tracer transport', *Adv. Geophys.*, **28A**, 301-323.

Mahlman, J. D., D. G. Andrews, D. L. Hartmann, T. Matsuno, and R. G. Murgatroyd (1984) 'Transport of trace constituents in the stratosphere', *Dynamics of the Middle Atmosphere*, J. R. Holton and T. Matsuno (Eds.), Terra Scientific Publishing, Tokyo, Japan, 387-416.

Matsuno, T. (1980) 'Lagrangian motion of air parcels in the stratosphere in the presence of planetary waves', *Pure Appl. Geophys.*, **118**, 189-216.

McIntyre, M. E., and Palmer (1983) 'Breaking planetary waves in the stratosphere', *Nature*, **305**, 593-600.

Newman, P. A., M. R. Schoeberl, and R. A. Plumb (1986) 'Horizontal mixing coefficients for two-dimensional chemical models calculated from National Meteorological Center data', *J. Geophys. Res.*, **91**, 7919-7924.

Plumb, R. A. (1979) 'Eddy fluxes of conserved quantities by small-amplitude waves', *J. Atmos. Sci.*, **36**, 1699-1704.

Plumb, R. A., and J. D. Mahlman (1987) 'The zonally averaged transport characteristics of the GFDL general circulation/transport model', *J. Atmos. Sci.*, **44**, 298-327.

Tung, K. K. (1982) 'On the two-dimensional transport of stratospheric trace gases in isentropic coordinates', *J. Atmos. Sci.*, **39**, 2330-2355.

Tung, K. K. (1986) 'Nongeostrophic theory of zonally averaged circulation. Part I: Formulation', *J. Atmos. Sci.*, **43**, 2600-2618.

Tung, K. K. (1987) 'A coupled model of zonally averaged dynamics, radiation and chemistry', in *Transport Processes in the Middle Atmosphere*, G. Visconti and R. R. Garcia (Eds.), D. Reidel.

Yang, H., and K. K. Tung (1990a) 'Nongeostrophic theory of zonally averaged circulation, Part II: Eliassen-Palm flux divergence and isentropic mixing coefficient', *J. Atmos. Sci.*, **47**, 215-241.

Yang, H., E. P. Olaguer and K. K. Tung (1990b) 'Simulation of the present-day chemical composition of the stratosphere using a coupled 2-D model in isentropic coordinates', *J. Atmos. Sci.*, accepted.

FUTURE OBSERVATIONS OF THE MIDDLE ATMOSPHERE

John C. Gille
National Center for Atmospheric Research
P. O. Box 3000
Boulder, CO 80303

ABSTRACT Future prospects for observation of the middle atmosphere over the next decade are reviewed. Continuations of present measurements, and operational satellite data are discussed, followed by ground based measurements, the Upper Atmosphere Research Satellite (UARS), and the Earth Observing System (Eos). The amount of data, number of quantities measured, and quality of data will allow many new problems to be addressed.

1. Introduction

A number of recent developments have lead to considerable optimism about future observations of the middle atmosphere. Plans are being made for the acquisition of more extensive data sets than in the past, with measurements of additional quantities or constituents, at improved resolution. Taken together there is an anticipation that the limitations of the present will be alleviated. Here, an attempt will be made to look ahead for one solar cycle, to the year 2000. A difficulty in discussing the future is that many of these plans are only on paper, and readily subject to change. Consequently, a prospective view such as this is subject to becoming outdated rather soon.

Here we will concentrate on major new systems, which entail significant expenditures and long range plan planning. However, important advances have often come from smaller, more individual instruments, which appear quickly and with little advance notice. By their nature, these are not subject to anticipation.

Before discussing the measurement systems themselves, it is worth asking what such measurements will be used for. Initially emphasis will probably be on mechanisms, in order to continue to improve our understanding of basic atmospheric processes that determine the state and evolution of the middle atmosphere, and that lead to improvements in our ability to model its future.

Simultaneously, increasing emphasis will be placed on the use of these measurements for monitoring, or determining the extent of any changes taking place in the atmosphere, and determining their causes. The connection between the two objectives is the belief that, if we understand the basic mechanisms, it will be sufficient later to monitor key variables (such as temperature and ozone), and the inputs that may be changing, (such as CFC's, methane, nitrous oxide, solar UV, etc.) in order to observe and understand change. There will also be increasing emphasis on the relationship of the middle atmosphere to the complete Earth system and other aspects of global change. In addition, some measurements will be acquired for purposes of making improved weather forecasts, and for space operations.

Measurements of the middle atmosphere have been made from the ground, aircraft, balloons, rockets, and satellites. Here we will concentrate on ground-based and satellite techniques, as these appear to be where most of the future "global" observations will be obtained. The following discussion begins with present observations that are continuing, and progresses on into the near and more distant future.

A. O'Neill (ed.),
Dynamics, Transport and Photochemistry in the Middle Atmosphere of the Southern Hemisphere, 217–225.
© 1990 *Kluwer Academic Publishers.*

2. Ongoing Research Observations

There are several research observations now being obtained that will extend into the future. The longest series of present satellite measurements are those from the Total Ozone Mapping Spectrometer (TOMS) on Nimbus 7, which is still operating on Nimbus 7 11 years (or one solar cycle) after launch in October, 1978. The instrument is described by Heath *et al.* (1978) and WMO (1990). It measures the backscattered solar radiation and the solar output in the near ultraviolet (300-380 nm); from the resulting albedo the total ozone may be derived. The spatial resolution is about 50 km at nadir, and because the instrument scans across track, complete global coverage is obtained for all latitudes for which the sun is a few degrees above the horizon-i.e. to 65 degrees latitude in the winter hemisphere during solstice months. TOMS measurements have played a major role in the analysis of trends in total ozone, and in studies of the Antarctic ozone hole. Because it has far exceeded its design life, the future length of this data record is uncertain. However, NASA has plans to fly 3 TOMS instruments during the 1990's, on Soviet, US, and Japanese satellites. Recent developments in understanding the drift of the instruments in flight now allow these data to stand on their own, as an independent source of information on global trends in total ozone. In addition, they allow one instrument to be inter-related to the next.

The Stratospheric Aerosol Monitor II (SAM II) is a second instrument on Nimbus 7 that is still functioning. The instrument has been described by McCormick *et al.* (1979) The SAM II measures the attenuation of sunlight through the atmosphere to determine the amount of aerosols in the high latitude stratosphere. Again, it has been in operation since 1978, and has provided extensive data on the aerosol content of the polar stratosphere, including the presence of polar stratospheric clouds and their role in polar ozone depletion.

The Stratospheric Aerosol and Gas Experiment II (SAGE II) experiment on the ERBE satellite similarly uses the solar occulatation principle to derive the vertical distributions of ozone, nitrogen dioxide, water vapor and aerosols from 65 N to 65 S, and from 20 km to 55 km altitude. The experiment has been described by Mauldin *et al.* (1985) and WMO (1990). These data have also been used in studies of trends in the vertical distribution of ozone.

3. NOAA Operational Data

Operational temperature data obtained by the U.S. National Oceanic and Atmospheric Administration (NOAA) have also been used extensively in middle atmospheric research. NOAA makes use of sounders on two spacecraft in low Earth orbit (~ 800- 1000 km altitude). The High-resolution Infrared Radiation Sounder (HIRS), the Microwave Sounding Unit (MSU) and the Stratospheric Sounding Unit (SSU) provided by the United Kingdom make up the sounding system. The latter has weighting functions that peak up to 2 mb (~45 km altitude). These measurements have been obtained since the launch of TIROS N in late 1978 and the subsequent NOAA spacecraft. For all of these instruments, the weighting functions are of the order of 2 scale heights wide, and true vertical resolution of the retrieved temperatures is about 1 scale height.

Beginning with NOAA K, planned for launch in about 1993 or 1994, the SSU and MSU will be replaced by the Advanced Microwave Sounding Unit (AMSU),which has weighting functions that are slightly narrower than the infrared weighting functions, but for which the highest weighting function peaks below 2 mb. There is concern that from then until the advent of a new system in the late 1990's there will be no operational data covering the mesosphere. (NOAA is considering an additional module, termed AMSU-C, to provide operational coverage above the 2 mb level, but no decision has been made at this time.)

In addition to the temperature sounders, since 1984 the spacecraft that obtains afternoon data has carried a second version of the Solar Backscatter Ultraviolet Spectrometer (SBUV/2) for obtaining operational measurements of total ozone and the vertical ozone profile. The weighting functions are similar to those of the SBUV on Nimbus 7 (WMO 1990). Analysis in the latter reference indicates that the ozone profile data contain independent information between 1 and 16 mb, or roughly 28 to 48 km. Unfortunately, there have been problems with both of the instruments now in orbit, and no data have yet been released for scientific study. Should these problems be resolved, the data will add greatly to the long term data record, and the ability to detect long term trends in the mean ozone profile.

NOAA plans to launch its own free-flier (spacecraft) to obtain afternoon measurements in the late 1990's (perhaps around 1998). This platform will carry a suite of operational instruments, including an operational version of the TOMS for total ozone measurements.

4. Ground-Based Measurements

In the past many important measurements of the middle atmosphere have been made from the ground. The Dobson spectrophotometers, which can measure total ozone as well as its vertical distribution (using the Umkehr technique) were the first of these. Ground-based spectrometers have provided measurements of total column amounts, and occasionally some information on the vertical distribution, of trace species above a few measurement sites. More recently radar techniques, including the measurements by the Mesosphere-Stratosphere-Thermosphere (MST) radars, have provided data on the winds in the lower stratosphere and upper mesosphere. Now lidar measurements are allowing frequent determinations of the vertical distribution of temperature, ozone, and winds, again above a few locations.

Most of the present ground-based observations of the middle atmosphere will continue.However, plans are now underway to collect some of the most powerful modern methods at a few ground based observatories, and to group these into a Network for the Detection of Stratospheric Change (NDSC). The objectives of such a network are:
1. To provide the earliest possible detection of changes in the stratosphere, and the means to understand them;
2. To permit scientific studies, such as studies of the temporal and spatial variability of atmospheric composition and structure; and
3. To provide the basis for validation and complementary measurements to use in conjunction with satellite systems.

A number of measurements have been identified as being of the highest priority to such a network, in the following order of importance:

Column ozone
Vertical ozone profile (0-70 km)
Temperature (0-70 km)
Vertical profile of ClO
Vertical profile of H_2O
Vertical distribution of aerosols
Vertical profile or column of NO_2
Stratospheric column of HCl
Vertical profiles of long-lived tracers CH_4 and N_2O
Other species (HNO_3, OH, $ClONO_2$)

Such a network is now in the initial stages of organization. The first locations will probably be in Europe, and in Hawaii, with additional stations in the Southern Hemisphere and tropics to be selected later.

5. Upper Atmosphere Research Satellite (UARS)

The UARS is a mission specifically designed to improve our understanding of the atmosphere above the tropopause, with primary emphasis on the stratosphere and mesosphere, regions where there is expected to be susceptibility to change by natural and anthropogenic perturbations.

The goals of the UARS project are (Reber, 1985) "to understand the mechanisms that control upper atmosphere structure and variability, the response of the upper atmosphere to natural and anthropogenic perturbations, and to define the role of the upper atmosphere in climate and climate variability."

During the definition of this mission, several scientific areas were identified as critical to achieving these goals. They include studies of energy input and loss, global photochemistry, dynamics, and the couplings among these processes, and among atmospheric regions. Based on these questions, the measurement requirements called out measurements of energy inputs as well as the vertical distributions of temperature, species concentrations, and, for the first time, directly observed winds. The measurements were required to be global in extent, with a latitudinal spacing of no more than 500 km, and a vertical resolution of half a scale height.

The UARS Observatory is scheduled to be launched by the space shuttle in the second half of 1991 into a 600 km orbit with a 57° inclination. Unlike the orbits of the satellites described previously, this orbit is not sun-synchronous, but precesses with a period of about 33 days. For thermal stability, and to orient the solar panels that provide the power, the spacecraft will be rotated about its yaw axis every 33 days, so one side of the observatory will always be away from the sun. This "cold" side is the preferred location for many of the instruments. A result is that a limb sounder that looks perpendicular to the velocity axis on the dark side will view to 80° latitude in one hemisphere, but only to 34° in the other hemisphere. Coverage of latitudes between 34° and 80° will therefore not be continuous.

The complement of instruments is listed in Table 1. It includes two instruments for the measurement of the solar ultraviolet irradiance and one instrument for the measurement of X-ray and particle energy inputs, four instruments for the measurement of species and temperature profiles, and two instruments for the measurement of winds.

For the instruments measuring the solar UV, one of the most difficult problems is that of accurate calibration, and maintenance of that calibration in orbit. The laboratory calibrations of SOLSTICE and SUSIM will be similar, and will employ standard sources that have been measured by several groups. The approach to long-term stability is rather different, however. SUSIM will carry calibration lamps, while SOLSTICE will compare its observations with an average of a large number of constant UV stars. The expectation is that this average will be stable over much longer periods than the life of SOLSTICE.

All the sounders view the limb, a necessity to achieve the high vertical resolution required by UARS. The ISAMS, as its name implies, is developed from the SAMS on Nimbus 7. It is an Infrared limb scanner that for the first time will make use of a small mechanical refrigerator to cool the detectors to achieve the high sensitivity needed. The refrigerator is expected to provide a longer lifetime than has been achieved with solid cryogens on previous limb-scanners. In addition, only ISAMS can observe on two opposite sides of the space-craft, improving the coverage of the observations.

The Cryogenic Limb Array Etalon Spectrometer (CLAES) stares at the infrared limb, using a tilting Fabry-Perot etalon to resolve individual spectral lines of the gases of interest. It is cooled by a two-stage solid cryogen in order to cool the detectors to about 14 K. Its lifetime is limited by the life of the solid cryogen, which is required to be at least 18 months.

The Microwave Limb Sounder (MLS) is similar in principle to these limb emission instruments, but measures the signal in single lines in the microwave region of the spectrum. A 1.6 meter antenna is required to obtain the half scale height vertical resolution.

Table 1.
UARS MEASUREMENTS[*]

Instrument	Description and Primary Measurements
Energy Input Measurements	
SOLSTICE--Solar-Stellar Intercomparison Experiment	Full disk solar irradiance spectrometer incorporating stellar comparison
	Solar spectral irradiance: 115-440 nm
SUSIM--Solar Ultraviolet Spectral Irradiance Monitor	Full disk solar irradiance spectrometer incorporating onboard calibration
	Solar spectral irradiance: 120-400 nm
PEM--Particle Environment Monitor	X-ray, proton, and electron spectrometers *In situ* energetic electrons and protons; remote sensing of electron energy deposition
Species and Temperature Measurements	
CLAES--Cryogenic Limb Array Etalon Spectrometer	Cooled interferometer sensing atmospheric infrared emissions
	$T, CF_2Cl_2, CFCl_3, ClONO_2, CH_4, O_3, NO_2, N_2O, HNO_3$, and H_2O
ISAMS--Improved Stratospheric and Mesospheric Sounder	Mechanically cooled radiometer sensing atmospheric infrared emissions $T, O_3, NO, NO_2, N_2O, HNO_3, H_2O, CH_4$, and CO
MLS--Microwave Limb Sounder	Microwave radiometer sensing atmospheric emissions ClO and H_2O_2
HALOE--Halogen Occultation Experiment	Gas filter/radiometer sensing sunlight occulted by the atmosphere HF and HCl
Wind Measurements	
HRDI--High Resolution Doppler Interferometer	Fabry-Perot spectrometer sensing atmospheric emission and scattering Two-component wind: 10-110 km
WINDII--Wind Imaging Interferometer	Michelson interferometer sensing atmospheric emission and scattering Two-component wind: 80-110 km

[*] Adapted from Reber (1985)

The third infrared instrument is the Halogen Occultation Experiment (HALOE), which, as its name suggests, measures the absorption of solar radiation by the atmosphere as the sun, as seen from the spacecraft, rises or sinks behind the atmosphere. Gas cells are used in 4 channels to enhance the sensitivity to weak absorption features in the atmosphere.

A major new type of measurement is the High Resolution Doppler Imager (HRDI), which measures the Doppler shift of spectral lines in the visible spectrum. At low altitudes (<~35km) the lines are seen as absorption features, while at high levels (>~60 km) they are emission (scattering) lines. An individual view gives the line-of-sight velocity; in order to determine the two-dimensional wind field, measurements are made looking twice at the same volume of air from two perpendicular directions.

A second wind measuring instrument, the Wind Imaging Interferometer (WINDII) also measures winds by measuring Doppler shifts of airglow features with a Michelson Interferometer, but these winds are generally in the thermosphere.

The quantities observed and their altitude ranges are indicated in Fig. 1.

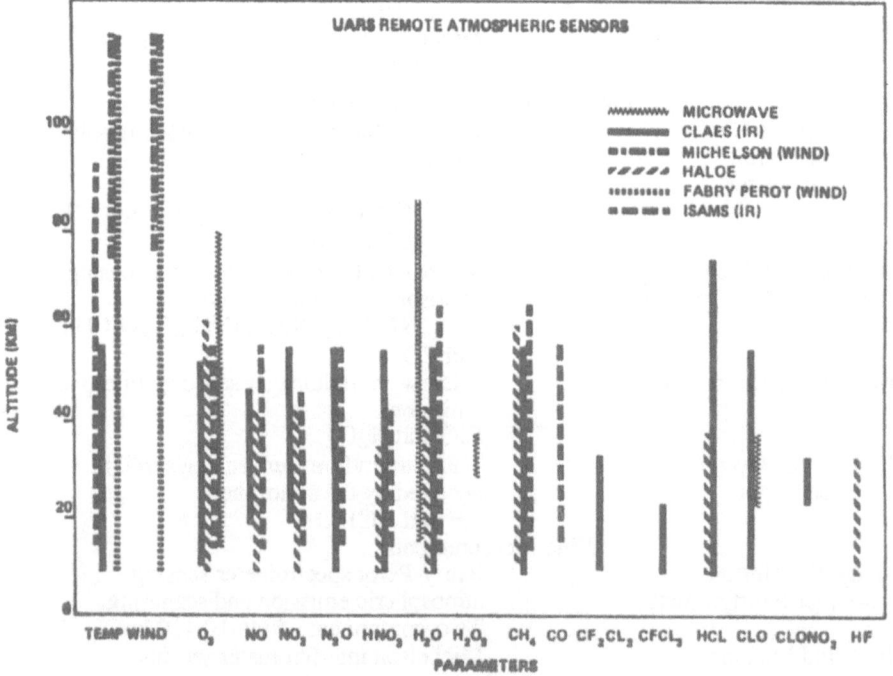

Figure 1. Altitude coverage of the atmospheric parameters measured by the UARS remote sensors. (From Reber, 1985).

6. Earth Observing System (Eos)

It is now clear that the Earth must be regarded as a large, interactive system consisting of major subsystems such as the atmosphere, the oceans, the cryosphere, the biosphere, etc. This view is spelled out very clearly, for instance, in Earth System Science (NASA,1988). This approach is embodied in the planning for the international Earth Observing System (Eos), which has as its goals the determination of whether the earth system is changing, as the result of natural or anthropogenic effects, and the observation and ultimately the understanding of the processes taking place, in order to improve our capability to model them. One of the ultimate goals is the improvement of our ability to predict future global changes.

To achieve these goals, the U.S., through NASA, the European Space Agency, and Japan will launch large, multi-instrument satellites beginning in the late 1990's The NASA satellites will be very large compared to previous spacecraft. Two series of three NASA spacecraft are planned, with each spacecraft having a five year lifetime, for a total observing time of 15 years for each series. The first series is planned to begin with the launch of Eos-A in 1997, and the second series with the launch of Eos-B in 2000.

Clearly, the focus of the Eos is much broader than the Middle Atmosphere, but considerable attention is being given to this critical region of the Earth system, since it is a region in which we know that there are long term changes in the ozone layer due to anthropogenic effects. Planning is now in an early stage, so only preliminary indications can be given at this time.

The first platform is largely devoted to the observation of the Earth's surface, with associated observations of tropospheric aerosols (whose effects must be removed from the surface measurements) and tropospheric sounding. It is also expected that an advanced limb sounder entitled the High Resolution Dynamics Limb Sounder (HIRDLS) will fly on these platforms. This instrument, a joint US-UK development, has measurement goals of $4°$ latitude by 4° longitude horizontal resolution, 1 km vertical resolution, and and the ability to sound the upper troposphere and low stratosphere. These are necessary to observe processes in the atmosphere that take place on much smaller scales than have been observable before, and also the critical region near the tropopause, in order to obtain a much improved understanding of the exchange of mass and constituents between the troposphere and stratosphere. Measurements of temperature, O_3, H_2O, CFC11, CFC12, CH_4, N_2O, NO_2, N_2O_5, HNO_3, and aerosols will be obtained.

An improved version of the SAGE is slated to fly as an attached payload on the space station, obtaining observations of aerosols, NO_2, H_2O, and other species from roughly 40°S to 40°N.

The second platform is likely to have a major emphasis on the middle atmosphere, including instruments such as an advanced Microwave Limb Sounder (MLS), which will improve on the UARS instrument by adding measurements of HCl, HOCl, HO_2, BrO, and OClO in the polar spring, as well as a number of other species and isotopic variants. Thus, key molecules in all the known ozone destruction cycles are measured.

A second instrument for the measurement of trace species is the Spectroscopy of the Atmosphere using Far InfraRed Emission (SAFIRE) experiment. Its goals include the measurement of vertical profiles of key O_y, HO_y, NO_y, ClO_y, and BrO_y constituents, including O_3, OH, HO_2, NO_2, HNO_3, HCl, HOCl, and HBr. This ambitious instrument will require a two-stage refrigerator to cool its far infrared detectors to a temperature near 10 K.

Direct measurement of winds will be obtained by the Stratospheric Wind Infrared Limb Sounder (SWIRLS) instrument, which will determine line-of-sight winds by measuring the Doppler shift of atmospheric emission lines. Again, it is necessary to measure winds in two orthogonal directions in order to determine the complete horizontal wind field. This technique promises to overcome two of the limitations of the HRDI on UARS; by measuring emission, observations can be obtained day and night, and measurements can be obtained in the 35-60 km altitude region which is inaccessible to HRDI.

Finally, it is anticipated that a later version of SOLSTICE will fly on the second platform, providing data on the spectral and temporal variations of solar UV that drive stratospheric photochemistry. In addition, there will probably be measurements of incoming energetic particles.

7. Concluding Remarks

The measuring systems discussed above are summarized in Fig. 2. We can briefly summarize some of the previous information by asking what completely new information we can look forward to obtaining. UARS will provide the first global measurements of middle atmospheric winds, and of some of the chlorine species. Subsequently, Eos will provide the first measurements of small longitudinal and vertical scales and the tropopause region, first measurements of some of the HO_y and BrO_y species, and extend or initiate the acquisition of long term data sets.

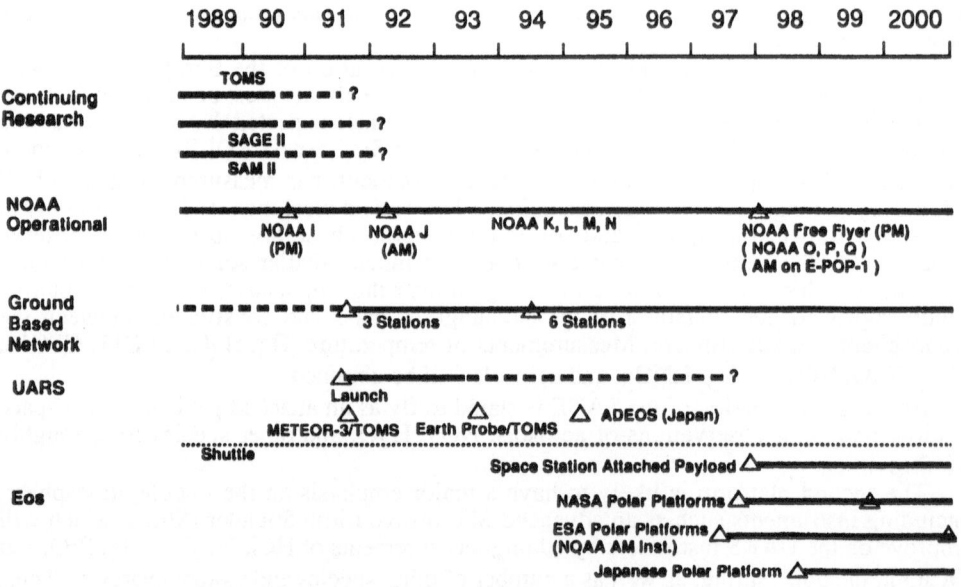

Figure 2. Future middle atmosphere measuring systems.

The remarks above have extended our survey to the year 2000, and in fact some of the Eos instruments will continue until 2015. It should be emphasized that crystal balls are always somewhat cloudy, especially far into the future. Past experience suggests several points that will probably also be true in the future:

> schedules will slip;
> some instruments won't work as well as expected;
> some instruments will be much more useful than expected;
> there will be new problems and phenomena, leading to new
> measurement requirements and applications.

However, certain tendencies in the 1990's appear likely. Satellite instruments will be bigger and more complex, more expensive, more flexible, and provide more data. Fortunately, future data systems will provide increased and easier access to the data. The future will see greater international collaboration and participation in these programs, aided by these data systems.

Scientifically, there will be greater emphasis on interactions, both dynamical and chemical, between the middle atmosphere and the troposphere. Improved instrument calibration and stability will allow greater attention to long-term change.

In conclusion, the next decade promises to provide a rich supply of data for future research.

Acknowledgments

This work was supported in part by the National Aeronautics and Space Administration under Grant W-16215. The National Center for Atmospheric Research is sponsored by the National Science Foundation.

References:

Heath, D., A.J. Krueger, and H. Park, The Solar Backscatter Ultraviolet (SBUV) and Total Ozone Mapping Spectrometer (TOMS) Experiment, Sec. 7, Nimbus 7 Users Guide, C. Madrid, Ed., Goddard Space Flight Center, Greenbelt, Maryland, 1978.

McCormick, M.P., P. Hamill, T.J. Pepin, W.P. Chu, T.J. Swissler and L.R. McMaster, Satellite studies of the stratospheric aerosol, Bull. Am. Met. Soc., 60, 1038, 1979.

Mauldin, L.E., N.H. Zaun, M.P. McCormick, J.H. Guy and W.R. Vaughn, Stratospheric Aerosol and Gas Experiment II Instrument: A functional description, Opt. Eng., 24, 307, 1985.

NASA, Earth System Science, a closer view, Report of the Earth System Sciences Committee, NASA, Washington, D.C., 1988.

Reber, C.A., Upper Atmosphere Research Satellite (UARS) Mission, NASA, Goddard Space Flight Center Document 430-1003-001, Greenbelt, Maryland, 1985.

WMO, The Ozone Trends Panel Report, R. Watson, Ed., in press, 1990.

MODELLING AND OBSERVATIONS : THE USE OF DIAGNOSTICS

DAVID G. ANDREWS
Meteorological Office Unit
Hooke Institute
Clarendon Laboratory
Parks Road
Oxford OX1 3PU
U.K.

ABSTRACT. This paper discusses some of the uses and limitations of dynamical diagnostics in the study of atmospheric models and observational data, and in the intercomparison of the two. Some of the problems of interpreting diagnostics are mentioned, with particular attention to the difficulties of extracting causal information. The need for controlled experiments with models for the latter purpose is stressed.

1. Introduction

The ultimate practical reason for studying the atmosphere is the need to predict its future behaviour on timescales of hours, days or years. To this end it is first necessary to pursue a seemingly more academic goal: that of understanding the basic physical, chemical and even biological processes occurring in the atmosphere. In general, the longer and more detailed the forecasts required, the greater the depth to which these processes must be explored.

How can these basic processes be identified? The 'scientific method' of the laboratory experimenter – observing phenomena, formulating hypotheses, testing them by controlled experiments, formulating revised hypotheses, and so on – cannot be used, since controlled experiments cannot generally be performed on the large-scale atmosphere itself.*

Instead, after the initial stages of discovering an atmospheric phenomenon by sifting through the data, a common approach is to construct models, each containing

---

* A possible exception to this rule is that advantage can occasionally be taken of cases, such as stratospheric sudden warmings or the recent Antarctic ozone depletion, where 'nature kindly performs for us' something like a controlled experiment: see, e.g., McIntyre (1982, 1989). Interhemispheric differences may perhaps be exploited in a similar way.

A. O'Neill (ed.),
Dynamics, Transport and Photochemistry in the Middle Atmosphere of the Southern Hemisphere, 227–235.
© 1990 *Kluwer Academic Publishers.*

a selection of processes that are hypothesized to be significant for that phenomenon. If a hypothesis seems equally applicable to the atmosphere and to a model, then the latter is a useful surrogate for the former (Lorenz, 1967), especially if it disproves the hypothesis. The model is usually represented by means of a closed set of mathematical equations, which must normally be solved numerically. Occasionally, a laboratory apparatus may provide a useful atmospheric model. The performance of each model, and thus the appropriateness of the hypothesized set of atmospheric processes, is usually judged by comparing its behaviour with that of the observed atmosphere.

A valuable way to optimise this method is to employ a whole *hierarchy* of models, starting with simple ones that contain a minimum of atmospheric processes and adding further processes in turn, attempting to reproduce the observed atmosphere to higher and higher accuracy and at the same time developing a conceptual framework for the interpretation of atmospheric and model behaviour (Hoskins, 1983a). The simplest models may be investigated analytically and their workings clearly understood: they may thus provide a basic 'physical intuition', which can then be applied to the interpretation of the more complicated models and which can help guard against serious errors in their formulation and coding. An understanding of the more complex models may, in turn, help in the interpretation of the observed atmosphere: as models simulate the atmosphere with greater and greater fidelity, it is generally assumed that they contain more and more of the relevant atmospheric processes and their interactions. However, it should be emphasized that this assumption may not always be valid: a deficient model can sometimes produce what appears to be a good simulation because of mutually-compensating errors, because it has been artificially constrained in some way or because its behaviour has not been compared in sufficient detail with that of the atmosphere.

Controlled experiments *can* be performed on models, for instance to explore the relative importance of different processes (some of which may be poorly understood) and to assess the possible impact of perturbations to the climate system. Examples include the use of complex general circulation models (GCMs) to study the response of the global atmosphere to imposed sea surface temperature anomalies, or the climatic effects of increasing carbon dioxide (see, e.g., Palmer and Mansfield, 1986, Schlesinger and Mitchell, 1987, and references). Ideally, these model experiments would suggest strategies for further observational campaigns, which in turn would lead to the formulation and testing of more refined hypotheses.

In addition to their role in the interpretation of observational data, models can also assist in the data-analysis process itself. For example, numerical weather prediction models are routinely employed in 'data assimilation', in which the constraints implied by the model's dynamics and physics are used to partially compensate for the inevitable sparsity and ambiguity of the observations (see, e.g., Lorenc, 1986). There may also be benefits from incorporating conceptual models – such as of atmospheric frontal structures – in this process, and taking advantage in other ways of the human brain's enormous computing power by employing a suitable 'man-machine mix', rather than using a fully-automated assimilation procedure

(B.J. Hoskins, personal communication, 1988; McIntyre, 1988). In these cases the 'analysed data' are to some extent model-dependent: great care is therefore needed if the same models are to be used for interpreting these data.

2. Diagnostics

We now turn to the question of how best to compare the atmosphere with a model, or one model with another. One possibility might be to compare the various velocity, temperature and constituent fields at each point in space and time. However, the sheer volume of data involved would make this an extremely complicated and time-consuming procedure, and very difficult to interpret – especially if the agreement is bad! More compact and meaningful ways of representing the atmospheric and model states, and the differences between them, are essential. Of course, numerous measures of these states are in use, but many of them are rather arbitrary or undiscriminating, and may tell us little about atmospheric mechanisms. For example the eddy geopotential amplitude, a popular measure of the deviation of the state from zonal symmetry, tends to be proportional to the local zonal-mean zonal wind, and its height and latitude variations may reflect mainly the local jet structure rather than any specific eddy property (McIntyre, 1982).

Ideally, we seek 'diagnostics' that compress and organise large volumes of data into manageable portions and also carry information about atmospheric processes. One meaning of the word 'diagnostic' is 'a distinctive symptom or characteristic' (Shorter Oxford English Dictionary); one would wish atmospheric diagnostic quantities to convey at least qualitative 'signatures' of processes such as baroclinic instability, Rossby-wave propagation and breaking, and so on. Better still, they should also give meaningful and discriminating quantitative measures of these processes, to allow sensitive comparisons between models and observations, or between one model and another. Of course, some ambiguity in the interpretation of diagnostics is likely to occur in practice, since distinct processes can have rather similar signatures in terms of any given diagnostic: a variety of different diagnostics will generally be needed to distinguish between different processes. For example, certain types of orographically-induced stratospheric planetary waves in shear are found to exhibit the same 'positive baroclinic energy conversion' (from 'mean' to 'eddy' available potential energy) as in models of baroclinic instability. Analysis with more refined diagnostics and (more conclusively) the use of suitable models show that – contrary to suggestions in the literature – these planetary waves are not in any way maintained by baroclinic instability in the stratosphere: see, e.g., Plumb (1983).

There has recently been a healthy tendency for a greater variety of diagnostics to be used in the analysis of atmospheric and model data. Each diagnostic has its own strengths and weaknesses. Consider, for example, the Eliassen-Palm (EP) flux and its divergence. As suggested by Edmon, Hoskins and McIntyre (1980) and Hoskins (1983b), the flux may be interpreted as giving an indication of the direction of net Rossby-wave propagation in the meridional plane; its divergence provides some feel for the regions of generation of such waves by nonlinear baroclinic instability, the regions of breaking in the subtropics, and a measure of the magnitudes of these sources and sinks of wave activity. The flux divergence also indicates the strength and location of the 'effective zonal body force due to the presence of the waves' that is felt by the zonal-mean flow. Weaknesses of the EP flux as a diagnostic include the facts that it cannot on its own distinguish between dynamically-different types of wave (e.g. Rossby waves and equatorial Kelvin waves) and that it only gives a measure of net propagation when waves are superimposed; see §3 for further discussion.

It has been assumed here that the required diagnostics can be calculated with acceptable accuracy from the available data. This may not always be the case in practice: for example, diagnostics involving the Lagrangian displacements of air parcels may be impossible to calculate directly from atmospheric data with any accuracy, and difficult to obtain even in models. Other diagnostics, such as the EP flux divergence or the potential vorticity, involve two or three differentiations, and may sometimes be prone to quite large uncertainty when calculated from atmospheric observations.

3. Physical Interpretation

The question of what constitutes a 'meaningful' diagnostic is not easy to answer. Possible candidates include quantities that occur in simple, idealised, yet self-consistent theoretical models, and to which a clear physical meaning can be attached within the context of such models. It is then (usually tacitly) supposed that a similar physical interpretation carries through to the more complex models and the real atmosphere. The simpler models can thus be thought of as providing a 'theoretical framework' for organising the data and interpreting aspects of the behaviour of the more complicated models and the atmosphere itself.

Consider for example the EP flux, described above. The interpretation of this flux as indicating the net direction of Rossby-wave propagation is a direct generalization from ideas of 'group velocity'. Thus for the simple quasi-geostrophic model of linear Rossby waves in a uniform zonal flow $\bar{u}$ on a beta-plane, one can apply classical 'wave-packet' ideas to show that modulations of the packet envelope travel with the group velocity c_g, given by $\partial \omega / \partial k$, where ω is the local frequency and k the local wavenumber. But in the same model it can be shown that c_g also equals $\mathbf{F}/A$, where $\mathbf{F}$ is the EP flux and A the 'wave-activity density' (see, e.g., Andrews, Holton and Leovy, 1987, page 188). In this model, therefore, the interpretation of the group velocity carries over to $\mathbf{F}/A$ and so the waves transfer information in a direction parallel or anti-parallel to the local direction of $\mathbf{F}$, depending on whether A is positive or negative.

Using 'WKBJ' techniques, one can extend this physical interpretation to a linear quasi-geostrophic model in which $\bar{u}$ varies slowly with position (i.e. on scales much greater than a wavelength). However, in a linear quasi-geostrophic model in which $\bar{u}$ is not slowly-varying, WKBJ methods fail, ω and k are not generally well-defined, and nor is c_g. On the other hand, $\mathbf{F}$ and A are still defined, and satisfy a *conservation law*

$$\partial A/\partial t + \nabla \cdot \mathbf{F} = 0, \qquad (1)$$

if the waves are adiabatic and frictionless. A is a 'conservable measure of wave-activity', with flux $\mathbf{F}$, and it seems sensible to regard $\mathbf{F}/A$ as a generalization of the group velocity concept to this model, at least in regions where $A \neq 0$. (Similar remarks apply to linear, primitive-equation models.) Note, however, that rapid variations in the mean flow can give significant reflection of wave-trains, and a calculation of $\mathbf{F}$ for a wave-field will give no indication, on its own, of the relative amounts of 'incident' and 'reflected' wave.

In yet more complex, nonlinear, models, relations of the form (1) can still be found, and diabatic or frictional wave effects can be incorporated by including a 'nonconservative' term on the right-hand side: see, e.g., Haynes (1988). But when nonlinear motions are allowed, it may be difficult to decide whether the processes under consideration should be regarded as 'wave-like' – as opposed to 'vortex-like', for example – and although $\mathbf{F}$ remains *some* kind of quantitative measure of the departures from zonal symmetry, its physical interpretation may be obscure.

There is a need for a set of controlled experiments with the 'linear, non-slowly-varying' model and the nonlinear models, to test how far the ideas about the physical interpretation of the EP flux in terms of wave-propagation can be pushed. The same is true for the interpretation of the EP flux divergence, mentioned above: for sufficiently small-amplitude waves the divergence of the EP flux can be directly related to non-conservative wave generation and dissipation processes, and also to wave transience. The usefulness of thinking in terms of a 'wave-induced' zonal force has been demonstrated in simple models of small-amplitude waves interacting with the zonal-mean flow. Experiments with numerical models suggest that these ideas carry over, to some extent, to finite-amplitude disturbances, but more systematic tests are desirable.

These remarks emphasize that one must exercise caution in extending the use of diagnostics from simple models to more complex ones: there is always the possibility that the original model may be too simple, and may omit some crucial aspect of the real atmosphere that negates the use of the diagnostic, without modification, in the real situation.

A well-known example of the latter difficulty is that associated with the concepts of 'wave-energy' and 'wave-energy flux'. While these quantities may be useful for describing small-amplitude waves propagating in a fluid that is basically at rest or in uniform motion, they may become difficult to interpret when such waves are superimposed on a mean shear flow, since energy can then be transferred from the waves to the mean flow, or vice versa. (In the latter case the EP diagnostics are arguably more useful since 'EP wave-activity', unlike wave-energy, is conservable even in the presence of mean shear flows. However, as mentioned above, EP diagnostics may in turn have their own shortcomings.) This example also illustrates the danger of assuming that seemingly familiar and well-established physical concepts – in this case 'wave-energy' – may help in the interpretation of a complex fluid-dynamical system, without checking whether this assumption is borne out in a fully self-consistent mathematical model of the system. This danger is increased when the diagnostic quantity is given a physically-inappropriate name in the first place.

4. Causal Mechanisms

Diagnostics can sometimes provide useful hints about possible causal mechanisms in the atmosphere. However, they can never give definitive information about such mechanisms and, if used without due caution, can be positively misleading. (An example is the naive assumption, mentioned above, that a baroclinic energy conversion from 'mean' to 'eddy' available potential energy necessarily implies that the waves are generated by baroclinic instability.) In a nonlinear, interactive system like the atmosphere, specific mechanisms of this kind may be very difficult to isolate from the complex web of cause-and-effect relationships. Careful controlled experiments with suitable models are always necessary to validate the ideas about causality suggested by simple diagnostics.

In discussing cause-and-effect relationships, it may be useful to distinguish between two cases. First there is the time-evolving state in which an imposed perturbation at some time t_0 can be shown to lead to certain consequences at a later time t_1. The simplest way of establishing the relationship is to perform a pair of model experiments, with and without the change at time t_0, and examine the different behaviour of the two experiments at later times. (More than two model runs may sometimes be necessary, to establish statistical significance.) In some cases the causal mechanism may have a clear physical interpretation, for example in terms of waves transferring information about the perturbation. However, the interpretation may be more elusive if the perturbation is of large amplitude, or if a long time has elapsed since t_0, when information about the initial perturbation may have spread into large regions of physical space and into many scales of motion. The perturbed state may then be so different from the unperturbed that any kind of comparison may be difficult.

The second case concerns the effect of different processes on a quasi-steady or time-averaged state. For instance, we may talk about 'the effect of the planetary waves on the climatological state of the winter stratosphere'. In this case, it is common to attempt to quantify cause-and-effect relationships using diagnostics based on certain terms in the time-averaged budgets of momentum, energy or wave-activity, for example. However, it must be emphasised that such an approach will normally supply no more than a check on the mutual consistency of different processes, and on its own will say nothing about causal relationships. (Thus the time-averaged, zonal-mean zonal momentum equation implies a certain balance between eddy-flux and mean-flow terms, which can sometimes be verified from data, but which on its own says nothing about causality.)

The most satisfactory way of identifying cause and effect in this second case is probably again by means of controlled model experiments, such as a control run including the waves and another run in which the waves are artificially suppressed, leaving other conditions as nearly fixed as possible. For example, Andrews, Mahlman and Sinclair (1983) strongly damped the upward-propagating, tropospherically-forced planetary waves in the lower stratosphere of a general circulation model, so as to exclude them from most of the model middle atmosphere (see their Section 6). The resulting changes to the zonal-mean flow were consistent with those suggested by an examination of the EP flux divergence, using insights from simpler models. (On the other hand, they contradicted earlier ideas involving heuristic arguments about the momentum flux divergence but not based on any self-consistent model.) This method of suppressing the eddies was quite drastic, but had the advantage of giving an unequivocal result. As an alternative, it may in some cases be preferable to consider the differential 'effect' resulting from a smaller change in the 'cause', if the signal-to-noise ratio is large enough (S.B. Fels, personal communication, 1983).

In both the time-evolving and the time-averaged examples, 'adjoint sensitivity theory' (Hall, 1986) may turn out to be an appropriate technique for quantifying some cause-and-effect relationships, by providing a systematic method of calculating the responses (assumed small) to small perturbations. In cases when the solutions of the adjoint equation can be identified as wave modes, this theory may also supply a physical interpretation of the ways in which information about the perturbation spreads throughout the system. However, as pointed out by M.E. McIntyre (personal communication 1988), the method may be less useful in the more interesting, sensitive, examples in which small causes produce large effects. In any case, our ideas of causality may need careful re-consideration when we are considering fully chaotic systems, which display strong sensitivity to initial conditions.

5. Conclusion

This paper has not attempted to make any definitive statements on how atmospheric and model data should be interpreted. Its purpose has been, rather, to stimulate a re-examination by atmospheric modellers and data analysts of some of our basic preconceptions concerning the uses to which data can be put, and to encourage further exploration of the ways in which the integrated study of observations, theories and hierarchies of models can help us improve our understanding of the atmosphere. Many atmospheric scientists are already well aware of the problems raised here: it is to be hoped that a continuing debate within the community will help resolve some of these difficult issues.

Acknowledgements

I am grateful to Alan O'Neill for pressing me, over a period of several years, to get to grips with the problems of the interpretation of data, and for questioning many of my glib statements on this issue. Brian Hoskins, Keith Shine, John Stanford, and especially Michael McIntyre made a number of useful comments on earlier drafts of the paper, and some of their suggestions have been incorporated here.

References

Andrews, D.G., Holton, J.R. and Leovy, C.B. (1987). *Middle Atmosphere Dynamics*. Academic Press, Orlando, Florida.

Andrews, D.G., Mahlman, J.D. and Sinclair, R.W. (1983). Eliassen-Palm diagnostics of wave, mean-flow interaction in the GFDL "SKYHI" general circulation model. *J. Atmos. Sci.*, **40**, 2768–2784.

Edmon, H.J., Jr., Hoskins, B.J. and McIntyre, M.E. (1980). Eliassen-Palm cross sections for the troposphere. *J. Atmos. Sci.*, **37**, 2600–2616. (Also Corrigendum, 1981, *J. Atmos. Sci.*, **38**, 1115.)

Hall, M.C.G. (1986). Application of adjoint sensitivity theory to an atmospheric general circulation model. *J. Atmos. Sci.*, **43**, 2644–2651.

Haynes, P.H. (1988). Forced, dissipative generalizations of finite-amplitude wave-activity conservation relations for zonal and nonzonal basic flows. *J. Atmos. Sci.*, **45**, 2352–2362.

Hoskins, B.J. (1983a). Dynamical processes in the atmosphere and the use of models. *Quart. J. Roy. Met. Soc.*, **109**, 1–21.

Hoskins, B.J. (1983b). Modelling of the transient eddies and their feedback on the mean flow. In *Large-scale dynamical processes in the atmosphere*, (eds. B.J. Hoskins and R.P. Pearce), pp. 169–199, Academic Press, Orlando, Florida.

Lorenc, A.C. (1986). Analysis methods for numerical weather prediction. *Quart. J. Roy. Met. Soc.*, **112**, 1177–1194.

Lorenz, E.N. (1967). *The nature and theory of the general circulation of the atmosphere*. World Meteorological Organization, Geneva.

McIntyre, M.E. (1982). How well do we understand the dynamics of stratospheric sudden warmings? *J. Meteor. Soc. Japan*, **60**, 37–65.

McIntyre, M.E. (1988). Numerical weather prediction: a vision of the future. *Weather*, **43**, 294–298.

McIntyre, M.E. (1989). On the Antarctic Ozone Hole. *J. Atmos. Terrest. Phys.*, **51**, 29–43.

Palmer, T.N. and Mansfield, D.A. (1986). A study of wintertime circulation anomalies during past El Niño events, using a high resolution general circulation model. I: Influence of model climatology. *Quart. J. Roy. Met. Soc.*, **112**, 613–638.

Plumb, R.A. (1983). A new look at the energy cycle. *J. Atmos. Sci.*, **40**, 1669–1688.

Schlesinger, M.E. and Mitchell, J.F.B. (1987). Climate model simulations of the equilibrium response to increased carbon dioxide. *Revs. Geophys.*, **25**, 760–798.

ANTARCTIC OZONE DEPLETION AND POTENTIAL EFFECTS ON THE GLOBAL OZONE BUDGET

W. L. GROSE, R. S. ECKMAN, R. E. TURNER, AND W. T. BLACKSHEAR
Atmospheric Sciences Division
NASA Langley Research Center
Hampton, Virginia 23665-5225
U.S.A.

ABSTRACT. In addition to the dramatic reductions in polar ozone observed in the springtime Antarctic stratosphere during the past decade, data from the Total Ozone Mapping Spectrometer (TOMS) instrument also provide evidence of a reduction in total columnar ozone extending into middle latitudes of the Southern Hemisphere (e.g. Watson et al., 1988). It has been suggested that dilution of the mid-latitude air by export of ozone-poor air from polar regions following breakup of the vortex would create a deficit which might persist for a long period because of the slow chemical replacement time (months to a year in the lower stratosphere) for ozone (e.g. Sze et al., 1989). If the deficit is maintained until the next springtime depletion episode, the effect might be cumulative, with a permanent reduction in the global ozone budget and, hence, a possible explanation for the mid-latitude reductions of ozone seen in the TOMS data. A study of the so-called "dilution effect" has been conducted using a three-dimensional chemistry/transport model. The results of the model simulations reveal a small, but significant, residual deficit in the total ozone in the Southern Hemisphere 1 year following the formation of an ozone hole in the polar regions. The results are supportive of the dilution concept being a possible explanation for the Southern Hemisphere mid-latitude changes in the TOMS data over the past decade. However, the conclusions must be qualified by noting that other factors which influence the ozone distribution have not been taken into account in this study.

A brief synopsis of the key results from this article is contained in World Meteorological Organization Report No. 20, Scientific Assessment of Stratospheric Ozone: 1989.

A. O'Neill (ed.),
Dynamics, Transport and Photochemistry in the Middle Atmosphere of the Southern Hemisphere, 237–251.
© 1990 *Kluwer Academic Publishers.*

1. Introduction

The observations reported by Farman et al. (1985) focused attention upon depletion of the ozone column during springtime in the polar regions. Subsequent examination of both satellite and Dobson ozone data revealed that a rapid decline in ozone occurred over the past decade each September after sunlight returned to the polar region. Reductions of as much as 50 percent were observed [WMO (1990), and references cited therein]. There is now substantial evidence [JGR (1989a and 1989b) special issues devoted to results from the Airborne Antarctic Ozone Experiment; also WMO (1990) and references cited therein] that is supportive of theories [e.g. Solomon et al. (1986) and McElroy et al. (1986)] attributing the ozone depletion to chlorine (and bromine, to a lesser extent) compounds.

In contrast with the decline in the ozone column resulting from chemical processes, recovery of the column is essentially due to dynamical processes [WMO (1990), and the references cited therein]. Replenishment of the polar ozone column occurs rapidly after breakup of the polar vortex. Ozone-poor air is exchanged with air masses with relatively higher ozone content from lower latitudes by large-scale, quasi-horizontal transport. Evidence for this large-scale, latitudinal mass exchange can be seen in the potential vorticity analyses of Newman (1986), which were carried out for a number of years. Atkinson et al. (1989) concluded that anomalously low total ozone values observed in Australia and New Zealand during December 1987 were also the result of ozone-poor air from the polar regions being transported to lower latitudes after breakup of the polar vortex. Their conclusions are supported by potential vorticity analyses and the strong correlation with the evolution of the total ozone observed in the TOMS data.

Gridded total ozone data from the TOMS instrument is shown in Fig. 1. The variation by month of the difference between the average of the 1987-1988 TOMS data and the average for 1979-1980 is displayed as a function of latitude. The 2-year averages for 1987-1988 and 1979-1980 have been used to suppress the effect of the quasi-biennial oscillation [e.g. Garcia and Solomon (1987) and Lait et al. (1989)]. The development of the ozone hole is readily evident from the large reductions seen at south polar latitudes during September and October. Reductions of 4 percent and greater are also apparent between 50-60S for most of the year. Reductions of 4 percent are also seen at 40S from March through July. Significant reductions extending toward mid-latitudes are seen to occur in November and December at the time when large lateral mixing would be expected to take place after vortex breakup. Reductions extending toward mid-latitudes are also seen in April and May when a surge in wave activity typically occurs in the Southern Hemisphere lower stratosphere [e.g. Hirota et al. (1983) and Randel (1988)].

The reductions at Southern Hemisphere mid-latitudes seen in the TOMS data may plausibly result from dilution of the mid-latitude air by export of the ozone-poor air from the polar region following onset of the final warming and breakup of the westerly vortex [Sze et al. (1989)]. Any resulting deficit in ozone created by such an effect

TOMS OZONE

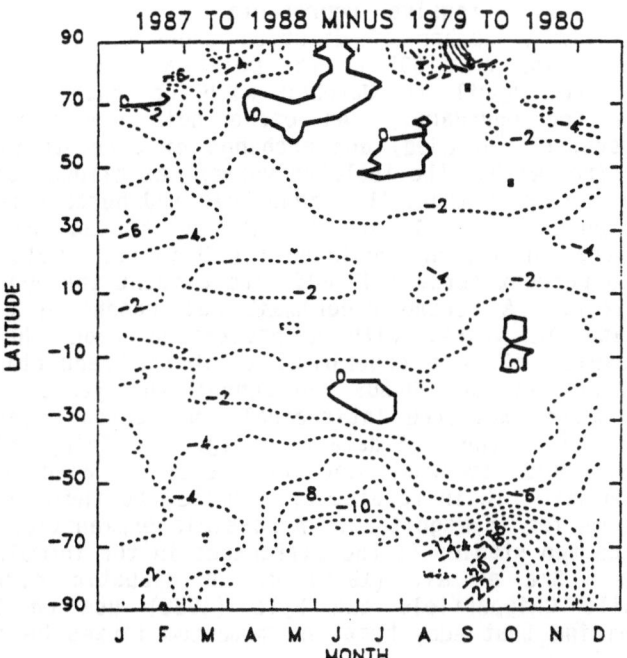

Figure 1. The percentage change in the total columnar ozone between the average for 1987-1988 and 1979-1980 from the TOMS instrument (figure supplied courtesy of R. Stolarski, NASA GSFC).

could then be maintained for a lengthy period because the characteristic time for chemical replacement of odd-oxygen in the lower stratosphere is relatively long (months to a year, depending upon latitude). If an ozone deficit at mid-latitudes lasts until formation of a polar ozone hole the following spring, the effect might be cumulative and have consequences for the global ozone budget. However, it is well recognized [e.g. WMO (1990); Watson et al. (1988) and the references cited therein], that year-to-year variations in the global distribution of ozone can occur from such diverse influences as dynamical variability, injection of volcanic aerosols, the solar cycle, and possibly even heterogeneous chemical processes outside the polar regions [Hofmann and Solomon (1989)]. Therefore, until comprehensive and realistic modeling studies are combined with further analysis of the existing data record, it will not be possible to unambiguously determine the exact cause (or combination of causes) of the aforementioned reductions of the mid-latitude ozone column.

240

Several model studies have been conducted to study the consequences of dilution. Sze et al. (1989) used a two-dimensional model [Ko et al. (1985)] with an extensive representation of the gas-phase chemistry to perform a number of dilution experiments. In their experiments, the region from 63S to the pole was assumed to be an isolated vortex from April to October with no exchange of air permitted across the boundary. An ozone hole was imposed (no heterogeneous processes included) and exchange of extra-vortical air masses with that from within the isolated vortex was allowed beginning in November. In one experiment, the ozone hole (80 percent reduction from 63S to the pole between 14 and 22 km) was imposed only in the first year with the simulation continued for 3 years. Column ozone reductions of 2-6 percent between 45-60S were evident throughout most of the second year. A second experiment was conducted with the simulation lasting 12 years, with a prescribed ozone hole each spring. The results of that experiment exhibited reductions of 3 percent and larger between 50-60S throughout the year. Broadly speaking, the results exhibited latitudinal and seasonal variations similar to those seen in the TOMS data (see Fig. 1). Chipperfield and Pyle (1988) performed somewhat similar studies, also using a two-dimensional model. Their results are similar to those of Sze et al. (1989), but smaller in magnitude. A possible explanation for the differences in the two studies is the difference in the formulation of the two models. Sze et al. (1989) use a diabatic circulation formulation, while Chipperfield and Pyle (1988) use an Eulerian formulation requiring that eddy heat and momentum fluxes be included explicitly.

Prather et al. have also modeled the dilution effect using a three-dimensional off-line tracer model and adopting a simplified, linearized ozone photochemistry. In this study, an ozone hole was imposed instantaneously on October 1 in the region from 70.5S to the pole between the 22 and 200 mb levels. A 2-year simulation was then conducted (again with an ozone hole imposed on October 1). At the end of the first year, column reductions of 2 percent were present as far equatorward as 40S. At the end of the second year, the average reduction in the column was approximately 20 percent larger than at the end of the first year. These results also indicated that most of the deficit that existed 1 year after imposition of the hole was in the upper troposphere. Cariolle et al. (1990) also conducted studies with a three-dimensional model. Their model also adopted a linearized formulation for the ozone chemistry, but was interactive in the general circulation model in contrast to the off-line approach used by Prather et al. (1990). The results of Cariolle et al. (1990) are generally similar, however.

In the present study, a three-dimensional chemistry/transport model was used to study some aspects of the dilution effect and possible consequences. The model study essentially differs from those of Prather et al. (1990) and Cariolle et al. (1990), in that the formulation of the chemistry is comprehensive (gas-phase chemistry only) and is not based upon linearization. In subsequent sections, the model is described, an outline of the experiment is presented, and the results are discussed.

2. Description of the Model

A three-dimensional, global, general circulation model with a comprehensive chemistry scheme is employed in this study. Previous descriptions of the model may be found in Blackshear et al. (1987), Grose et al. (1987), and Grose et al. (1989). On-going improvements warrant an updated description at this time. The model consists of three components: a general circulation (GCM) module, a transport module, and a chemistry module.

The general circulation module employs a spectral, primitive equation model extending from the surface to approximately 60 km. The effects of orography and land-ocean thermal contrast are represented. A sigma coordinate system (Phillips, 1957) is used with 12 vertical levels. The time integration is accomplished with a semi-implicit technique with a 30-minute time step.

An "off-line" approach [see e.g. Mahlman and Moxim (1978)] is used to perform the transport simulations. In this case, the time-dependent wind and temperature fields generated by the GCM are used as input to the set of mass continuity equations for the tracers. The formulation of the transport scheme is essentially identical to that of the GCM. A scale-selective, biharmonic diffusion term is included to parameterize horizontal, sub-grid scale processes. Following Mahlman and Moxim (1978), we do not include vertical sub-grid diffusion. Special care is taken in the treatment of negative mixing ratios when they occur. A non-linear diffusion term is applied locally to convert the finite difference approximation of the vertical advection term from central to upwind differencing.

The chemistry of the middle atmosphere exhibits a wide range of temporal scales. To avoid numerical instabilities, a "family" chemistry scheme is adopted. Species are grouped into families where the time scale for production or loss is much greater than that of the member constituents [e.g. Brasseur and Solomon (1984)]. This approach allows for the use of longer time steps than would otherwise be necessary if a coupled set of differential equations explicitly transporting each species were employed. Nevertheless, relatively large expenditures of CPU time are required to run annual simulations of the model. For example, on the NASA/Ames Numerical Aerodynamic Simulation (NAS) Cray Y-MP, approximately 4.5 CPU minutes per model day are presently required to run the transport/chemical modules.

The present chemical scheme considers 39 species using 115 chemical and photolytic reactions. Six families and species are explicitly transported: O_x (O, O(^{1}D), O_3), NO_y (N, NO, NO_2, NO_3, HNO_4, $ClONO_2$), Cl_x (Cl, ClO, HCl, HOCl, $ClONO_2$, ClO_2), HNO_3, N_2O_5 and, H_2O_2. Odd hydrogen (H, OH, HO_2) is assumed to be in photochemical equilibrium. The methane oxidation cycle is included by considering the CH_x family. Partitioning of the remaining individual species is performed at each time step making use of the relevant equilibrium relationships.

Long-lived species (H_2O, H_2, CH_4, N_2O, CH_3Cl, $CFCl_3$, CF_2Cl_2, CH_3CCl_3, and CO) are specified as a function of latitude, altitude, and season based on observations, where available, or from two-dimensional model results [Garcia and Solomon, (1983)].

The photolysis rate calculation makes use of look-up tables to reduce computational time. The rates are stored as a function of overhead absorber column density and, where appropriate, temperature. A comparison with detailed photolysis calculations shows satisfactory agreement with the approximate method employed here. The effects of multiple scattering are not accounted for in the present version of the model. Absorption cross sections were obtained from DeMore et al. (1987) and Baulch et al. (1982). The ultraviolet solar irradiances are derived from the recommendations of Frederick et al. (1986) based on recent satellite and rocket measurements. Kinetic rates are taken from the compilation of DeMore et al. (1987). Unlike the photolysis rate calculation, the bi- and termolecular rate constants are calculated at each time step.

Polar night chemistry presents a special problem in stratospheric models as the chemical lifetimes of many species become quite long compared with the time step used and it, therefore, becomes difficult to partition species under these conditions. The present version of the model allows for a simplified representation of O_x, NO_y, and N_2O_5 chemistry in darkness. Other families and species are transported, but no chemistry is allowed to occur. This approach leads to these individual family members remaining indeterminant in darkness.

3. Description of the Experiment

The results discussed in subsequent sections of this paper were extracted from two model simulation runs. The first run was initialized on January 1 and then integrated for 32 months (through August of year 3) with standard gas-phase chemistry as mentioned in the previous section on model description. The last 12 months of this run are designated as the control simulation. The second model run, designated as the perturbed simulation, was initialized on September 1 with conditions from year 2 of the control simulation. During the month of September, a linear destruction rate was imposed such that 90 percent of the ozone between 27 and 112 mb from 69S to the Pole is destroyed by the end of the month. From October 1 onward, the simulation was continued through the following August with only the standard gas-phase chemistry.

4. Discussion of Results

Some comparisons of model simulated fields (winds, temperatures, and constituents) with observations have been described in earlier publications [e.g. Blackshear et al. (1987), Grose et al. (1987), and Grose et al. (1989)]. For the purposes of this paper, only the ozone distribution is discussed. The zonal mean distribution of ozone for model day September 1 is shown in Fig. 2. The peak mixing ratio of 12 ppmv occurs at 10 mb near 15N. Examination of the seasonal evolution of the zonal mean ozone distribution revealed that the location of peak mixing ratio traversed the Equator during the annual cycle. This distribution is generally similar to that obtained from satellite data for similar periods [e.g. the SBUV data in Russell (1986)] and exhibits differences typical of those seen in other models [e.g. WMO (1990) and Jackman et al. (1989)]. Values of the ozone mixing ratio in the upper stratosphere suffer from the well-known discrepancy of being somewhat lower than observations. For example, the calculated ozone mixing ratio at 1 mb in the Tropics is about 10 percent lower than observed.

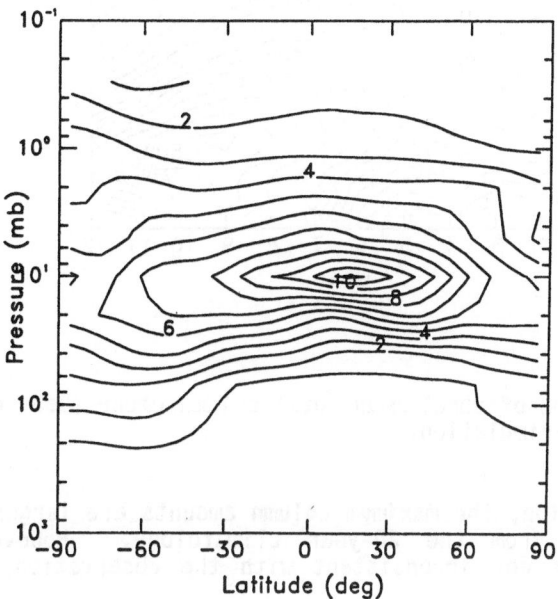

Figure 2. Zonal mean ozone (ppmv) cross sections from simulation for model day September 1.

The evolution of the zonal mean column ozone during an annual cycle from the control simulation is shown in Fig. 3. The most prominent features are the spring high-latitude maximum in the Northern Hemisphere, the broad equatorial minimum, and the maximum during austral spring near 60S. These general features are in reasonable agreement with the 20-year average distribution presented by London (1980), but differ in some details. As an example, the Southern Hemisphere spring maximum shows a pronounced poleward progression with time. The model results display only a slight tendency in this

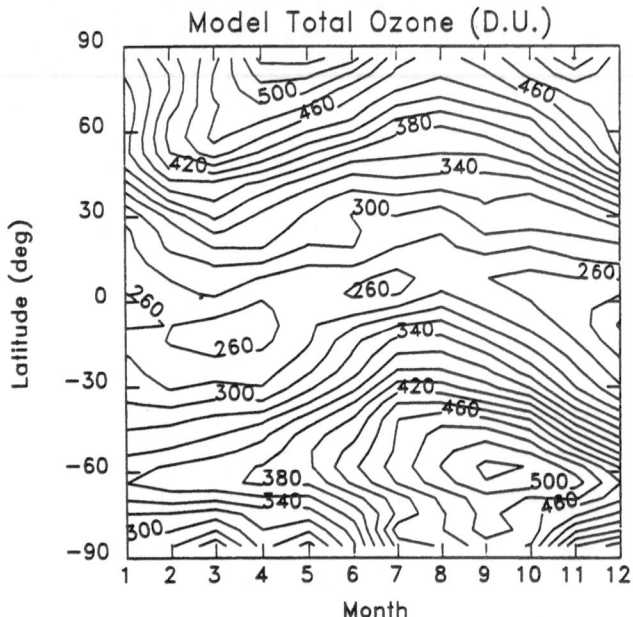

Figure 3. Variation of zonal mean total column ozone (Dobson units) from model control simulation.

respect. In addition, the maximum column amounts are larger than the equivalent values from the 20-year climatology. However, those maximum values are not inconsistent with the observations for some individual years.

The percentage difference in the total ozone between the control and perturbed simulation 1 month (model day October 1) after initiation of the perturbed simulation is shown in Fig. 4. Differences of approximately 50 percent at the pole, 10 percent at 60S, and 2 percent at 45S are apparent. Part of the deficit outside the region where the ozone depletion occurred (69S to the pole) is non-physical, resulting from inadequate horizontal resolution for resolving the steep latitudinal gradients existing at the perimeter of the "hole." By the end of December, the differences amount to approximately 12 percent at the pole, 8 percent at 60S, and 5 percent at 45S. Analysis of the ozone tendency equation on an isentropic surface (450K) confirms that horizontal eddies are acting to redistribute the ozone deficit during

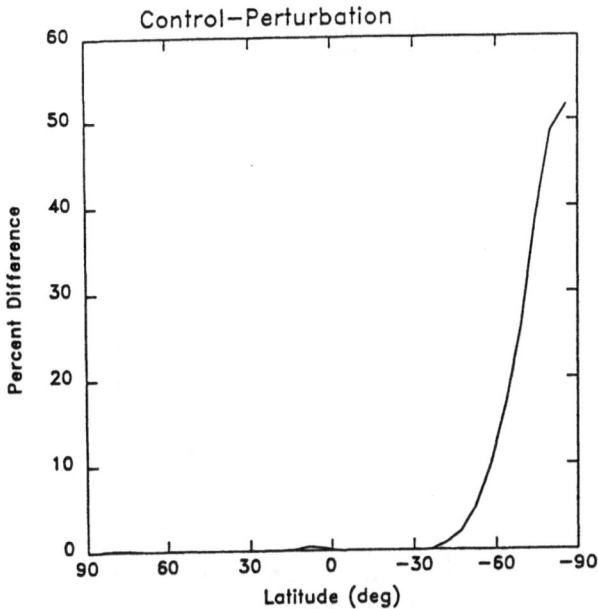

Figure 4. Percent difference in column ozone between control and perturbed simulations 1 month after initiation of perturbed simulation.

this period, decreasing the column between 65 and 80S. By mid-April, a maximum difference of 6 percent in the column between the control and perturbation simulation occurs near 60S. A difference of 2 percent occurs as far equatorward as 30S. The column differences at the end of the simulation 1 year later are shown in Fig. 5. By this time, a maximum difference of approximately 2 percent is seen to occur at 60-70S (solid curve). The difference in the column above 100 mb is shown in the dashed curve. Clearly the majority of the deficit now resides below the 100-mb level. In this respect, the results are consistent with those of Prather et al. (1990) and Cariolle et al. (1990).

246

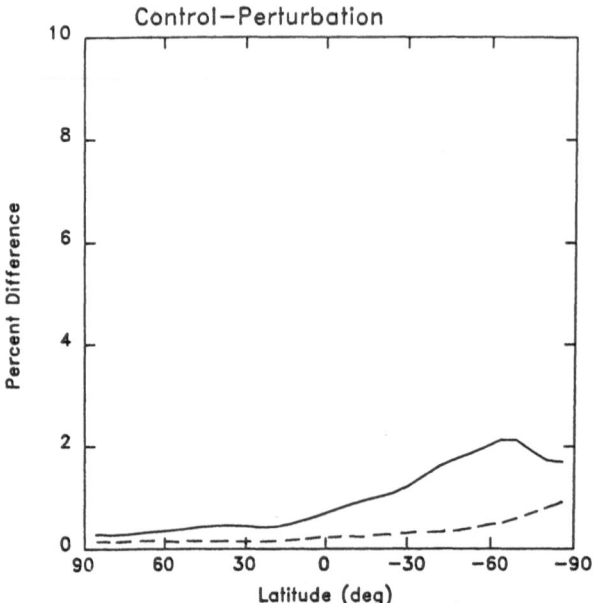

Figure 5. Percent difference in column ozone between control and perturbed simulations 1 year after initiation of perturbed simulation. Solid curve corresponds to total column. Dashed curve corresponds to column above 100 mb.

Finally, the integrated total column ozone (global and by hemisphere) differences between the control and perturbed simulations are shown as a function of time in Fig. 6. One month after initiation of the perturbed simulation, a deficit of approximately 5.75 percent (relative to the control simulation) of the total ozone in the Southern Hemisphere is present. The residual deficit existing 1 year later is approximately 1.3 percent of the total ozone in the Southern Hemisphere. Again, the majority of the deficit resides below the 100-mb level in the model.

The results of the study suggest that there is some possibility of a small, but cumulative effect on the total ozone budget. However, major qualifications on these results must be kept in mind. As mentioned in the introduction, year-to-year variations in the ozone distribution may result from dynamical variability, the injection of volcanic aerosols, the solar cycle, and the possibility of ozone

247

depletion occurring as a result of heterogeneous chemical processes outside of polar regions. These factors have not been addressed in the present study. In fact, the status of current atmospheric models, as well as inadequate data on all these factors, precludes a comprehensive model simulation study of their contributions to changes in the ozone distribution relative to those which might occur as a result of the dilution effect. Another factor which we have not

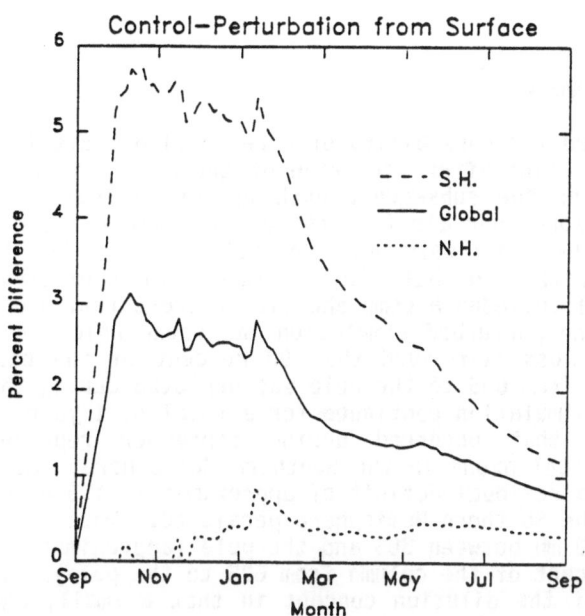

Figure 6. Variation in percent differences in integrated total column ozone (global and hemisphere) between control and perturbed simulations.

addressed in this study is the fact that not only ozone-poor air, but denitrified air as well will be exported to lower latitudes after breakup of the polar vortex. This factor could also influence the ozone distribution. We also commented earlier about the inadequate horizontal resolution of the model (and all other three-dimensional models including chemistry, to our knowledge) for resolving the extremely sharp horizontal gradients at the edge of the ozone hole. Inadequate vertical resolution may, as well, influence our results and should be borne in mind when evaluating the results.

In spite of these qualifying remarks, the most remarkable aspect of this and the other studies of the dilution effect cited earlier, is the degree to which the results agree with one another. There are substantial differences among the various models, each having its own peculiar shortcomings. Nevertheless, although differences in detail exist, a general conclusion is common to all of the studies. The conclusion is that dilution of mid-latitude ozone by transport of ozone-poor air from polar latitudes after breakup of the depleted Antarctic polar vortex may well be a contributing factor to the observed decadal trend in total ozone at mid-latitudes of the Southern Hemisphere.

5. Concluding Remarks

We have examined the possibility of a residual deficit in the global ozone budget resulting after formation of the Antarctic ozone hole in austral spring and the subsequent exchange of air masses from lower latitudes with ozone-poor air from the polar regions after breakup of the vortex and the final warming. Control and perturbed simulations were conducted with a global, three-dimensional chemistry/transport model. The model included a comprehensive representation of gas-phase chemistry. In the perturbed simulation, an ozone hole was imposed by adding a linear loss term such that 90 percent of the ozone in the region 27-112 mb from 69S to the pole was depleted during the month of September. The simulation continued for a total of 1 year.

The depletion that occurred during September represented 5.75 percent of the total ozone in the Southern Hemisphere. At the end of the simulation, a residual deficit of approximately 1.3 percent of the total ozone in the Southern Hemisphere persisted. Most of the deficit resided below 100 mb between 30S and the pole (approximately 1 percent at 30S and 2 percent of the column from 60S to the pole). The results are supportive of the dilution concept in that a small, and possibly cumulative, effect on the total ozone budget occurred as a result of the combined effect of the relatively fast (order of weeks) quasi-horizontal transport of ozone-poor air to lower latitudes after vortex breakdown and the relatively slow (order of month to a year) chemical replacement of odd-oxygen in the lower stratosphere.

Obviously, further studies (both model and observational) must be conducted to definitively assess the impact of dilution on the global ozone budget. An extension of this present study is underway to remedy some of the shortcomings which are mentioned earlier. In the study currently underway, the vertical resolution of the model has been enhanced by doubling the number of levels. In addition, heterogeneous chemical processes have been incorporated into the model. Also, the HCl and $ClONO_2$ have been removed from the Cl_x family and are treated explicitly. The results of the study will be reported in the near future.

6. Acknowledgments

The authors wish to thank R. Stolarski for furnishing the figure displaying the TOMS data. We also are grateful for helpful discussions with A. O'Neill, J. Mahlman, A. Plumb, and M. Prather. Thanks are also due to D. Miller and S. Johnson for preparation of the manuscript.

7. References

Atkinson, R. J., Matthews, R. A., Newman, P. A, and Plumb, R. A (1989) "Evidence of the Mid-Latitude Impact of Antarctic Ozone Depletion", Nature, 340, 290-294.

Baulch, D. L. et al. (1982) "Evaluated Kinetic and Photochemical Data for Atmospheric Chemistry: Supplement 1", J. Physical and Chemical Reference Data, 11, 327-496.

Blackshear, W. T., Grose, W. L., and Turner, R. E. (1987) "Simulated Sudden Stratospheric Warming: Synoptic Evolution", Quarterly Journal of the Royal Meteorological Society, 113, 815-846.

Brasseur, G. and Solomon, S. (1984) Aeronomy of the Middle Atmosphere, D. Reidel, Dordrecht, Holland.

Cariolle, D., Lasserre-Bigorry, Royer, J.-F., and Geleyn, J.-F. (1990) "A GCM Simulation of the Springtime Antarctic Ozone Decrease and its Impact on Mid-Latitudes", Journal of Geophysical Research, 95, 1883-1898.

Chipperfield, M. P. and Pyle, J. A. (1988) "Two-Dimensional Modeling of the Antarctic Lower Stratosphere", Geophysical Research Letters, 15, 875-878.

DeMore, W. B. et al. (1987) "Chemical Kinetics and Photochemical Data for Use in Stratospheric Modeling, Evaluation No. 8", JPL Publication 87-41, Jet Propulsion Laboratory, Pasadena, CA, 196 pp.

Farman, J. C, Gardiner, B. G, and Shanklin, J. D. (1985) "Large Losses of Total Ozone in Antarctica Reveal Seasonal ClO_x/NO_x Interaction", Nature, 315, 207-210.

Frederick, J. E. (1986) "Radiative processes", in Atmospheric Ozone 1985, WMO Report 16, Washington, D.C., 349-392.

Garcia, R. and Solomon, S. (1983) "A Numerical Model of the Zonally Averaged Dynamical and Chemical Structure of the Middle Atmosphere", J. Geophysical Research, 88, 1379-1400.

Garcia, R. R. and Solomon S. (1987) "A Possible Relationship Between Interannual Variability in Antarctic Ozone and the Quasi-Biennial Oscillation", Geophysical Research Letters, 14, 848.

Grose, W. L., Nealy, J. E., Turner, R. E., and Blackshear, W. T. (1987) "Modeling the transport of chemically active constituents in the stratosphere", in G. Visconti and R. Garcia (eds.), Transport Processes in the Middle Atmosphere, D. Reidel and Co., 229-250.

Grose, W. L., Eckman, R. S., Turner, R. E, and Blackshear, W. T. (1989) "Global modeling of ozone and trace gases", in T. Schneider et al. (eds.), Atmospheric Ozone Research and Its Policy Implications, Elsevier Publishers, Amsterdam, 1021-1035.

Hirota, I., Hirooka, T., and Shiotani, M. (1983) "Upper Stratosphere Circulation in the Two Hemispheres", Quarterly Journal of the Royal Meteorological Society, 109, 443.

Hoffmann, D. J. and Solomon, S. (1989) "Ozone Destruction Through Heterogeneous Chemistry Following the Eruption of El Chichon", Journal of Geophysical Research, 94, 5029-5041.

Jackman, C. H., Seals, R. K., Jr., and Prather, M. J. (eds.), (1989), "Two-dimensional intercomparison of stratospheric models", NASA Conference Publication 3042.

Journal of Geophysical Research (1989a), 94, No. 9.

Journal of Geophysical Research (1989b), 94, No. 14.

Ko, M. K. W., Tung, K. K., Weisenstein, and Sze, N. D. (1985) "A Zonal Mean Model of Stratospheric Tracer Transport in Isentropic Coordinates: Numerical Simulations for Nitrous Oxide and Nitric Acid", Journal of Geophysical Research, 90, 2313-2329.

Lait, L. R., Schoeberl, M. R, and Newman P. A. (1989) "Quasi-Biennial Modulation of the Antarctic Ozone Depletion", Journal of Geophysical Research, 94, 11559-11571.

London, J. (1979) "The observed distribution and variations of total ozone", in M. Nicolet and A. Aikin (eds.), Proceedings of the NATO Advanced Study Institute on Atmospheric Ozone, U.S. Department of Transportation, 31-44.

Mahlman, J. D. and Moxim, W. J. (1978) "Tracer Simulation Using a Global General Circulation Model: Results From a Mid-Latitude Instantaneous Source Experiment", J. Atmospheric Sciences, 35, 1340-1374.

McElroy, M. B., Salawitch, R. J., Wofsy, S. C, and Logan, J. A. (1986) "Antarctic Ozone: Reductions Due to Synergistic Interactions of Chlorine and Bromine", Nature, 321, 759-762.

Newman, P., (1986) "The Final Warming and Polar Vortex Disappearance During the Southern Hemisphere Spring" Geophysical Research Letters, 13, 1228-1231.

Prather, M., Garcia, M. M., Suozzo, R., and Rind, D. (1990) "Global Impact of the Antarctic Ozone Hole: Dynamical Dilution with a 3-D Chemical Transport Model", Journal of Geophysical Research, 95, 3449-3472.

Randel, W. (1988) "The Anomalous Circulation in the Southern Hemisphere Stratosphere During Spring 1987", Geophysical Research Letters, 5, 911-914.

Russell, J. M. III (1986) MAP Handbook, 22.

Solomon, S., Garcia R. R., Rowland F. S., and Wuebbles D. J. (1986) "On the Depletion of Antarctic Ozone", Nature, 321, 755-758.

Sze, N. D. et al. (1989) "Antarctic Ozone Hole: Possible Implications for Ozone Trends in the Southern Hemisphere" Journal of Geophysical Research, 94, 11521-11528.

Watson, R. T. and Ozone Trends Panel, M. J. Prather and Ad Hoc Theory Panel, and M. J. Kurylo and NASA Panel for Data Evaluation (1988) "Present State of Knowledge of the Upper Atmosphere 1988: An Assessment Report", NASA Reference Publication 1208.

World Meteorological Organization (1990) "Scientific Assessment of Stratospheric Ozone: 1989", Report No. 20, Vol.1.

ALPHABETICAL INDEX OF AUTHORS

SUBJECT INDEX

Japanese Antarctic Research Expedition 145

257